Birkhäuser

Modern Birkhäuser Classics

Many of the original research and survey monographs in pure and applied mathematics published by Birkhäuser in recent decades have been groundbreaking and have come to be regarded as foundational to the subject. Through the MBC Series, a select number of these modern classics, entirely uncorrected, are being re-released in paperback (and as eBooks) to ensure that these treasures remain accessible to new generations of students, scholars, and researchers.

David Spring

Convex Integration Theory

Solutions to the h-principle in geometry
and topology

Reprint of the 1998 Edition

 Birkhäuser

David Spring
Department of Mathematics
Glendon College
2275 Bayview Avenue
Toronto, Ontario M4N 3M6
Canada
dspring@glendon.yorku.ca

2010 Mathematics Subject Classification 58C35, 57R99

ISBN 978-3-0348-0059-4   e-ISBN 978-3-0348-0060-0
DOI 10.1007/978-3-0348-0060-0

© 1998 Birkhäuser Verlag
Originally published under the same title as volume 92 in the Monographs in Mathematics series by
Birkhäuser Verlag, Switzerland, ISBN 978-3-7643-5805-1
Reprinted 2010 by Springer Basel AG

Cover design: deblik, Berlin

Printed on acid-free paper

Springer Basel AG is part of Springer Science+Business Media

www.birkhauser-science.com

ACKNOWLEDGEMENTS

I should like to thank the many colleagues for their personal support expressed over the years for this book project. In particular, I thank M. Gromov and Y. Eliashberg for their sound advice and for their encouragement and support for my work. I should like to thank Kam Siu-Man for his assistance with the illustrations. I also gratefully acknowledge the financial support that I have received from NSERC over several years in support of my research project for this book.

CONTENTS

CHAPTER 1

INTRODUCTION

§1. Historical Remarks

Convex Integration theory, first introduced by M. Gromov [17], is one of three general methods in immersion-theoretic topology for solving a broad range of problems in geometry and topology. The other methods are: (i) Removal of Singularities, introduced by M. Gromov and Y. Eliashberg [8]; (ii) the covering homotopy method which, following M. Gromov's thesis [16], is also referred to as the method of sheaves. The covering homotopy method is due originally to S. Smale [36] who proved a crucial covering homotopy result in order to solve the classification problem for immersions of spheres in Euclidean space.

These general methods are not linearly related in the sense that successive methods subsumed the previous methods. Each method has its own distinct foundation, based on an independent geometrical or analytical insight. Consequently, each method has a range of applications to problems in topology that are best suited to its particular insight. For example, a distinguishing feature of Convex Integration theory is that it applies to solve closed relations in jet spaces, including certain general classes of underdetermined non-linear systems of partial differential equations. As a case of interest, the Nash-Kuiper C^1-isometric immersion theorem can be reformulated and proved using Convex Integration theory (cf. Gromov [18]). No such results on closed relations in jet spaces can be proved by means of the other two methods. On the other hand, many classical results in immersion-theoretic topology, such as the classification of immersions, are provable by all three methods.

In this context it would of interest to have an historical account of immersion-theoretic topology as it has developed during the past several decades. The history of immersion theory is rich and complex, with contributions from many leading topologists in different countries. To date, the literature contains no general overview of the history of immersion theory. For a brief account of Convex Integration theory and its relation to the early history of immersion theory, cf. Spring [39].

Gromov's treatise [18] serves as a milestone in immersion-theoretic topology. In this book, Gromov reformulates and reflects on the basic methods and

D. Spring, *Convex Integration Theory: Solutions to the h-principle in geometry and topology*, Modern Birkhäuser classics, DOI 10.1007/978-3-0348-0060-0_1, © Springer Basel AG 2010

results of immersion theory. In particular, the above three methods have been rewritten and generalized, with special attention to open problems and to research areas of contemporary interest such as symplectic topology. The book contains an enormous amount of material, expressed throughout in a very concise fashion (cf. McDuff [29], for a review of Gromov's book). Gromov's reformulation of Convex Integration theory [18, Chapter 2.4] generalizes considerably his original paper, Gromov [17], where only open and closed relations in spaces of 1-jets are treated. Convex Integration theory is decidedly more sophisticated than the other two theories. The theory of iterated convex hull extensions over higher order jet spaces is quite subtle, involving both the topology and geometry of jet spaces (Chapters VI, VII) and the $C^\perp$-approximation theorem in analysis (cf. Chapter III). However, Gromov's presentation of the basic results of Convex Integration theory is difficult to follow since important details of the proofs are not provided explicitly in his text; these details are left to the reader, often without accompanying hints. For example, iterated convex hull extensions in the case of non-ample relations are not discussed although they are required in the proof of the Nash-Kuiper C^1-isometric immersion theorem. The technique for constructing global parametrized families of strictly surrounding paths (cf. C-structures in Chapter II) is not mentioned.

For these and other reasons, it is useful to have an independent exposition of Convex Integration theory which, because it concentrates on this one topic, can proceed at a measured pace, with detailed attention to the special techniques and proof procedures that pertain to the theory. Furthermore, for those who might wish to learn enough of the theory in order to understand how it works in the case of relations in spaces of 1-jets, an important case both historically and in practice, a separate proof of the h-principle in the 1-jet case is desirable. This is accomplished in Chapter IV. The aims of this book may be summarized as follows:

- To provide a detailed exposition of the basic theorems of Convex Integration theory for relations over jet spaces which are both open and ample (Chapters I–VIII).

- To generalize the theory to relations over jet spaces which are open but not necessarily ample (Chapters VII, VIII).

- To provide a detailed exposition of the applications of Convex Integration theory to the complementary case of closed relations in jet spaces, in particular to underdetermined non-linear systems of PDEs (Chapter IX).

- To introduce the connection between Convex Integration theory and Filippov's Relaxation Theorem in classical Optimal Control theory (Chapter X).

The book is divided into two parts: Part I, Chapters I to IV, contains all the background constructions to the general theory including the $C^\perp$-approximation theorem in Chapter III. Chapter IV applies this theory to the prove the C^0-dense h-principle for open ample relations in spaces of 1-jets. Part II, Chapters V to X, provides the general theory of convex hull extensions for open relations over higher order jet spaces, with applications to PDE theory (Chapter IX), and to proving the C^{r-1}-dense h-principle (Chapter VIII). In Chapter V the background analytic and topological constructions that appear in Part I are generalized to the microfibration case. The book is self-contained as possible. Part I may be read independently for the basic underlying theory; applications involving higher order jet spaces and PDE theory require the theory in Part II.

Chapter IX on PDEs contributes to the topological literature on non-linear PDE theory. Other topological approaches in the literature, based on the co-homology of exterior differential systems, include the work of Bryant *et al.* [6]. The distinctive feature of the convex integration approach to solving PDEs is the large scale topological and analytic displacement of higher order jets, based on the $C^\perp$-Approximation Theorem, III 3.8. Chapter X on the Relaxation Theorem is included in part for historical reasons. For the past 20 years leading researchers in the separate fields of differential topology and of control theory were unaware that the fundamental analytical approximation problem treated in the Relaxation Theorem is essentially the same analytic approximation problem treated in Convex Integration theory, even though the contexts and motivations for studying this analytic problem are different in these respective fields. Filippov's Relaxation Theorem first came to this author's attention during a survey of the literature in preparation for an earlier manuscript of this book. In Chapter X we prove a type of C^r-Relaxation Theorem for partial differential inclusions of order $r \geq 1$.

Gromov's book [18] records over 15 years of his research in immersion-theoretic topology. His beautiful theory of convex hull extensions, with applications to solving non-linear systems of PDEs, is presented in his book mainly as a series of concisely stated theorems with some proof sketches. Many interpretations of the theory are therefore possible. Our approach, which generalizes the theory to the case of non-ample relations, with applications to non-ample systems of PDEs, is based on a detailed analysis of convex hull extensions (Chapter VII). With some exceptions, the detailed proofs of the main theorems appear here for the first time. Nevertheless not all of the topics in Gromov's treatment of Convex Integration theory appear in these pages.

Gromov's presentation of Convex Integration theory in his treatise has been a principal source of inspiration for this book. We have profited much from his formulation of the theory, and also from the clarifications received from him through personal communications over a period of many years concerning these and related topics.

§2. Background Material

The background material for this book consists mainly of basic topics in differential topology at the graduate school level: the theory of smooth manifolds and smooth fiber bundles; the theory of jet spaces, and the theory of smooth approximations of maps between manifolds, including smooth approximations of sections of smooth bundles. Convenient references for this material include Bredon [5], Hirsch [22], and also Steenrod [38]. Algebraic topology and homotopy theory are employed only secondarily in the theory in order to identify obstructions to the construction of sections of specific relations over jet spaces, just as in the case of classical immersion-theoretic topology.

1.2.1. We review some basic facts about jet spaces and we introduce some notation that is used throughout this book. Let $\mathcal{H}_r(n, q)$ denote the real vector space of homogeneous polynomial maps of total degree r from $\mathbf{R}^n$ to $\mathbf{R}^q$. $\mathcal{H}_1(n, q) = \mathcal{L}(\mathbf{R}^n, \mathbf{R}^q)$, the vector space of linear maps from $\mathbf{R}^n$ to $\mathbf{R}^q$; $\mathcal{H}_0(n, q) = \mathbf{R}^q$. For each $r \geq 1$,

$$\mathcal{H}_r(n, q) = \oplus_1^q \mathcal{H}_r(n, 1); \quad \dim \mathcal{H}_r(n, 1) = \frac{n(n+1)\cdots(n+r-1)}{r!}. \tag{1.1}$$

Let $f \in C^r(U, W)$; $U \subset \mathbf{R}^n$, $W \subset \mathbf{R}^q$ are open. Then the rth order derivative $D^r f \in C^0(U, \mathcal{L}^r_{\text{sym}}(\mathbf{R}^n, \mathbf{R}^q))$, where $\mathcal{L}^r_{\text{sym}}(\mathbf{R}^n, \mathbf{R}^q)$ is the space of symmetric r-linear maps form $\mathbf{R}^n$ to $\mathbf{R}^q$. In what follows we consider (same notation) $D^r f \in C^0(U, \mathcal{H}_r(n, q))$: for all $(x, h) \in U \times \mathbf{R}^n$, $D^r f(x)(h) = D^r f(x)(h, \ldots, h) \in \mathbf{R}^q$. Let $v = (v_1, v_2, \ldots, v_n)$ be a basis of $\mathbf{R}^n$. With respect to coordinates $(u_1, \ldots, u_n)$ in the basis v ($\partial_j = \partial/\partial u_j$; $\partial_v^\alpha = \partial_1^{p_1} \circ \partial_2^{p_2} \circ \cdots \circ \partial_n^{p_n}$) then explicitly, for all $(x, h) \in U \times \mathbf{R}^n$ ($r \geq 1$),

$$D^r f(x)(h) = (h_1 \partial_1 + h_2 \partial_2 + \cdots + h_n \partial_n)^r f(x)$$

$$= \sum_{|\alpha|=r} \frac{r!}{\alpha!} h^\alpha \, \partial_v^\alpha f(x) \in \mathcal{H}_r(n, q). \tag{1.2}$$

where $h = (h_1, h_2, \ldots, h_n)$ in the basis v of $\mathbf{R}^n$; $h^\alpha = h_1^{p_1} h_2^{p_2} \cdots h_n^{p_n}$; $|\alpha| = p_1 + p_2 \cdots + p_n$; $\alpha! = p_1! \ldots p_n!$. If $p \in \mathcal{H}_r(n, q)$ then $D^r p = r!p$. We adopt the convention that the vector space coordinates of a polynomial $g \in \mathcal{H}_r(n, q)$ is the sequence obtained by weighting the $\mathbf{R}^q$-coefficients of the polynomial g with the corresponding factor $\alpha!$ in (1.2): $g(h) = \sum_{|\alpha|=r} h^\alpha c_\alpha$ ($c_\alpha \in \mathbf{R}^q$) has coordinates $(\alpha! c_\alpha) \in \mathbf{R}^{q d_r}$, where $d_r = \dim \mathcal{H}_r(n, 1)$. In these weighted coordinates, for all $x \in U$,

$$\frac{1}{r!} D^r f(x) = (\partial_v^\alpha f(x))_{|\alpha|=r}$$

$$\left(\frac{\partial^r f}{\partial u_{i_1} \ldots \partial u_{i_r}}(x) : 1 \leq i_1 \leq \cdots \leq i_r \leq n \right) \in \mathbf{R}^{q d_r}. \tag{1.3}$$

Let $J^r(U, W) = U \times W \times \prod_{s=1}^{r} \mathcal{H}_s(n, q)$, $r \geq 1$; $J^0(U, W) = U \times W$. $J^r(U, W)$ is a manifold of dimension $n + q + q \left(\sum_{s=1}^{r} d_s \right)$. Employing the product $J^r(U, W) = J^s(U, W) \times \prod_{j=s+1}^{r} \mathcal{H}_j(n, q)$, there is a natural projection map $p_s^r \colon J^r(U, W) \to J^s(U, W)$, $0 \leq s \leq r - 1$.

There is a continuous map, $\pi_r \colon C^r(U, W) \to C^0(U, J^r(U, W))$, $f \mapsto j^r f$, the r-jet extension of f, such that for all $x \in U$,

$$j^r f(x) = (x, f(x), Df(x), \frac{1}{2!} D^2 f(x), \dots, \frac{1}{r!} D^r f(x)); \quad j^0 f(x) = (x, f(x)).$$

For $h = x - x_0$, $j^r f(x_0)(h)$ is the Taylor expansion of f of order r at x_0. From (1.3), the coordinates of $j^r f(x)$ are given by the coordinates of the successive homogeneous polynomials $\frac{1}{s!} D^s f(x) \in \mathcal{H}_s(n, q)$ $(f = (f_1, \dots, f_q) \colon U \to \mathbf{R}^q)$:

$$j^r f(x) = \left(x, f_\mu(x), \frac{\partial f_\mu}{\partial u_{i_1}}(x), \frac{\partial^2 f_\mu}{\partial u_{i_1} \partial u_{i_2}}(x), \dots, \frac{\partial^r f_\mu}{\partial u_{i_1} \dots \partial u_{i_r}}(x) \right) \qquad (1.4)$$
$$1 \leq \mu \leq q; \quad 1 \leq i_1 \leq i_2 \dots \leq i_p \leq n; 1 \leq p \leq r.$$

There are projection maps $s \colon J^r(U, W) \to U$ (source map); $\tau \colon J^r(U, W) \to W$ (target map) for which $s \circ j^r f(x) = x \in U$; $\tau \circ j^r f(x) = f(x) \in W$. $s \colon J^r(U, W) \to U$ is a product bundle, fiber $W \times \prod_{s=1}^{r} \mathcal{H}_s(n, q)$. Let $0 \leq s \leq r$ be integers. The natural projection map $p_s^r \colon J^r(U, W) \to J^s(U, W)$ satisfies $p_s^r \circ j^r f(x) = j^s f(x)$, for all $x \in U$, $0 \leq s \leq r - 1$.

Employing homogeneity, if $p \in \mathcal{H}_s(n, q)$ then for all $j \neq s$,

$$D^j p(x - x_0)(x_0) = 0; \quad D^s p(x - x_0)(x_0) = s! p, \ s \geq 1.$$

The projection map, $p_{r-1}^r \colon J^r(U, W) \to J^{r-1}(U, W)$, is a product vector bundle, fiber $\mathcal{H}_r(n, q)$, whose fiber over $j^{r-1} f(x_0)$, $x_0 \in U$, consists of all the r-jet extensions $j^r(f + p(x - x_0))(x_0)$, $p \in \mathcal{H}_r(n, q)$.

Note that the map, $U \times C^r(U, W) \to J^r(U, W)$, $(x, f) \mapsto j^r f(x)$, is surjective. Indeed, let $z = (x_0, y_0, p_s) \in J^r(U, W)$, where $p_s \in \mathcal{H}_s(n, q)$, $1 \leq s \leq r$. Thus $j^r f(x_0) = z$, where $f(x) = y_0 + \sum_{s=1}^{r} p_s(x - x_0)$.

Let $\sim$ be the relation on $U \times C^r(U, W)$: $(x, f) \sim (y, g)$ if and only if $x = y$ and $j^r f(x) = j^r g(x)$ i.e. f, g have the same derivatives up to order r at x (a relation on germs of functions at $x \in U$). $(x, f) \sim (y, q)$ is an equivalence relation, whose equivalence classes are denoted $[f]_r(x)$, and for which the map $[f]_r(x) \mapsto j^r f(x)$ is well-defined, injective, hence a bijection from the set of equivalence classes to $J^r(U, W)$. Thus $J^r(U, W)$ is naturally identified as the space of r-jets of germs of C^r-maps from U to W.

The Manifold $X^{(r)}$. Let $p: X \to V$ be a smooth fiber bundle, $\dim V = n$, with fiber F, $\dim F = q$. $\Gamma^r(X)$ denotes the space of C^r-sections of the bundle $p: X \to V$ in the compact-open C^r-topology; $\Gamma(X) = \Gamma^0(X)$. For each $x \in V$ sections $f, g \in \Gamma^r(X_U)$, U a neighbourhood of x in V, have the same r-jet at $x \in V$ if in some (hence all) local coordinates f, g have the same derivatives up to order r at x. This r-jet relation at x is an equivalence relation, for which the set of all equivalence classes $[f]_r(x)$, $x \in V$, is denoted by $X^{(r)}$; $X^{(0)} = X$. There are natural projection maps $s: X^{(r)} \to V$, $[f]_r(x) \mapsto x$ (source map); $\tau: X^{(r)} \to X$, $[f]_r(x) \mapsto f(x)$ (target map). Clearly the r-jet relation on $\Gamma^r(X)$ is compatible with the s-jet relation on $\Gamma^s(X)$ for all s, $0 \leq s \leq r$. Hence there is a natural projection map, $p_s^r: X^{(r)} \to X^{(s)}$, $[f]_r(x) \mapsto [f]_s(x)$, $0 \leq s \leq r$ (the case $s = 0$ is the target map).

Suppose the base manifold V and the fiber F are manifolds without boundary. Locally the bundle X is a product bundle: $p: X_U = U \times F \to U$ where $U \equiv \mathbf{R}^n$ is a chart on V. We identify $\Gamma^r(X_U) = C^r(U, F)$. If $f \in \Gamma^r(X)$ and $f(U) \subset W$, $W \equiv \mathbf{R}^q$ is a chart on F, then $f|_U \in C^r(U, W)$, and $[f]_r(x) \in X^{(r)}$ is represented locally by $j^r(f|_U)(x) \in J^r(U, W)$ in the coordinates (1.4).

The bundle $p: X \to V$ comes equipped with transition maps $g = g_{AB}: A \cap B \to \mathrm{Diff}(F)$ that are associated to overlapping open charts A, B in the base manifold V. Let $f \in \Gamma^r(X)$. In the above local coordinates (charts on V, F are identified to the corresponding open sets respectively in $\mathbf{R}^n, \mathbf{R}^q$), let $f|_A \in C^r(A, W_1)$, $f|_B \in C^r(B, W_2)$. For all $x \in A \cap B$:

$$f|_B(x) = g(x) \circ f|_A(x); \quad g(x) \in \mathrm{Diff}(W_1, W_2).$$

Consequently, the coordinates (1.4) for the r-jet extensions $j^r f|_A$, $j^r f|_B$ are related by the following change of coordinates formula: for all $x \in A \cap B$ (we identify $g \equiv \mathrm{ev}\, g: (A \cap B) \times W_1 \to W_2$),

$$j^r f|_B(x) = j^r g \circ (\mathrm{id}, f|_A)(x) \in J^r(B, W_2). \tag{1.5}$$

Employing the Chain Rule, $D^k f|_B$ is a sum of terms involving all of the derivatives $D^j f|_A$, $D^j g$, $0 \leq j \leq k$, $1 \leq k \leq r$. For example, in the case $k = 1$,

$$D f|_B(x) = D_1\, g(x, f|_A(x)) + D_2\, g(x, f|_A(x)) \circ D f|_A(x) \tag{1.5a}$$

where D_1, D_2 are derivatives respectively in the $\mathbf{R}^n$- and $\mathbf{R}^q$-variables.

Thus $X^{(r)}$ is naturally topologized as a smooth manifold, the manifold of r-jets of germs of C^r-sections of the bundle $p: X \to V$, whose charts are equivalent to $J^r(U, W)$ above, where $U \subset \mathbf{R}^n$, $W \subset \mathbf{R}^q$ are open sets. $[f]_r(x) \in X^{(r)}$ is represented in these chart coordinates by the coordinates (1.4) for $j^r(f|_U)(x) \in J^r(U, W)$. The change of coordinates map for overlapping charts $J^r(A, W_1)$,

$J^r(B, W_2)$ on $X^{(r)}$ is induced by the Chain Rule calculation in formula (1.5) for $[f]_r(x) \in J^r(A, W_1) \cap J^r(B, W_2)$.

The source map $s\colon X^{(r)} \to V$, $[f]_r(x) \mapsto x$, is smooth, and is equivalent in local coordinates to the projection source map $s\colon J^r(U, W) \to U$ defined above in the Euclidean space case. Similarly the projection map $p_s^r\colon X^{(r)} \to X^{(s)}$ is smooth, $0 \leq s \leq r$, and is equivalent in local coordinates to the corresponding natural projection map $p_s^r\colon J^r(U, W) \to J^s(U, W)$.

Also $s\colon X^{(r)} \to V$ is a smooth fiber bundle, fiber $G = F \times \prod_{s=1}^r \mathcal{H}_s(n, q)$, that is associated to the smooth bundle $p\colon X \to V$ as follows: Let A, B be overlapping open charts in V. The transition map $h_{AB}\colon A \cap B \to \mathrm{Diff}(G)$ for the bundle $s\colon X^{(r)} \to V$ is the function of the derivatives $D^j g_{AB}$, $0 \leq j \leq r$, that is induced from the change of coordinates formula (1.5). For example in case $r = 1$, the fiber of $s\colon X^{(1)} \to V$ is $G = F \times \mathcal{L}(\mathbf{R}^n, \mathbf{R}^q)$. Let $(y, A) \in F \times \mathcal{L}(\mathbf{R}^n, \mathbf{R}^q)$ be a point in the fiber over $x \in V$. In local coordinates, employing (1.5a),

$$h_{AB}(y, A) = D_1\, g_{AB}(x, y) + D_2\, g_{AB}(x, y) \circ A.$$

In case $r > 1$, h_{AB} is obtained by successive differentiations in local coordinates for $j^r f|_B$ in formula (1.5). Similar considerations show that the projection $p_{r-1}^r\colon X^{(r)} \to X^{(r-1)}$ is a smooth bundle (discussed below), and hence by composition of projections that the projection $p_s^r\colon X^{(r)} \to X^{(s)}$ is a smooth bundle.

$\Gamma(X^{(r)})$ denotes the space of continuous sections, in the compact-open topology, of the bundle $s\colon X^{(r)} \to V$. Let $\alpha \in \Gamma(X^{(r)})$ and let $f = p_0^r \circ \alpha \in \Gamma(X)$ be the 0-jet component of α. Explicitly, employing the equivalence in local coordinates of the source map to the projection $s\colon J^r(U, W) \to U$, where $U \equiv \mathbf{R}^n$ is a chart on V and $f(U) \subset W$, for all $x \in U$,

$$\alpha(x) = \big(x, f(x), (\varphi_\beta(x))_{|\beta| \leq r}\big) \in J^r(U, W),$$

where the functions $\varphi_\beta \in C^0(U, \mathbf{R}^q)$ are continuous for all multi-indices β, $1 \leq |\beta| \leq r$ (for each $x \in U$, the coordinates $(\varphi_\beta(x) \in \mathbf{R}^q)_{|\beta|=p}$ represent the polynomial $\sum_{|\beta|=p}(h^\beta/\beta!)\varphi_\beta(x)$ in $\mathcal{H}_p(n, q)$, $1 \leq p \leq r$).

For each $f \in \Gamma^r(X)$ there is a continuous map $j^r f \in \Gamma(X^{(r)})$, $x \mapsto [f]_r(x)$, the r-jet extension of the section f, such that in local coordinates in $X^{(r)}$ over a chart U in V, for all $x \in U$, $[f]_r(x) = j^r(f|_U)(x) \in J^r(U, W)$ where $f(U) \subset W$ i.e. locally, $(j^r f)|_U \in \Gamma(X_U^{(r)})$ has coordinates (1.4). The smooth projection map $p_s^r\colon X^{(r)} \to X^{(s)}$ satisfies $p_s^r \circ j^r f = j^s f \in \Gamma(X^{(s)})$, $0 \leq s \leq r$.

The affine bundle $p_{r-1}^r\colon X^{(r)} \to X^{(r-1)}$. For each $r \geq 1$ the projection map $p_{r-1}^r\colon X^{(r)} \to X^{(r-1)}$ is an affine bundle, fiber $\mathcal{H}_r(n, q)$. Explicitly, employing local coordinates in a chart $J^r(U, \mathbf{R}^q)$ on $X^{(r)}$, the fiber over $j^{r-1}h(x_0) \in X^{(r-1)}$,

$h \in \Gamma^r(X)$, $x_0 \in U$, consists of all the r-jet extensions, $j^r(h + p(x - x_0))(x_0)$, $p \in \mathcal{H}_r(n, q)$ i.e. the fibers are equivalent to $\mathcal{H}_r(n, q)$. Applying the change of coordinates formula (1.5) to $f = g_{AB} \circ (h + p(x - x_0))$ it follows from the homogeneity of polynomials in $\mathcal{H}_r(n, q)$ that,

$$j^{r-1} f(x_0) = j^{r-1}(g_{AB} \circ h)(x_0);$$
$$D^r f(x_0) = D^r g_{AB} \circ (h + p(x - x_0))(x_0)$$

for all $p \in \mathcal{H}_r(n, q)$. Thus, employing the Chain Rule, a change of coordinates on charts of $X^{(r)}$ induces an affine transformation of the $\mathcal{H}_r(n, q)$-fiber over the base point $j^{r-1}h(x_0)$ of the bundle $p_{r-1}^r \colon X^{(r)} \to X^{(r-1)}$. The "translation" term of the affine transformation is induced by the derivatives $D^j h(x_0)$, $0 \le j \le r - 1$ (included in the base point $j^{r-1}h(x_0)$) in the Chain Rule calculation for $D^r f(x_0)$. For example in case $r = 1$, employing (1.5a), the transformation on the fiber in $X^{(1)}$ over the base point $(x, y) \in U \times F \equiv X_U$ is given in local coordinates as follows,

$$T(A) = D_1\, g_{AB}(x, y) + D_2\, g_{AB}(x, y) \circ A \quad \text{for all } A \in \mathcal{L}(\mathbf{R}^n, \mathbf{R}^q),$$

which includes the translation term $D_1\, g_{AB}(x, y)$. Hence $p_{r-1}^r \colon X^{(r)} \to X^{(r-1)}$ is an affine bundle, fiber $\mathcal{H}_r(n, q)$. In case $r = 1$, $p_0^1 \colon X^{(1)} \to X^{(0)} = X$ is an affine bundle, fiber $\mathcal{H}_1(n, q) = \mathcal{L}(\mathbf{R}^n, \mathbf{R}^q)$. A similar analysis, employing the Chain Rule, shows that the projection map $p_s^r \colon X^{(r)} \to X^{(s)}$ is a smooth affine bundle, fiber $\prod_{j=s+1}^r \mathcal{H}_j(n, q)$, for all s, $0 \le s \le r - 1$.

A special case of interest is the product F-bundle, $X = V \times F \to V$. We identify $\Gamma^r(X) \equiv C^r(V, F)$. In this case the jet space $X^{(r)}$ is commonly denoted by $J^r(V, F)$ in the literature. Local representatives of $f \in C^r(V, F)$ on overlapping charts A, B of the base manifold V are ($g_{AB} = \mathrm{id}$):

$$f|_B = \psi^{-1} \circ f|_A \circ \varphi \in C^r(U, \mathbf{R}^q),$$

where $U \subset \mathbf{R}^n$ is open and ψ, φ are corresponding chart maps. Thus the change of coordinates map (1.5) for charts on $J^r(V, F)$ simplifies to the formula,

$$j^r f|_B(x) = j^r(\psi^{-1} \circ f|_A \circ \varphi)(x).$$

In case $r = 1$ note that $Df|_B = D\psi^{-1} \circ Df|_A \circ D\varphi$, from which it follows that the bundle $p_0^1 \colon X^{(1)} \to X$ specializes to a vector bundle, fiber $\mathcal{L}(\mathbf{R}^n, \mathbf{R}^q)$. Consequently a section $\alpha \in \Gamma(X^{(1)})$ corresponds to a tangent bundle morphism (same notation) $\alpha \colon T(V) \to T(F)$ which covers $g \in C^0(V, F)$ where g is the 0-jet component of the section α. This bundle morphism interpretation is discussed again at the end of §1.3.1.

Suppose now $p\colon X \to V$ is a vector bundle. In this case the transition maps are of the form $g_{AB}\colon A \cap B \to GL(F)$, where $GL(F)$ is the vector space of linear automorphisms of the vector space F. Employing (1.5) the zero section $z \in \Gamma(X)$ extends to the zero section $j^r z \in \Gamma(X^{(r)})$; hence all of the affine bundles $s\colon X^{(r)} \to V$, $p_s^r\colon X^{(r)} \to X^{(s)}$, $0 \le s \le r$, specialize to vector bundles.

In case V and the fiber F of $p\colon X \to V$ are smooth manifolds possibly with boundary, then $X^{(r)}$ is a smooth manifold, with corners in case both of these manifolds have non-empty boundary. For the purposes of the proofs of the main theorems of this book, which are carried out inductively over a covering of the base manifold V by a system of charts, it is convenient to adjoin if necessary a collar $\partial V \times [0,1)$ to ∂V, and a collar $\partial F \times [0,1)$ to ∂F. That is, without loss of generality we adopt throughout the *standing assumption* that V, F are manifolds without boundary i.e. charts in V, F that meet ∂V, ∂F respectively will extend into the above collars. For this purpose note that in case $\partial V \ne \emptyset$, the bundle $p\colon X \to V$ extends over the collar added to ∂V. A similar extension applies to microfibrations $\rho\colon \mathcal{R} \to X^{(r)}$ in case $\partial V \ne \emptyset$.

In addition to the compact-open C^r-topology on $\Gamma^r(X)$ there is also the Whitney fine or strong C^r-topology on $\Gamma^r(X)$ for which neighbourhoods are defined as follows. Let $f \in \Gamma^r(X)$ and let $\mathcal{N}$ be a neighbourhood of the image $j^r(f)(V) \subset X^{(r)}$. Let $\mathcal{N}_f = \{g \in \Gamma^r(X) \mid j^r(g)(V) \subset \mathcal{N}\}$. The Whitney fine C^r-topology on $\Gamma^r(X)$ is the topology generated by the family of all subsets $\mathcal{N}_f$. Informally, $f, g \in \Gamma^r(X)$ are close in the fine C^r-topology if their r-jet extensions are C^0-close in $X^{(r)}$ at all points of V.

Smooth Approximation Theorem 1.1. *Let $h \in \Gamma^k(X)$ and let $\mathcal{N}$ be a neighbourhood of the image $j^k h(V) \subset X^{(k)}$, $k \ge 0$. Suppose $K \subset V$ is closed such that h is C^r in a neighbourhood U of K, $r \ge k$. There is a homotopy rel K, $H\colon [0,1] \to \Gamma^k(X)$, $H_0 = h$, such that for all $t \in [0,1]$:*

(i) *The image $j^k H_t(V) \subset \mathcal{N}$.*

(ii) *On a smaller neighbourhood U_1 of K, $\mathrm{ev}\, H\colon [0,1] \times U_1 \to X$ is a C^r map; for all $w \in K$, $j^r H_t(w) = j^r h(w)$.*

(iii) *$H_1 \in \Gamma^r(X)$; H_1 is C^∞ on $V \setminus K$.*

Proof. The theorem is classical although not usually stated in this precise form. Some details follow. Up to a small homotopy rel W of h, W a neighbourhood of K such that $\overline{W} \subset U$, one may assume that h is a C^r-section. Let $\epsilon\colon X^{(r)} \to [0,\infty)$ be continuous such that $j^r h(K) = \epsilon^{-1}(0)$. Since $s\colon X^{(r)} \to V$ is a smooth bundle, employing the exponential map one can identify neighbourhoods of $B = j^r h(V)$ in $X^{(r)}$ with vertical disk bundles over B ($= \ker Ds$) with respect to $Ds\colon T(X^{(r)}) \to T(V)$. Let $\mathcal{M}$ be a vertical ϵ-disk neighbourhood of $B \setminus j^r h(K)$. Employing smooth approximation theory in the C^r-topology, there is a small homotopy rel K, $H\colon [0,1] \to \Gamma(X)$, $H_0 = h$, such that H_1 is smooth on $V \setminus K$ and such that $j^r H_t(V \setminus K) \subset \mathcal{M}$ for all t. Since $\epsilon = 0$ on $j^r h(K)$, for each $t \in [0,1]$ $\lim_{x \to w} j^r H_t(x) = j^r h(w)$ uniformly on compact sets for all $w \in K$. It follows that H_t is of class C^r for all $t \in [0,1]$ (cf. Hirsch [22] Theorem 2.5); hence $H_1 \in \Gamma^r(X)$. $\square$

§3. h-Principles

1.3.1. Let $\rho\colon \mathcal{R} \to X^{(r)}$ be a continuous map, referred to as a relation $\mathcal{R}$ over $X^{(r)}$, where $X^{(r)}$ is the space of r-jets associated to a smooth bundle $p\colon X \to V$. $\Gamma(\mathcal{R})$ is the space of continuous sections of the map $s \circ \rho\colon \mathcal{R} \to V$ in the compact-open topology. If $\alpha \in \Gamma(\mathcal{R})$ then $\rho \circ \alpha \in \Gamma(X^{(r)})$. A section $\alpha \in \Gamma(\mathcal{R})$ is *holonomic* if there is a C^r-section $f \in \Gamma^r(X)$ such that $\rho \circ \alpha = j^r f \in \Gamma(X^{(r)})$. Such a section f is unique since $p_0^r \circ \rho \circ \alpha = f \in \Gamma(X)$ i.e. f is the 0-jet component of the section $\rho \circ \alpha$. An important example is the case of the inclusion $i\colon \mathcal{R} \to X^{(r)}$ of a subset $\mathcal{R}$ of $X^{(r)}$, referred to as a *partial differential relation* in $X^{(r)}$. In this case $\alpha \in \Gamma(\mathcal{R})$ is holonomic if $\alpha = j^r f$ for a unique C^r-section $f \in \Gamma^r(X)$.

Relations over $X^{(r)}$ occur in various contexts. For example let $\mathcal{R} \subset X^{(s)}$ be a relation. The projection $p_r^s\colon \mathcal{R} \to X^{(r)}$ is a relation over $X^{(r)}$ for all $r < s$. In this case a holonomic section $\alpha \in \Gamma(\mathcal{R})$ consists of a C^r-section $f \in \Gamma^r(X)$ whose r-jet extension $j^r f \in \Gamma(X^{(r)})$ extends continuously to the section $\alpha \in \Gamma(X^{(s)})$. Thus the construction of f is "guided above" by the relation $\mathcal{R} \subset X^{(s)}$. As a particular case, the free map relation $\mathcal{R} \subset X^{(2)}$ is an over relation for the immersion relation in $X^{(1)}$ (cf. the discussion on the C^1-dense h-principle for free maps at the end of §1.3.2). The C^r-dense h-principle (below) applied to open, ample relations $\mathcal{R} \subset X^{(s)}$, $s > r$, leads to refined results in immersion-theoretic topology.

Let $\mathcal{R}$ be a relation over $X^{(r)}$. Employing the atlas of charts $J^r(U, W)$ on $X^{(r)}$ (cf. §1.2.1) a section $\alpha \in \Gamma(\mathcal{R})$, for which $f = p_0^r \circ \rho \circ \alpha \in \Gamma(X)$ is the 0-jet component, is represented locally as a section of the bundle $s\colon J^r(U, W) \to U$, where $f(U) \subset W$. Thus $\rho \circ \alpha|_U \in C^0(U, J^r(U, W))$; in local coordinates, for all $x \in U$,

$$\rho \circ \alpha(x) = \big(x, f(x), (\varphi_\beta(x))_{|\beta| \le r}\big) \in J^r(U, W), \qquad (1.6)$$

where $\varphi_\beta \in C^0(U, \mathbf{R}^q)$ is a continuous function for all multi-indices β, $1 \le |\beta| \le r$ (for each $x \in U$, the coordinates $(\varphi_\beta(x) \in \mathbf{R}^q)_{|\beta|=p}$ represent a polynomial map in $\mathcal{H}_p(n, q)$, $1 \le p \le r$). Employing the local coordinate representation (1.4), α is holonomic ($= j^r f$) if and only if locally over all charts $U \subset V$, the sequence of functions,

$$(\varphi_\beta)_{|\beta| \le r} = (\partial_v^\beta f)_{|\beta| \le r} = \left(\frac{\partial^p f}{\partial u_{i_1} \ldots \partial u_{i_p}} \in C^0(U, \mathbf{R}^q)\right) \qquad (1.7)$$

$$1 \le i_1 \le i_2 \le \cdots \le i_p \le n; \; 1 \le p \le r.$$

The main object of study in this book is the topology and geometry of open and closed relations $\mathcal{R} \subset X^{(r)}$, and in a more general context the topology and geometry of relations $\rho \colon \mathcal{R} \to X^{(r)}$ which are microfibrations (Part II), $r \geq 1$. Open relations in jet spaces occur naturally in differential topology. For example, the classical problem of immersing a manifold V into a manifold N, $\dim V = n$, $\dim N = q$, $n \leq q$, is analyzed in terms of an open relation of the 1-jet space $X^{(1)}$ of the product bundle $X = V \times N \to V$. Explicitly, let $\mathcal{R} \subset X^{(1)}$ be the open relation that is defined pointwise in the local coordinates (1.6) as follows: the vectors, $\varphi_i(x) \in \mathbf{R}^q$, $1 \leq i \leq n$, are linearly independent (an open condition). Employing (1.7), a section $\alpha \in \Gamma(\mathcal{R})$ is holonomic i.e. $\alpha = j^1 f$, if and only if over all charts $U \subset V$ the vectors $\partial f / \partial u_i(x) \in \mathbf{R}^q$, $1 \leq i \leq n$, are linearly independent for all $x \in U$, which is precisely the immersion condition on a map $f \in C^1(V, N)$.

Similarly, the free map relation is the open relation $\mathcal{R} \subset X^{(2)}$ of the above product bundle such that, pointwise in the local coordinates (1.6), the $n + n(n + 1)/2$ vectors, $(\varphi_\beta(x) \in \mathbf{R}^q)_{|\beta| \leq 2}$, are linearly independent. Employing (1.7), a section $\alpha \in \Gamma(\mathcal{R})$ is holonomic ($= j^2 f$) if and only if over all charts $U \subset V$, $\partial f / \partial u_i(x)$, $1 \leq i \leq n$, $\partial^2 f / \partial u_j \partial u_k(x)$, $1 \leq j \leq k \leq n$, are linearly independent vectors in $\mathbf{R}^q$ for all $x \in U$, which is precisely the free map condition on a map $f \in C^2(V, W)$ ($q \geq n + n(n + 1)/2$).

Note that linear independence of vectors is expressed in terms of the maximal rank of the corresponding matrix of components of these vectors. Further open relations in jet spaces $X^{(r)}$, $r \geq 1$, can be defined locally in terms of inequalities on the ranks of corresponding matrices whose entries are obtained from the jet space coordinates (1.6). Indeed, this jet space formalism is employed expressly to provide a general framework for the study of open relations that occur in differential topology.

Closed relations in jet spaces $X^{(r)}$ occur in the study of partial differential equations (PDEs). For example, let $F = (F_1, \ldots, F_k) \colon X^{(r)} \to \mathbf{R}^k$ be a continuous map. Thus $\mathcal{R} = F^{-1}(0) \subset X^{(r)}$ is closed and is expressed locally in the coordinates (1.6) by the system of k-equations:

$$F_\mu(x, f(x), (\varphi_\beta(x))_{|\beta| \leq r}) = 0; \quad 1 \leq \mu \leq k. \tag{1.8}$$

Employing the coordinates (1.7), a section $\alpha \in \Gamma(\mathcal{R})$ is holonomic i.e. $\alpha = j^r f$, if and only if over all charts $U \subset V$,

$$F_\mu\left(x, f(x), \frac{\partial f}{\partial u_i}(x), \quad , \frac{\partial^r f}{\partial u_{i_1} \ldots \partial u_{i_r}}(x)\right) = 0; \quad 1 \leq \mu \leq k. \tag{1.9}$$

Thus a holonomic section of the closed relation $\mathcal{R} = F^{-1}(0) \subset X^{(r)}$ consists of a C^r-section $f \in \Gamma^r(X)$ of the bundle $p \colon X \to V$ such that $F(j^r f(x)) = 0$ for all

$x \in V$ i.e. the section f solves the system (1.9) of k-PDEs globally over the base manifold V. In Chapter IX systems (1.9) are solved by the methods of convex integration where the closed relation $\mathcal{R} = F^{-1}(0)$ satisfies in addition certain nowhere flat conditions (II §1; IX Theorem 9.1) which are geometrical in nature and which are easily verified in practice. However, the scope of the results are necessarily less comprehensive than the corresponding results for open relations in jet spaces. Indeed, the nowhere flat hypotheses systematically exclude linear and quasi-linear systems of PDEs and apply only to underdetermined rth order systems (1.9), or to systems that are reducible to the underdetermined case, $r \geq 1$ (generically determined systems (1.9) are *not* amenable to the methods of convex integration). In particular, there is no uniqueness of solutions. However the space of solutions are dense in the C^{r-1}-fine topology. Furthermore only vector valued functions can be treated by the theory, in the following sense: if locally $f \in C^r(U, W)$ solves the system (1.9), where $U \subset \mathbf{R}^n$, $W \subset \mathbf{R}^q$, then significant results can be obtained only in the case $q \geq 2$. In particular, the Monge–Ampère equation and other non-linear systems of PDEs that involve real-valued functions $f \in C^r(U, \mathbf{R})$ cannot be solved by the methods of convex integration.

The real strength of the results of Chapter IX lies in the general theory of underdetermined systems of PDEs. Indeed, generically underdetermined systems (1.9) can be solved locally, and also globally if the bundle and nowhere flat hypotheses are valid globally. A different perspective on underdetermined systems of PDEs is developed in Chapter X. The Relaxation Theorem in Optimal Control theory, which solves first order differential inclusions in ODE theory up to C^0-approximation, is generalized, X Theorem 10.2, to solve partial differential inclusions of all orders $r \geq 1$ up to C^{r-1}-approximation.

Convex Integration theory is important since it is the only theory to date in immersion-theoretical topology that provides a unified approach for solving both open and closed relations in jet spaces. Furthermore it is more general in scope than the method of sheaves (covering homotopy method), Gromov [16], in the sense that the relations $\mathcal{R} \subset X^{(r)}$ that are subject to the method of convex integration are not required to be invariant under the pseudogroup of diffeomorphisms of the base space V of the bundle $p\colon X \to V$. For example, systems of PDEs (1.9) that are studied in Chapter IX do not generally satisfy the above invariance property.

We conclude this section with some historical remarks on relations $\mathcal{R}$ in spaces of 1-jets. Let V, N be smooth manifolds; $X = V \times N \to V$ is the product bundle. Since $p_0^1\colon X^{(1)} \to X^{(0)} = X$ is a vector bundle, fiber $\mathcal{L}(\mathbf{R}^n, \mathbf{R}^q)$, then employing the manifold structure of $X^{(1)}$, it follows that a section $\alpha \in \Gamma(X^{(1)})$ is equivalent to a vector bundle morphism $F\colon T(V) \to T(N)$ which covers $g\colon V \to N$, where $g \in C^0(V, N)$ is the 0-jet component of α. In local coordinates, for each $x \in V$, the linear map $F(x)$ on tangent spaces is the $q \times n$ matrix whose columns are the vectors $\varphi_i(x) \in \mathbf{R}^q$ of the local coordinates (1.6) for $\alpha(x)$ in the case $r = 1$. Thus specifying a relation $\mathcal{R} \subset X^{(1)}$ is equivalent to specifying conditions on the bundle map $F \colon T(V) \to T(N)$. For example, the immersion relation above corresponds to bundle monomorphisms $F \colon T(V) \to T(N)$. Similarly, the submersion relation in $X^{(1)}$ corresponds to bundle epimorphisms $F \colon T(V) \to T(N)$. In this way, the bundle morphism formulation of classical results in immersion-theoretic topology due to Smale [36], Hirsch [21], Feit [12], Phillips [31] has an equivalent formulation in terms of corresponding relations $\mathcal{R} \subset X^{(1)}$. This formulation of problems in immersion-theoretic topology in terms of relations in jet spaces was developed extensively in Gromov [16], [18].

1.3.2. Solving Relations over $\mathbf{X^{(r)}}$. Let $\rho \colon \mathcal{R} \to X^{(r)}$ be a relation over $X^{(r)}$. Solving the relation $\mathcal{R}$ means constructing a holonomic section $\beta \in \Gamma(\mathcal{R})$. This is done in two stages: In the first stage, we construct, or are given, a section $\alpha \in \Gamma(\mathcal{R})$. The construction of the section α is a topological problem which involves algebraic topology and is considered the easy part of the problem of constructing a holonomic section of $\mathcal{R}$. For example, in the case of the problem of immersing a manifold V into a manifold N, discussed above, a section $\alpha \in \Gamma(\mathcal{R})$ is equivalent to the existence of a bundle monomorphism $H \colon T(V) \to T(N)$. The construction of a bundle monomorphism is a topological problem that is analyzed in cohomology by obstruction theory. Similarly, the solution of the system of equations (1.8) requires the construction of a suitable sequence of continuous functions φ_β, for all $|\beta| \le r$. Again this is a topological problem which in practice can often be solved at least locally.

The main difficulty is the second stage of the problem: the passage from an arbitrary section $\alpha \in \Gamma(\mathcal{R})$ to a holonomic section. Following Gromov, the most optimistic expectation is expressed in the following general *h*-principle.

Homotopy Principle. The relation $\mathcal{R}$ over $X^{(r)}$ satisfies the *h-principle* (homotopy principle) if each section $\alpha \in \Gamma(\mathcal{R})$ is homotopic through sections $\alpha_t \in \Gamma(\mathcal{R})$, $t \in [0, 1]$, $\alpha_0 = \alpha$, to a holonomic section $\alpha_1 \in \Gamma(\mathcal{R})$ i.e. up to homotopy, every section of $\mathcal{R}$ is holonomic. There is also a relative notion: Suppose $\rho \circ \alpha = j^r f_0$ in a neighbourhood U of a closed subspace $K \subset V$ where $f_0 \in \Gamma(X)$ is C^r on U i.e. α is holonomic on U. The relative *h*-principle requires in addition that α be homotopic rel K to a holonomic section.

The *h*-principle applies in the cases of the classical open relations in jet spaces (immersions, Feit's Theorem, submersions of open manifolds) that were first studied in immersion-theoretic topology. In case $p \colon X \to V$ is a holomorphic bundle and V is a Stein manifold then the *h*-principle reduces to Oka's principle, which is in fact older than the *h*-principle, and is proved in great generality in Gromov [19].

We remark here that in general the above homotopy $\alpha_t \in \Gamma(\mathcal{R})$ is large i.e. the holonomic section α_1 in general is not a small perturbation of the given section $\alpha \in \Gamma(\mathcal{R})$. Indeed, the size of the perturbation is controlled by the diameters of the fiberwise surrounding paths in the relation $\mathcal{R}$, as explained in Chapter III. On the other hand, in case α is holonomic up to jets of order $r-1$ i.e. $p_{r-1}^r \circ \rho \circ \alpha = j^{r-1}g$, the homotopy α_t can be chosen to be a small perturbation of α, when restricted to jets of order $\leq r-1$: the sections $p_{r-1}^r \circ \rho \circ \alpha_t \in \Gamma(X^{(r-1)})$, $t \in [0,1]$, are C^0-close to the section $j^{r-1}g \in \Gamma(X^{(r-1)})$ (cf. the C^{r-1}-dense h-principle below). It is this controlled approximation on lower order jets that is employed in Chapter IX to solve underdetermined and related systems (1.9) of PDEs.

Parametric h-Principle. Let Z be an auxiliary smooth compact manifold. A map $g\colon Z \to \Gamma(X)$ is of class C^r if the evaluation map $\operatorname{ev} g\colon Z \times V \to Z \times X$, $(z,x) \mapsto (z, g_z(x))$ is of class C^r. The relation $\mathcal{R}$ over $X^{(r)}$ satisfies the *parametric h-principle* if each continuous map $h\colon Z \to \Gamma(\mathcal{R})$ is homotopic to a map $g\colon Z \to \Gamma(\mathcal{R})$ which is holonomic in the following sense: there is a continuous map $G\colon Z \to \Gamma^r(X)$ such that for all $z \in Z$, $\rho \circ g(z) = j^r(G_z) \in \Gamma(X^{(r)})$. In practice G is constructed to be of class C^r. There is also a relative principle in case $K \subset Z$ is closed and h is holonomic on $\mathfrak{Op}\,K$ (a neighbourhood of K in Z): the homotopy connecting h, g is constant on K.

Weak Homotopy Equivalence. A continuous map $f\colon A \to B$ is a *weak homotopy equivalence* if on the homotopy level f induces a one-to-one correspondence between the path components of A and of B, and if for each $x \in A$, $f_*\colon \pi_i(A,x) \to \pi_i(B, f(x))$ is a group isomorphism for all $i \geq 1$. Let $\mathcal{R} \subset X^{(r)}$ be a relation and let $\Gamma_{\mathcal{R}}(X)$ denote the subspace of $\Gamma^r(X)$ consiting of all sections h such that $j^r h \in \Gamma(\mathcal{R})$. The induced map $J\colon \Gamma_{\mathcal{R}}(X) \to \Gamma(\mathcal{R})$, $h \mapsto j^r h$, is continuous.

In general the algebraic topology of the space $\Gamma_{\mathcal{R}}(X)$ is difficult to analyze. The main work in immersion-theoretic topology over the past few decades has shown that, in many cases of geometrical and topological interest, the map J induces a weak homotopy equivalence. Consequently, on the homotopy level, the space $\Gamma_{\mathcal{R}}(X)$ is analyzed by studying the homotopy groups of the space $\Gamma(\mathcal{R})$. Since $\Gamma(\mathcal{R})$ is a space of continuous sections, the computations of $\pi_i(\Gamma(\mathcal{R}))$, $i \geq 0$, can be carried out in principle, and often in practice, by employing techniques from algebraic topology. Thus, from the viewpoint of immersion-theoretic topology, a weak homotopy equivalence theorem reduces the geometric analysis of the space $\Gamma_{\mathcal{R}}(X)$ to the equivalent homotopy-theoretic problem of determining the homotopy groups of the space $\Gamma(\mathcal{R})$. It is in this sense that one solves the relation $\mathcal{R} \subset X^{(r)}$. Note that the h-principle for $\mathcal{R} \subset X^{(r)}$ is equivalent to the surjectivity of the map J_* on path components. The induced isomorphism J_* on homotopy groups follows from the above parametric h-principle applied to the parameter space of spheres. The weak homotopy equivalence results are proved in IV Theorem 4.7 in case $\mathcal{R} \subset X^{(1)}$ is open and ample, and in VIII Theorem 8.15 in the general case of microfibrations $\rho \colon \mathcal{R} \to X^{(r)}$.

C^i-dense h-Principle. Let $\rho \colon \mathcal{R} \to X^{(r)}$ be a relation over $X^{(r)}$. Let $\alpha \in \Gamma(\mathcal{R})$, $f \in \Gamma^i(X)$, where $0 \leq i \leq r - 1$, such that $p_i^r \circ \rho \circ \alpha = j^i f \in \Gamma(X^{(i)})$ i.e. α is holonomic up to jets of order i. The relation $\mathcal{R}$ satisfies the h-*principle* C^i-*close to* f if for each neighbourhood $\mathcal{N}$ of the image $j^i f(V)$ in $X^{(i)}$ there is a homotopy $H \colon [0, 1] \to \Gamma(\mathcal{R})$, $H_0 = \alpha$, such that:

(i) For all $t \in [0, 1]$, the image $p_i^r \circ \rho \circ H_t(V) \subset \mathcal{N}$.

(ii) The section $H_1 \in \Gamma(\mathcal{R})$ is holonomic: $\rho \circ H_1 = j^r f_1$, $f_1 \in \Gamma^r(X)$.

In particular f_1 is a C^i-approximation of f in the Whitney fine C^i-topology. The relation $\mathcal{R}$ satisfies the C^i-*dense h-principle*, (equivalently: the h-principle for $\mathcal{R}$ is everywhere dense in the fine C^i-topology) if the relation $\mathcal{R}$ satisfies the h-principle C^i-close to f for all data (α, f) as above.

The above C^i-dense h-principle is sufficient for most applications and is employed in the case of open ample relations in jet spaces. A stronger form, employed in the study of short maps (Chapter VIII Theorem 8.12), asserts holonomicity up to jets of order i throughout the homotopy: there is a homotopy $(H_t, f_t) \in \Gamma(\mathcal{R}) \times \Gamma^i(X)$, $(H_0, f_0) = (\alpha, f)$, $t \in [0, 1]$, such that in addition,

(iii) For all $t \in [0, 1]$, $p_i^r \circ \rho \circ H_t = j^i f_t \in \Gamma(X^{(i)})$.

In case $i = 0$ the strong form of the C^0-dense h-principle is equivalent to the C^0-dense h-principle. Indeed, let $f_t \in \Gamma(X)$ be the 0-jet component of the homotopy $\rho \circ H_t \in \Gamma(X^{(r)})$, $t \in [0, 1]$. Since $j^0 f_t = f_t$, the pair (H_t, f_t) satisfies condition (iii). In case $i \geq 1$, this strong form of the h-principle is technically delicate and was an open question in the theory; it is mentioned without proof by Gromov only in the context of open relations $\mathcal{R}$ which are ample ([18] p.174). Proofs of the strong C^i-dense h-principles, Theorems 8.4, 8.14, are based on our strong version of the h-Stability Theorem 7.2.

To conclude this section it is instructive to note that the covering homotopy method, due originally to Smale [36], and developed extensively in Gromov's thesis [16], proves only a C^0-dense h-principle in general. That is, the C^i-dense h-principle is beyond the scope of the covering homotopy method for all $i \geq 1$. For example, with respect to the free map relation $\mathcal{R} \subset X^{(2)}$ discussed above in §1.3.1, let $\alpha \in \Gamma(\mathcal{R})$ be a section for which the 0-jet component $f \in C^1(V, N)$ is a C^1-immersion. If $q > n + n(n+1)/2$, then the free map relation $\mathcal{R}$ is ample and satisfies the strong C^1-dense h-principle (VIII Theorem 8.12). In particular there is a small homotopy f_t, $t \in [0, 1]$, $f_0 = f$, in the fine C^1-topology on $C^1(V, N)$ such that:

(iv) $f_1 \in C^2(V, N)$ is a free map.

(v) For all $t \in [0, 1]$, f_t is an immersion.

By contrast, the construction of the above homotopy f_t in the fine C^1-topology is beyond the scope of the covering homotopy method. Indeed, the best result for the free map relation that can be proved by the covering homotopy method is the existence of a homotopy f_t, $t \in [0, 1]$, $f_0 = f$, in the fine C^0-topology on $C^1(V, N)$ such that $f_1 \in C^2(V, N)$ is a free map.

The impossibility of proving C^i-dense h-principles, $i \geq 1$, by the covering homotopy method is traced to the basic covering homotopy construction, due originally to Smale [36], which consists of introducing a small "twisting" homotopy in the C^0-topology on compactly supported (internal) diffeomorphisms of the *domain* manifold, in order to carry out the required covering homotopy. This twisting homotopy of diffeomorphisms is necessarily large in the C^1-topology. By contrast, the method of convex integration constructs a controlled (external) homotopy of a C^r map in the *range* manifold, subject to suitably small approximations on all lower order derivatives, as detailed in the $C^\perp$-Approximation Theorem (III Theorem 3.8). It is the controlled approximations on all lower order derivatives that is the real strength of the method of convex integration.

§4. The Approximation Problem

1.4.1. We briefly review the analytic problem that underlies Convex Integration theory. In its simplest form it is expressed as the following approximation problem for curves in Euclidean space. Let $A \subset \mathbf{R}^q$ be open, connected, and let $f \in C^1([a, b], \mathbf{R}^q)$ be a C^1-curve such that for all $t \in [a, b]$ the derivative $f'(t) \in \operatorname{Conv} A$, the convex hull of A. The problem is to construct, for each $\epsilon > 0$, a curve $g = g_\epsilon \in C^1([a, b], \mathbf{R}^q)$ such that ($\| \ \|$ is the sup-norm on $C^0([a, b], \mathbf{R}^q)$),

(1) For all $t \in [a, b]$ the derivative $g'(t) \in A$.

(2) $\|g - f\| < \epsilon$.

Thus the problem is to C^0-approximate a curve whose derivative lies in the convex hull of $A \subset \mathbf{R}^q$ by a curve whose derivative lies in A. Stated in this way the above approximation problem is the C^1-version of a well-known problem in the literature of Optimal Control theory. Indeed, assuming that A is compact (not necessarily connected) and that f is absolutely continuous (so that $f(t) = f(a) + \int_a^t f'(s)\, ds$), Filippov's Relaxation Theorem [13] solves the above approximation problem for an absolutely continuous function g, where the derivative conditions $f'(t) \in \operatorname{Conv} A$, $g'(t) \in A$ are understood a.e. (almost everywhere). The literature in Optimal Control theory does not discuss the case that A is connected and the functions f, g are of class C^1 (the applications of the Relaxation Theorem involve functions with discontinuous derivatives). From this point of view the context of the above approximation problem is differential inclusions and 1st order ordinary differential equations (ODE theory). The solution to the above approximation problem is a type of C^1-Relaxation Theorem. In general the open connected set A depends continuously on parameters so that for all $t \in [a, b]$,

$$f'(t) \in \text{Conv } A(t, f(t)); \quad g'(t) \in A(t, g(t)),$$

where $g = g_\epsilon$ is required to satisfy the approximation (2). Filippov's Relaxation Theorem is formulated in terms of the above parametrized set-valued map A whose set values are compact in $\mathbf{R}^q$ and satisfy a Lipschitz property (cf. Chapter X).

For the purposes of applications to the topology of manifolds the above approximation problem in one variable is generalized in Convex Integration theory to the following approximation problem for partial derivatives of functions of several variables. Let $f \in C^r(I^n, \mathbf{R}^q)$, $r \geq 1$, where I^n is an n-rectangle in $\mathbf{R}^n$, $n \geq 1$, such that for all $x \in I^n$ ($A \subset \mathbf{R}^q$ is open, connected),

$$\partial_t^r f(x) \in \text{Conv } A,$$

where $(u_1, \ldots, u_{n-1}, t)$ are coordinates in $\mathbf{R}^n$ and $\partial_t^r = \partial^r/\partial t^r$. The problem is to construct, for each $\epsilon > 0$, a function $g = g_\epsilon \in C^r(I^n, \mathbf{R}^q)$ such that ($\| \ \|$ is the sup-norm on $C^0(I^n, \mathbf{R}^q)$),

(3) For each $x \in I^n$, $\partial_t^r g(x) \in A$.

(4) $\|\partial^\alpha(g - f)\| < \epsilon$ for all derivatives ∂^α, $|\alpha| \leq r$, such that $\partial^\alpha \neq \partial_t^r$.

Here $\partial^\alpha = \partial_1^{p_1} \circ \cdots \circ \partial_n^{p_n}$; $\partial_j = \partial/\partial u_j$, $1 \leq j \leq n$ ($u_n = t$); $|\alpha| = p_1 + \cdots + p_n$. The derivatives $\partial^\alpha \neq \partial_t^r$ are denoted as "$\perp$"-derivatives (complementary to the pure derivative ∂_t^r). Thus the problem is to approximate a C^r-function f whose derivatives $\partial_t^r f$ lie in the convex hull of A by a C^r-function g whose derivatives $\partial_t^r g$ lie in A, where the approximation is on all $\perp$-derivatives of f. This approximation problem, though not discussed in the literature on the Relaxation Theorem, may be considered as a problem of rth order partial differential inclusions, $r \geq 1$. In case $r = 1$ this approximation problem first appears in Gromov [17]. In general the open connected set A depends continuously on parameters so that for all $x \in I^n$ and all derivatives ∂^α such that $|\alpha| \leq r$ and $\partial^\alpha \neq \partial_t^r$ ($\partial^\alpha h = h$ if $|\alpha| = 0$),

$$\partial_t^r f(x) \in \text{Conv } A(x, \ldots, \partial^\alpha f(x), \ldots); \quad \partial_t^r g(x) \in A(x, \ldots, \partial^\alpha g(x), \ldots),$$

where $g = g_\epsilon$ is required to satisfy the approximation (4). The general solution to this problem is formulated as the $C^\perp$-Approximation Theorem, III Theorem 3.8, based on Gromov [18], Spring [35], with applications to proving a C^r-Relaxation Theorem in Chapter X.

The convex integration technique for proving the h-principle consists of applying the $C^\perp$-Approximation Theorem, III Theorem 3.8, discussed above, to the topological setting of a certain affine $\mathbf{R}^q$-bundle $p_\perp^r : X^{(r)} \to X^\perp$ (cf. Chapter VI). In local coordinates the base space $X^\perp$ corresponds to $\perp$-derivatives and the $\mathbf{R}^q$ fiber corresponds to pure rth order derivatives $\partial_t^r f$ of sections $f \in \Gamma^r(X)$. The bundle $p_\perp^r : X^{(r)} \to X^\perp$ appeared first in Gromov [18] in the context of the theory of convex hull extensions (details in Chapter VIII). The geometrical role of the manifold $X^\perp$ in the theory is described briefly as follows. Let $\mathcal{R} \subset X^{(r)}$ be a relation and let $\beta \in \Gamma(\mathcal{R})$. Suppose $f \in \Gamma^r(X)$ is a C^r-section such that $j^{r-1}f$ is the $(r-1)$-jet component of β i.e. β is holonomic up to jets of order $r-1$:

$$p_{r-1}^r \circ \beta = j^{r-1}f \in \Gamma(X^{(r-1)}).$$

Thus for all $x \in V$ the r-jets $\beta(x), j^r f(x)$ lie in the same affine fiber $X_w^{(r)}$ over the base point $w = j^{r-1}f(x)$ of the affine bundle $p_{r-1}^r : X^{(r)} \to X^{(r-1)}$.

Employing affine geometry one decomposes the affine fiber $X_w^{(r)}$ into certain families of parallel affine $\mathbf{R}^q$-planes, called principal directions in $X_w^{(r)}$. The construction of these principal directions employs a purely algebraic result, VI Proposition 6.4, which states that the vector space of homogeneous polynomials $\mathcal{H}_r(n, 1)$ admits a basis of monomials. Geometrically, for each $x \in V$, there is a piecewise linear path in the fibers $X_w^{(r)}$ that joins $j^r f(x)$ to $\beta(x)$ such that each component linear path lies in one of the affine $\mathbf{R}^q$-planes of a principal direction. Topologically each principal direction defines an affine $\mathbf{R}^q$-bundle $X^{(r)} \to X^\perp$ over a base manifold $X^\perp$ (cf. Chapter VI) which "intercalates" the affine bundle $p_{r-1}^r : X^{(r)} \to X^{(r-1)}$. If $\mathcal{R} \subset X^{(r)}$ is open and ample then, roughly speaking, the $C^\perp$-Approximation Theorem applies sequentially along successive principal directions to construct a sequence of homotopic sections $g_j \in \Gamma^r(\mathcal{R})$, $1 \le j \le m$, $g_1 = \beta$, $g_m \in \Gamma(\mathcal{R})$ is holonomic, and such that in local coordinates g_j, g_{j-1} satisfy given $C^\perp$-approximations.

In the case $r = 1$, the geometry of principal directions can be finessed and the h-principle can be proved directly. This is done in Chapter IV. However, for relations $\rho \colon \mathcal{R} \to X^{(r)}$, $r \ge 2$, the affine geometry of the bundles $p_\perp^r : X^{(r)} \to X^\perp$ is essential, and constitutes the main concentration of this book in Part II.

CHAPTER 2

CONVEX HULLS

In this Chapter special results in the geometry of convex hulls are developed which are required for the analytic approximation theory in Chapter III. The main result of this chapter is the Integral Representation Theorem 2.12 which, simply stated, represents a continuous function $f : B \to \mathbf{R}^q$, with values in the convex hull of a connected open set $X \subset \mathbf{R}^q$, as a Riemann integral whose integrand is a continuous function with values in X: for all $b \in B$,

$$f(b) = \int_0^1 h(t,b)\,dt,$$

where $h\colon [0,1] \times B \to X$ is continuous. Over each point $b \in B$, the map h is defined to be a suitable reparametrization of a contractible loop in X which strictly surrounds the point $f(b)$. A parametrized family of contractible loops (parametrized by the space B) together with a specific parametrized family of contractions is called a C-structure with respect to f. A key observation, Lemma 2.2, important for subsequent applications, is that the space of C-structures is itself a contractible space. Employing the lemma, one is able to glue together local C-structures in a neighbourhood of each point $b \in B$ to obtain a global C-structure over B, with respect to which one constructs the map h in the above Riemann integral.

§1. Contractible Spaces of Surrounding Loops

Let $\rho\colon X \to \mathbf{R}^q$, $q \geq 1$, be continuous where X is path connected. Denote by $\operatorname{Conv} X$ the convex hull of the ρ-image of X in $\mathbf{R}^q$: the subset of $\mathbf{R}^q$ consisting of all convex linear combinations $\sum_{i=1}^N p_i \rho(x_i)$, $N \geq 1$, where $x_i \in X$ and $\sum_{i=1}^N p_i = 1$, $p_i \in [0,1]$, $i = 1, 2, \ldots, N$. (A convenient reference for the principal properties of convex hulls is Hörmander [23]). Let $\operatorname{IntConv} X$ denote the *interior* (possibly empty) of $\operatorname{Conv} X$ in $\mathbf{R}^q$. In case X is not path connected and $x \subset X$, we employ the notation $\operatorname{Conv}(X,x)$, respectively $\operatorname{IntConv}(X,x)$, to denote the convex hull, respectively the interior of the convex hull, of the ρ-image of the path component in X to which x belongs. Note that in case $\rho\colon X \to \mathbf{R}^q$ is the inclusion of an open path connected subspace then the convex hull of X coincides with the interior of

D. Spring, *Convex Integration Theory: Solutions to the h-principle in geometry and topology*, Modern Birkhäuser classics, DOI 10.1007/978-3-0348-0060-0_2, © Springer Basel AG 2010

the convex hull of X. Indeed if $z \in \mathrm{Conv}\, X$ then there is an integer k, $0 \le k \le q$, and an affine k-simplex σ_k which contains z and whose vertices lie in X. Since X is open one easily constructs an affine q-simplex Δ_q whose vertices lie in X and which contains σ_k in its interior. More generally, if $\rho\colon X \to \mathbf{R}^q$ is a microfibration (V Lemma 5.7) then $\mathrm{Conv}(X, x) = \mathrm{IntConv}(X, x)$ for all $x \in X$.

Ample Sets. A continuous map $\rho\colon X \to \mathbf{R}^q$ is *ample* if for all $x \in X$, $\mathrm{Conv}(X, x) = \mathbf{R}^q$. That is, X is ample if the convex hull of the ρ-image of each path component of X is all of $\mathbf{R}^q$. Examples of open ample sets $X \subset \mathbf{R}^q$ which occur in the theory are $X = \mathbf{R}^q \setminus L$, where L is a smooth (or stratified) submanifold of codimension ≥ 2.

Nowhere Flat Sets. For the purposes of applying convex integration theory to the construction of solutions of non-linear systems of partial differential equations, the following affine notion is also required. A subset X in $\mathbf{R}^q$ is *nowhere flat* if for each affine $(q-1)$-dimensional hyperplane H of $\mathbf{R}^q$, the intersection $H \cap X$ is nowhere dense in X. For example, the unit sphere in $\mathbf{R}^q$ (in the Euclidean metric) is nowhere flat in $\mathbf{R}^q$, $q \ge 2$. Evidently, any affine subspace of $\mathbf{R}^q$ is flat i.e., not nowhere flat.

In case $X \subset \mathbf{R}^q$ is nowhere flat then $X \subset \bigcup_{x \in X} \overline{\mathrm{IntConv}(X, x)}$ in $\mathbf{R}^q$. Indeed, each $x \in X$ is a vertex of a non-degenerate affine q-simplex in $\mathbf{R}^q$, all of whose vertices are points in X.

It is useful for what follows to remark the obvious connection between Riemann integration and convex hulls. Let $f\colon [0,1] \to \mathbf{R}^q$ be continuous and let $X = \mathrm{im}\, f \subset \mathbf{R}^q$. Evidently a Riemann sum for the integral $\int_0^1 f(t)\, dt$ is a point of $\mathrm{Conv}\, X$. Furthermore, since X is compact, $\mathrm{Conv}\, X$ is compact and hence the integral $\int_0^1 f(t)\, dt \in \mathrm{Conv}\, X$. Geometrically, in case $[0,1]$ is parametrized by arc-length of the C^1-regular curve f, $\int_0^1 f(t)\, dt$ is the barycentre of X.

Surrounding Loops. Let $\rho\colon X \to \mathbf{R}^q$ be continuous; fix $x \in X$. Suppose $g\colon [0,1] \to X$ is a continuous loop at x: $g(0) = g(1) = x$. A point $z \in \mathrm{Conv}(X, x)$ is *surrounded*, respectively *strictly surrounded*, by the loop g if z lies in the convex hull, respectively the interior of the convex hull, of the composed path $\rho \circ g(t) \in \mathbf{R}^q$, $t \in [0,1]$. The main interest is in loops g in X that surround z and which are homotopically trivial i.e. there is a base point preserving homotopy of loops in X which connects g to the constant path C_x. Such loops are easily constructed as follows. Let $z \in \mathrm{Conv}(X, x)$, respectively $z \in \mathrm{IntConv}(X, x)$. There is a loop λ in X, based at x, such that $\mathrm{im}\, \rho \circ \lambda$ contains the vertices of a simplex that contains z, respectively contains the vertices of a q-simplex in $\mathbf{R}^q$ whose interior contains z. Then the product $g = \lambda * \lambda^{-1}$ is a contractible loop that surrounds z, respectively strictly surrounds z.

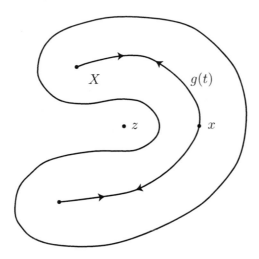

FIGURE 2.1

For each $z \in \mathrm{Conv}(X, x)$ let $X_x^z \subset C^0([0,1], X) \times C^0([0,1]^2, X)$ be the subspace, in the compact-open topology, of pairs of contractible loops at x which surround z, together with a contraction of the loop to C_x: pairs (g, G) where $g \colon [0,1] \to X$, $g(0) = g(1) = x$, g surrounds z and $G \colon [0,1]^2 \to X$ is a base point preserving homotopy which contracts the loop g to the constant path C_x. Thus, for all $(t, s) \in [0,1]^2$,

$$G(t, 0) = x; \; G(t, 1) = g(t); \; G(0, s) = G(1, s) = x. \tag{2.1}$$

Analogously, $\mathrm{int}\, X_x^z$ is the space of pairs (g, G) as above where g strictly surrounds z.

The space $\mathrm{int}\, X_x^z$ is an example of a space of C-structures (cf.§2) on an $\mathbf{R}^q$-bundle, in this case the bundle $\mathbf{R}^q \to *$ over a point, with respect to $z \in \mathrm{IntConv}(X, x)$, $x \in X$. Note that the loop g of the pair (g, G) is included for convenience; g is the map at time 1 of the homotopy G. The following elementary result is essential for the proof of the existence of C-structures in §2, Proposition 2.3.

Lemma 2.1. *For all* $(x, z) \in X \times \mathrm{Conv}(X, x)$, *respectively all* $(x, z) \in X \times \mathrm{IntConv}(X, x)$, *the space* X_x^z, *respectively* $\mathrm{int}\, X_x^z$ *is contractible.*

Proof. We prove the lemma for the space X_x^z; the proof for the space $\mathrm{int}\, X_x^z$ is similar and is omitted. We show that X_x^z deformation retracts in a canonical way to each point in X_x^z: there is a continuous map $\mathcal{D} \colon X_x^z \times X_x^z \times [0,1] \to X_x^z$ such that for all $u, v \in X_x^z$, the corresponding map $\mathcal{D}_u \colon X_x^z \times [0,1] \to X_x^z$,

$\mathcal{D}_u(v,t) = \mathcal{D}(u,v,t)$, satisfies the property:

$$\mathcal{D}_u(v,1) = v; \ \mathcal{D}_u(v,0) = u. \tag{2.2}$$

Thus for each $u \in X_x^z$ the map $\mathcal{D}_u$ is a deformation retract of X_x^z to u which depends continuously on $u \in X_x^z$.

Briefly, the map $\mathcal{D}$ is constructed canonically as follows. Let $u = (g,G), v = (h,H) \in X_x^z$ and let $*$ denote the product operator on the space of loops. Employing the contractions H, G one constructs a canonical homotopy of closed loops in X which connects h, g, and which, at each stage, consists of closed loops which surround z:

$$h \sim h * C_x \sim h * g \sim C_x * g \sim g. \tag{2.3}$$

Furthermore, again employing the contractions H, G, each of the surrounding loops in the above homotopy is canonically contractible to the constant path C_x. Specifically, with respect to the two middle homotopies in (2.3), the product of paths in the t variable, $H(t,u)*G(t,su), 0 \le u \le 1$, is a homotopy which connects C_x $(u = 0)$ to the loops which surround z, $h(t) * G(t,s), 0 \le s \le 1$, $(u = 1)$. Similarly $H(t,(1-s)u) * G(t,u), 0 \le u \le 1$, is a homotopy which connects C_x $(u = 0)$ to the loops which surround z, $H(t,(1-s)) * g(t), 0 \le s \le 1$, $(u = 1)$. With respect to the two end homotopies of (2.3), standard homotopies which connect $h, h * C_x$, respectively which connect $C_x, C_x * g$, are employed to construct a homotopy of contractions which connects $H_s, H_s * C_x$, respectively which connects $C_x * G_s, G_s, 0 \le s \le 1$. Precise details are left to the reader.
□

§2. C-Structures for Relations in Affine Bundles

Let $p: E \to B$ be an affine $\mathbf{R}^q$-bundle over a second-countable paracompact base space B (for example, a manifold B). In the applications $p: E \to B$ occurs mainly as the restriction over a submanifold $B \subset X^\perp$ of a naturally defined affine $\mathbf{R}^q$-bundle $p_\perp^r: X^{(r)} \to X^\perp$ associated to a smooth fiber bundle $p: X \to V$ and to a codimension 1 tangent hyperplane field τ on the base space V (cf. Chapter VI). The affine structure means that the transition functions of the bundle take their values in the group of affine transformations of $\mathbf{R}^q$. $\Gamma(E)$ denotes the space of continuous sections of the bundle E, in the compact-open topology. Let $\rho: \mathcal{R} \to E$ be a continuous map referred to as a *relation* over E. For example $\mathcal{R} \subset E$ and ρ is the inclusion map. $\Gamma(\mathcal{R})$ denotes the space of continuous sections of the map $p \circ \rho: \mathcal{R} \to B$ in the compact-open topology. If $\alpha \in \Gamma(\mathcal{R})$ then $\rho \circ \alpha \in \Gamma(E)$. $\Gamma_K(E), \Gamma_K(\mathcal{R})$ are the corresponding spaces of sections over the subspace $K \subset B$.

For each $b \in B$ let $E_b = p^{-1}(b)$, the $\mathbf{R}^q$-fiber over the base point b; $\mathcal{R}_b = \mathcal{R} \cap \rho^{-1}(E_b)$, the subspace of $\mathcal{R}$ lying over b (possibly empty). If $a \in \mathcal{R}_b$ then Conv$(\mathcal{R}_b, a)$ denotes the convex hull (in the fiber E_b) of the ρ-image of the path component of $\mathcal{R}_b$ to which a belongs (cf. §1).

Ample Relations. A relation $\rho \colon \mathcal{R} \to E$ is *ample* if for all pairs $(b, a) \in B \times \mathcal{R}_b$, Conv$(\mathcal{R}_b, a) = E_b$ (by convention this includes the case $\mathcal{R}_b$ is empty).

C-structures. Let $\rho \colon \mathcal{R} \to E$ be a relation over E. Suppose $f \in \Gamma(E)$, $\beta \in \Gamma(\mathcal{R})$ satisfy the property that for all $b \in B$,

$$f(b) \in \mathrm{IntConv}(\mathcal{R}_b, \beta(b)).$$

A *C-structure* $(C = \text{contractible})$ over a subset $K \subset B$, with respect to f, β, is a pair consisting of a contractible loop of sections in $\Gamma_K(E)$, based at β_K, which fiberwise strictly surrounds the section f_K together with a contraction of the loop to β_K: a pair (g, G) where $g \colon [0, 1] \to \Gamma_K(\mathcal{R})$ is continuous, $g(0) = g(1) = \beta_K$ (the restriction of β to K) such that for all $b \in K$ the path $g_b \colon [0, 1] \to \mathcal{R}_b$, $g_b(t) = g(t, b)$, strictly surrounds $f(b)$; $G \colon [0, 1]^2 \to \Gamma_K(\mathcal{R})$ is a (fiberwise) base point preserving contraction of g to β_K: for all $t, s \in [0, 1]$,

$$G(t, 1) = g(t); \ G(t, 0) = \beta_K; \ G(0, s) = G(1, s) = \beta_K. \tag{2.4}$$

The set of all C-structures (g, G) over K, with respect to f, β, is topologized as a subspace of $C^0\left([0, 1], \Gamma_K(\mathcal{R})\right) \times C^0\left([0, 1]^2, \Gamma_K(\mathcal{R})\right)$, in the compact-open topology. In this bundle theoretic context, if $\rho \colon X \to \mathbf{R}^q$ is continuous then the space int X_x^z (§1, Surrounding Paths) is precisely the space of C-structures with respect to $x \in X$, $z \in \mathrm{IntConv}(X, x)$, for the trivial bundle over a point $\mathbf{R}^q \to *$. The proof of Lemma 2.1 obviously admits a parametric version which is stated as the following lemma.

Lemma 2.2. *For each $K \subset B$, the space of C-structures over K with respect to f, β is contractible.*

Proof. The proof is analogous to the proof of Lemma 2.1, for which all surrounding maps and contracting homotopies carry an additional K-space of parameters. Indeed from Lemma 2.1, int X_x^z is canonically contractible, via the deformation retract $\mathcal{D}$, to each point of int X_x^z. To prove the lemma one applies $\mathcal{D}$ fiberwise to the relation $\mathcal{R}$ over E: for each $b \in K$ the deformation retract $\mathcal{D}$ of int X_x^z is applied to the case $X = \mathcal{R}_b$ with respect to $z = f(b) \in E_b$, $x = \beta(b) \in \mathcal{R}_b$. In this way $\mathcal{D}$ induces a canonical deformation retract $\mathcal{D}_K$ of the space of C-structures over K with respect to f, β to each C-structure in the space. A useful, though schematic, description of the deformation $\mathcal{D}_K$ is described succinctly as follows (cf. (2.3)). Full details are left to the reader. If (h, H), (g, G), are C-structures over K with respect to f, β then,

$$h \sim h * C_\beta \sim h * g \sim C_\beta * g \sim g.$$

where $C_\beta \colon [0,1] \to \Gamma_K(\mathcal{R})$ is the constant section β. In particular, let (g_0, G_0), (g_1, G_1) be C-structures over K with respect to f, β. There is a homotopy of C-structures (g_t, G_t), $t \in [0,1]$, over K with respect to f, β which connects (g_0, G_0), (g_1, G_1). $\qquad\qquad\square$

The principal result in this section is the following proposition which establishes the existence of C-structures over each $K \subset B$ in case $\mathcal{R} \subset E$ is open. In case $\mathcal{R} \subset E$ is open then for each $b \in B$ and $a \in \mathcal{R}_b$, $\mathrm{Conv}(\mathcal{R}_b, a) = \mathrm{IntConv}(\mathcal{R}_b, a)$. Since $f(b) \in \mathrm{IntConv}(\mathcal{R}_b, \beta(b))$, it follows that pointwise there is a C-structure over each point $b \in B$ with respect to $f(b) \in E_b$, $\beta(b) \in \mathcal{R}_b$ i.e. there is a contractible loop g in $\mathcal{R}_b$ based at $\beta(b)$ that strictly surrounds $f(b)$ in the fiber E_b. The following proposition constructs these strictly surrounding contractible loops continuously over the base space B.

Proposition 2.3. *Let $\mathcal{R} \subset E$ be open. Suppose $\beta \in \Gamma(\mathcal{R})$ (a fiberwise base point map) and $f \in \Gamma(E)$ satisfy the property that for all $b \in B$, $f(b) \in \mathrm{Conv}(\mathcal{R}_b, \beta(b))$. There is a C-structure (ψ, H) globally defined over B with respect to f, β. Explicitly, there is a continuous map $\psi \colon [0,1] \to \Gamma(\mathcal{R})$, $\psi(0) = \psi(1) = \beta$, such that for all $b \in B$, the path $\psi_b \colon [0,1] \to \mathcal{R}_b$, $\psi_b(t) = \psi(t, b)$, strictly surrounds $f(b)$. Furthermore, there is a base point preserving homotopy $H \colon [0,1]^2 \to \Gamma(\mathcal{R})$ which contracts ψ to the constant path of sections C_β: For all $t, s \in [0,1]^2$,*

$$H(t, 1) = \psi(t); \quad H(t, 0) = \beta; \quad H(0, s) = H(1, s) = \beta.$$

Proof. The local existence of C-structures in a neighbourhood of each point of B is proved in Lemma 2.4 below. The contractibility of the space of C-structures, Lemma 2.2, is employed in an essential way to patch together these local constructions of C-structures to obtain a globally defined C-structure over all of B. The details are as follows.

Throughout this text we employ the convenient notation, due to Gromov [18]: if $Z \subset Y$ then $\mathfrak{Op}\, Z$ denotes an open neighbourhood (i.e., an "opening") of Z in Y. $\mathfrak{Op}\, Z$ is employed in the sense of a germs: with no change of notation, $\mathfrak{Op}\, Z$ may be replaced by a smaller neighbourhood, if necessary, during the course of a proof.

Lemma 2.4. *For all $b \in B$, each C-structure over $\{b\}$ with respect to f, β extends to a C-structure over $\mathfrak{Op}\, b$ with respect to f, β.*

Proof. Fix $b \in B$. By hypothesis, $f(b) \in \text{IntConv}(\mathcal{R}_b, \beta(b))$. Let (g, G) be a C-structure over b with respect to $f(b), \beta(b)$ (as noted above C-structures exist over each point $b \in B$). Employing the hypothesis that $\mathcal{R} \subset E$ is open, one extends as follows the C-structure (g, G) over the point b to a C-structure (h, H) with respect to f, β, over a suitably small neighbourhood $\mathfrak{Op}\, b$. Since we are working locally near b we assume that the bundle $p \colon E \to B$ is the product bundle: $p \colon E = B \times \mathbf{R}^q \to B$.

Let $L = \operatorname{im} g = \operatorname{im} G$, a compact set in $\mathcal{R}_b$. Since $\mathcal{R}$ is open there is a neighbourhood $\mathfrak{Op}\, b$ such that $\mathfrak{Op}\, b \times L \subset \mathcal{R}$. For each $y \in \mathfrak{Op}\, b$ let $(g(y), G(y))$ be the translate of the C-structure (g, G) over $\{b\}$ to the fiber $\mathcal{R}_y$. For $\mathfrak{Op}\, b$ sufficiently small, for all $y \in \mathfrak{Op}\, b$ the loop $g(y)$ strictly surrounds $f(y)$, and also the line segment $\ell(y)$ in E_y that joins $\beta(y)$ to $(y, \beta(b))$ satisfies $\ell(y) \subset \mathcal{R}_b$.

Let $h \colon [0, 1] \to \Gamma_{\mathfrak{Op}\, b}(\mathcal{R})$, $h(0) = h(1) = \beta_{\mathfrak{Op}\, b}$, be the family of loops obtained by conjugating the translated loop $g(y)$ with the parametrized line segment $\ell(y)$: for all $y \in \mathfrak{Op}\, b$, $t \in [0, 1]$,

$$h(t, y) \;=\; \ell(y)(t) * g(y)(t) * \ell(y)^{-1}(t), \tag{2.5}$$

Similarly, let $H_0 \colon [0, 1]^2 \to \Gamma_{\mathfrak{Op}\, b}$ be the contraction of h obtained by conjugating $G(y)$ with $\ell(y)$: for all $y \in \mathfrak{Op}\, b$, $(t, s) \in [0, 1]^2$,

$$H_0(t, s, y) \;=\; \ell(y)(t) * G(y)(t, s) * \ell(y)^{-1}(t), \tag{2.6}$$

followed by a contraction of the segment $\ell(y)$ to the base point $\beta(y)$. Evidently the pair (h, H) is a C-structure over $\mathfrak{Op}\, b$ with respect to f, β.   □

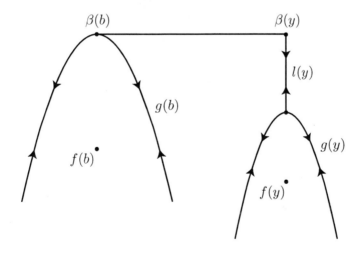

FIGURE 2.2

Employing Lemma 2.2, one patches together the local C-structures constructed above over $\mathfrak{Op}\, b$ for all $b \in B$. The inductive step for this process is as follows.

Lemma 2.5. *Let* $K, L \subset B$ *be closed and let* (g_0, G_0), (g_1, G_1) *be* C-*structures over* $\mathfrak{Op}\, K$, $\mathfrak{Op}\, L$ *respectively with respect to* f, β. *There is a* C-*structure* (h, H) *over* $\mathfrak{Op}\,(K \cup L)$ *with respect to* f, β *such that* $(h, H) = (g_0, G_0)$ *over (a smaller)* $\mathfrak{Op}_1 K$ *and* $(h, H) = (g_1, G_1)$ *over* $\mathfrak{Op}\,(L \setminus \mathfrak{Op}\,(K \cap L))$.

Proof. In case $K \cap L = \emptyset$ then one may assume $\mathfrak{Op}\, K \cap \mathfrak{Op}\, L = \emptyset$ in which case the lemma is trivial. Suppose now $M = K \cap L \neq \emptyset$. Applying Lemma 2.2, there is a homotopy of C-structures (g_s, G_s), $s \in [0, 1]$, over $\mathfrak{Op}\, M$ with respect to f, β which connects the C-structures (g_0, G_0) to (g_1, G_1) on $\mathfrak{Op}\, M$. Furthermore since $K \cap (L \setminus \mathfrak{Op}\, M) = \emptyset$, there are disjoint neighbourhoods $\mathfrak{Op}_1 K$, $\mathfrak{Op}_2(L \setminus \mathfrak{Op}\, M)$ in the paracompact space B.

Let $\lambda \colon B \to [0, 1]$ be continuous such that $\lambda = 0$ on $\mathfrak{Op}_3 K$ and $\lambda = 1$ on $B \setminus \mathfrak{Op}_1 K$, where $\overline{\mathfrak{Op}_3 K} \subset \mathfrak{Op}_1 K$. Let $H \colon [0, 1]^2 \to \Gamma(\mathcal{R})$ be the homotopy such that: (i) $H = G_0$ on $\mathfrak{Op}_3 K$; (ii) $H = G_1$ on $\mathfrak{Op}_2(L \setminus \mathfrak{Op}\, M)$; (iii) for all $(s, t, b) \in [0, 1]^2 \times \mathfrak{Op}\, M$, $H(s, t, b) = G_{\lambda(b)}(s, t, b)$. One verifies that H is well defined hence continuous. Setting $h(t, b) = H(t, 1, b)$ it follows that (h, H) is a C-structure over $\mathfrak{Op}\,(K \cup L)$. $\square$

Returning to the proof of the proposition, according to Lemma 2.4 there is a C-structure over $\mathfrak{Op}\, b$ for each $b \in B$. Hence there are countable locally finite open covers $\{W_i\}$, $\{U_i\}$ of the base space B such that for all i, $\overline{W}_i \subset U_i$, and there is a C-structure over U_i. Inductively on n, let $K = \bigcup_{i=1}^{i=n} \overline{W}_i$, a closed set in B, and suppose (ψ_n, G_n) is a C-structure over $\mathfrak{Op}\, K$. Applying Lemma 2.5 to the closed set K and to $\overline{W}_{n+1}$, there is a C-structure (ψ_{n+1}, G_{n+1}) over $\mathfrak{Op}\,(K \cup \overline{W}_{n+1})$ such that $(\psi_{n+1}, G_{n+1}) = (\psi_n, G_n)$ over $\mathfrak{Op}\, K$. Since the covers $\{W_i\}$, $\{U_i\}$ are locally finite, it follows that the maps,

$$\psi = \lim_{n \to \infty} \psi_n \colon [0, 1] \to \Gamma(\mathcal{R}); \quad G = \lim_{n \to \infty} G_n \colon [0, 1]^2 \to \Gamma(\mathcal{R}),$$

are well-defined and continuous. Consequently, (ψ, G) is a C-structure over B with respect to $f \in \Gamma(E)$, $\beta \in \Gamma(\mathcal{R})$. $\square$

Proposition 2.3 admits several refinements which are required for the full development of convex integration theory. These refinements are stated below as a series of complements to the above proposition. In all these complements, $\mathcal{R} \subset E$ is open.

Complement 2.6 (Relative Theorem). *Let* N *be a neighbourhood of* $\beta(B)$ *in* E. *Suppose there is a closed subspace* K *of* B *such that* $f = \beta$ *on* K. *There is a*

C-structure (h, H) *over* B *with respect to* f, β, *such that over* $\mathfrak{Op}\, K$, *the image of* H *lies in* N:

$$H\left([0,1]^2 \times \mathfrak{Op}\, K\right) \subset N.$$

Proof. The point here is that in case $b \in K$ the *C*-structure (h, H) that is constructed over $\mathfrak{Op}\, b$ in Lemma 2.4 may be chosen to satisfy the additional property that $\operatorname{im} H \subset N$. Passing to a locally finite subcover of the open cover $\{\mathfrak{Op}\, b\}_{b \in K}$ of K, the inductive proof procedure of Proposition 2.3 applies to construct a *C*-structure (h_1, H_1) over $\mathfrak{Op}\, K$ such that the image of H_1 lies in N. Applying Lemma 2.5 to the *C*-structure (h_1, H_1) over $\mathfrak{Op}\, K$ and to any *C*-structure over B one obtains a *C*-structure (h, H) over B which equals (h_1, H_1) over (a smaller) $\mathfrak{Op}\, K$. $\qquad\square$

Remark 2.7. If one replaces "strictly surrounds" by "surrounds" in the definition of a *C*-structure then the relative theorem above can be improved to state that, in case $f = \beta$ on $\mathfrak{Op}\, K$, then one can choose H to be the constant homotopy equal to the base point section β over (a smaller) $\mathfrak{Op}\, K$. However the strictly surrounding property for *C*-structures is indispensable for the general theory, for example in the proof of the Integral Representation theorem below.

Suppose $p \colon E \to B$ is a smooth (i.e., C^∞) affine $\mathbf{R}^q$-bundle. Since $\mathcal{R}$ is open in E and since the "strictly surrounds" property of a *C*-structure is an open condition, the standard smooth approximation theorems apply to prove the following smooth refinement of Proposition 2.3.

Complement 2.8 (C^∞-Structures). *Suppose, in addition, $p \colon E \to B$ is a smooth affine $\mathbf{R}^q$-bundle. Let $f \in \Gamma(E)$, $\beta \in \Gamma(\mathcal{R})$ satisfy the hypothesis of Proposition 2.3, where, in addition, β is a smooth section. There is a C-structure (h, H) over B, with respect to f, β, such that H (and hence h) is a smooth map i.e., the evaluation map $H \colon [0,1]^2 \times B \to E$ is a smooth map.*

Complement 2.9 (Parameters). *Let P be a compact Hausdorff space (a parameter space). Let $f \colon P \to \Gamma(E)$, $\beta \colon P \to \Gamma(\mathcal{R})$ be continuous maps such that for all $(p, b) \in P \times B$, the following convex hull property obtains:*

$$f(p, b) \in \operatorname{IntConv}(\mathcal{R}_b, \beta(p, b)). \tag{2.7}$$

There is P-parameter family of C-structures (h, H) over B with respect to f, β. That is, there are continuous maps,

$$h \colon P \times [0,1] \to \Gamma(\mathcal{R}); \quad H \colon P \times [0,1]^2 \to \Gamma(\mathcal{R}),$$

such that, for each $p \in P$, (h_p, H_p) is a C-structure over B with respect to the sections $f(p), \beta(p)$. (One employs the notation, $h_p(t) = h(p, t)$; $H_p(t, s) = H(p, t, s)$.)

Proof. Let id $\times p\colon P \times E \to P \times B$ be the pullback of the affine bundle $p\colon E \to B$ along the projection map onto the second factor, $\pi\colon P \times B \to B$. Then $P \times \mathcal{R}$, the pullback of $\mathcal{R}$, is open in $P \times E$, and the maps f, β induce obvious sections (same notation) in $\Gamma(P \times E)$, respectively $\Gamma(P \times \mathcal{R})$. Employing (2.7), the hypothesis of Proposition 2.3 is satisfied with respect to the maps f, β above, from which the Complement follows. $\qquad\square$

Corollary 2.10 (Ample Relations). *Let $\mathcal{R}$ be open and ample in E, and suppose $\mathcal{R}$ admits a section $\beta \in \Gamma(\mathcal{R})$. For each section $f \in \Gamma(E)$ there is a C-structure (h, H) globally defined over B, with respect to f, β.*

The point here is that the convex hull hypothesis of Proposition 2.3 is automatically satisfied for any $f \in \Gamma(E)$. Indeed, since $\mathcal{R}$ is ample in E, it follows by definition that, for each $b \in B$, the convex hull of each path component of $\mathcal{R}_b$ is E_b.

§3. The Integral Representation Theorem

Let $X \subset \mathbf{R}^q$, $x \in X$, and let $z \in \mathrm{IntConv}(X, x)$. The space $\mathrm{int}\, X_x^z$ is precisely the space of C-structures with respect to $z \in \mathbf{R}^q$, $x \in X$, for the trivial bundle over a point, $\mathbf{R}^q \to *$. In what follows, C-structures in $\mathrm{int}\, X_x^z$ are employed to obtain a representation of the point z as a Riemann integral whose integrand is a function with values in X. The main result of this section, the Integral Representation Theorem 2.12, establishes this Riemann integral representation continuously over the base space, in the context of an affine $\mathbf{R}^q$-bundle.

Proposition 2.11. *Let $X \in \mathbf{R}^q$, $x \in X$, and let $z \in \mathrm{IntConv}(X, x)$. Each C-structure $(g, G) \in \mathrm{int}\, X_x^z$ can be reparametrized to a C-structure (h, H) such that $z = \int_0^1 h(t)\, dt$.*

Proof. Let $(g, G) \in \mathrm{int}\, X_x^z$ be a C-structure and let $0 = s_0 < s_1 \cdots < s_{n+1} = 1$ be a partition of the interval $[0, 1]$ such that z is contained in the interior of the convex hull of the points $g(s_i)$, $1 \leq i \leq n$. For each i, $1 \leq i \leq n$, let $d\mu_i$ be a positive measure on $[0, 1]$ such that $\int_0^1 d\mu_i = 1$, and $d\mu_i \approx \delta(s - s_i)$. ($\delta$ is the Dirac delta function). For example, $d\mu_i$ is represented by a positive, continuous density function f_i on $[0, 1]$ such that $\int_0^1 f_i(s)\, ds = 1$, and $f_i(s)$ is concentrated near s_i, $1 \leq i \leq n$.

Let $b_i = \int_0^1 g\, d\mu_i$, $1 \leq i \leq n$. Let $\epsilon > 0$. If each $d\mu_i$ is a sufficiently close approximation to $\delta(s - s_i)$ then, with respect to a norm $\|\ \|$ on $\mathbf{R}^q$,

$$\|b_i - g(s_i)\| < \epsilon, \quad 1 \leq i \leq n, \tag{2.8}$$

Since z is in the interior of the convex hull of the points $g(s_i)$, $1 \leq i \leq n$, it follows from (2.8) that for $\epsilon > 0$ sufficiently small, $\mathfrak{Op}\, z$ is contained in the convex hull of the points b_i, $1 \leq i \leq n$. One may therefore assume,

$$z = \sum_{i=0}^{n} p_i \, b_i, \quad \text{where} \sum_{i=1}^{n} p_i = 1, \; p_i \in [0,1], \; 1 \leq i \leq n. \tag{2.9}$$

We remark here that one may choose $n = q+1$ and $g(s_i) \in \mathbf{R}^q$, $1 \leq i \leq q+1$, to be the vertices of a $q+1$-simplex, from which it follows that the above barycentric coordinates $(p_i)_{1 \leq i \leq q+1}$ are unique and strictly positive. This is important for proving the continuity of these coordinates in the parametric (bundle) version Theorem 2.12 below.

The continuous measure $d\mu = \sum_{i=1}^{n} p_i \, d\mu_i$ on the interval $[0,1]$ is positive, $\int_0^1 d\mu = 1$, and with respect to $d\mu$ one has the following integral representation:

$$\int_0^1 g \, d\mu = \sum_{i=1}^{n} p_i \int_0^1 g \, d\mu_i$$
$$= \sum_{i=1}^{n} p_i \, b_i = z. \tag{2.10}$$

Employing a simple change of coordinates, one obtains the integral representation (2.10) with respect to Lebesgue measure on $[0,1]$. Explicitly, let $\lambda(t) = \int_0^t d\mu$. Evidently, $\lambda(0) = 0$, $\lambda(1) = 1$, and $d\lambda/dt > 0$ on $[0,1]$. Define $h = g \circ \lambda^{-1}$. Clearly $h(0) = h(1) = x$; h strictly surrounds z and, employing the change of coordinates $s = \lambda(t)$, it follows from (2.9) that,

$$\int_0^1 h(s) \, ds = \int_0^1 g \circ \lambda^{-1}(s) \, ds = \int_0^1 g \, d\mu = z. \tag{2.11}$$

Let $H \colon [0,1]^2 \to X$ be the reparametrization, $H(t,s) = G(\lambda^{-1}(t), s)$. Hence (h, H) is a C-structure in int X_x^z for which the integral representation (2.11) obtains. $\qquad\square$

Theorem 2.12 (Integral Representation). *Let $p \colon E \to B$ be an affine $\mathbf{R}^q$-bundle over a second-countable paracompact space B. Let $\mathcal{R} \subset E$ be open and suppose $\beta \in \Gamma(\mathcal{R})$ and $f \in \Gamma(E)$ satisfy the property that, for all $b \in B$,*

$$f(b) \in \mathrm{Conv}(\mathcal{R}_b, \beta(b)).$$

Each C-structure (g, G) over B with respect to f, β can be reparametrized to a C-structure (h, H) such that for all $b \in B$ (recall $h \colon [0,1] \to \Gamma(\mathcal{R})$), $f(b) = \int_0^1 h(t, b) \, dt$.

Proof. Applying Proposition 2.3, there exists a C-structure (g, G) over B with respect to the sections $f \in \Gamma(E)$, $\beta \in \Gamma(\mathcal{R})$. The proof consists of suitably reparametrizing a C-structure (g, G) to obtain a C-structure (h, H) over B for which the above integral representation of f obtains. The map $g \colon [0, 1] \to \Gamma(\mathcal{R})$ satisfies the property that $g(0) = g(1) = \beta$, and for each $b \in B$ the corresponding path $g_b \colon [0, 1] \to \mathcal{R}_b$ strictly surrounds $f(b)$ in the fiber E_b. Consequently (since "strictly surrounding" is an open condition), for each $b \in B$ there is a neighbourhood U of b and a partition $0 < s_1 < s_2 \cdots < s_{q+1} < 1$ of the interval $[0, 1]$, such that for all $y \in U$ the sequence of points $g_y(s_1), g_y(s_2), \ldots, g_y(s_{q+1})$, spans an affine q-simplex $\Delta(y)$ in the fiber E_y, and $f(y)$ is an interior point of $\Delta(y)$. In particular the barycentric coordinates of $f(y) \in \Delta(y)$ are strictly positive continuous functions on U (cf. the remark following (2.9) above).

Recall, Proposition 2.11, the positive, continuous measures $d\mu_i$ on $[0, 1]$, $\int_0^1 d\mu_i = 1$, and such that $d\mu_i \approx \delta(s - s_i)$, $1 \leq i \leq q + 1$. For each $y \in U$ let $b_i(y) = \int_0^1 g_y(s) \, d\mu_i$, $1 \leq i \leq q + 1$. If $d\mu_i$ is a sufficiently close approximation to $\delta(s - s_i)$, $1 \leq i \leq q + 1$, and if U is a sufficiently small neighbourhood of b, then, for all $y \in U$, the sequence of points $b_1(y), b_2(y), \ldots, b_{q+1}(y)$ also spans an affine q-simplex $\Delta'(y)$ in the fiber E_y, and $f(y)$ is an interior point of $\Delta'(y)$. In particular, the barycentric coordinates of $f(y)$ in the q-simplex $\Delta'(y)$ are strictly positive continuous functions of $y \in U$. Consequently, there is a neighbourhood $W \equiv W(b)$ of b, $\overline{W} \subset U$, and globally defined continuous functions, compactly supported in U, $p_i \colon B \to [0, 1]$, $1 \leq i \leq q + 1$, such that the sequence of functions (p_i) on W are the barycentric coordinates of the section f: for all $y \in W$,

$$f(y) = \sum_{i=1}^{q+1} p_i(y) \, b_i(y); \quad \sum_{i=1}^{q+1} p_i(y) = 1. \qquad (2.12)$$

Let $\{W_j\}$ be a countable locally finite subcover of the above open cover $\{W(b)\}_{b \in B}$ of the base space B. Thus, for each index j, (2.12) applies to the section f over W_j. Explicitly, in the above notation (the index j corresponds to W_j) for all $y \in W_j$,

$$f(y) = \sum_{i=1}^{q+1} p_i^j(y) \, b_i^j(y); \quad \sum_{i=1}^{q+1} p_i^j(y) = 1.$$

Let $\{q_j \colon B \to [0, 1]\}_{j \geq 1}$ be a partition of unity subordinate to the cover $\{W_j\}$. Also, for each index j, one employs the notation, $d\mu_i^j$, $1 \leq i \leq q + 1$, to denote the measures above on $[0, 1]$, with respect to the open set W_j. Thus $b_i^j(y) = \int_0^1 g(s, y) \, d\mu_i^j$, $1 \leq i \leq q + 1$. Let $d\mu$ be the B-parameter family of measures on the interval $[0, 1]$ defined as follows. For each $b \in B$,

$$d\mu(b) = \sum_{j=1}^{\infty} \sum_{i=1}^{q+1} q_j(b) \, p_i^j(b) \, d\mu_i^j. \qquad (2.13)$$

Evidently, $d\mu(b)$ is a positive measure on $[0,1]$, continuous in the parameter $b \in B$, such that for all $b \in B$, $\int_0^1 d\mu(b) = 1$. Furthermore, employing (2.13), $f \in \Gamma(E)$ has the following integral representation with respect to the B-parameter of measures $d\mu(b)$. For each $y \in B$,

$$
\begin{aligned}
\int_0^1 g(s,y) \, d\mu(y) &= \sum_{j=1}^{\infty} q_j(y) \sum_{i=1}^{q+1} p_i^j(y) \int_0^1 g(s,y) \, d\mu_i^j \\
&= \sum_{j=1}^{\infty} q_j(y) \sum_{i=1}^{q+1} p_i^j(y) \, b_i^j(y) = \sum_{j=1}^{\infty} q_j(y) \, f(y) = f(y).
\end{aligned}
\tag{2.14}
$$

We now change coordinates to obtain an integral representation for the section f with respect to Lebesgue measure. Let $\lambda \colon [0,1] \times B \to [0,1]$, be the continuous function, $\lambda(t,b) = \int_0^t d\mu(b)$. Thus for all $b \in B$, $\lambda(0,b) = 0$, $\lambda(1,b) = 1$, and the derivative $\partial\lambda/\partial t \colon [0,1] \times B \to [0,1]$ is a continuous, positive function. Consequently, for each $b \in B$, the inverse function $\lambda^{-1}(t,b)$ exists and $\lambda^{-1} \colon [0,1] \times B \to [0,1]$ is a continuous function.

Let $h \colon [0,1] \to \Gamma(\mathcal{R})$ be the map, $h(s,b) = g(\lambda^{-1}(s,b),b)$. Evidently, h is continuous, $h(0) = h(1) = \beta$, and employing the change of coordinates $s = \lambda(t,b)$, it follows from (2.14) that for each $b \in B$,

$$
\begin{aligned}
\int_0^1 h(s,b) \, ds &= \int_0^1 g(\lambda^{-1}(s,b),b) \, ds \\
&= \int_0^1 g(t,b) \, d\mu(b) = f(b).
\end{aligned}
\tag{2.15}
$$

Let $H \colon [0,1]^2 \to \Gamma(\mathcal{R})$ be the corresponding reparametrization, $H(t,s,b) = G(\lambda^{-1}(t,b),s,b)$. Hence (h,H) is a C-structure over B which satisfies the integral representation (2.15) for the section $f \in \Gamma(\mathcal{R})$. □

The Integral Representation Theorem 2.12 admits a series of complements which are derived from the corresponding Complements 2.6 to 2.10. Again, $\mathcal{R} \subset E$ is open for all these complements.

Complement 2.13 (Relative Theorem). *Let N be a neighbourhood of $\beta(B)$ in E. Suppose there is a closed subspace K of B such that $f = \beta$ on K. There is a C-structure (h,H) over B such that,*

(i) *For each $b \in B$, $f(b) = \int_0^1 h(t,b) \, dt$.*
(ii) *Over $\mathfrak{Op}\, K \subset B$, the image of H lies in N:*

$$
H([0,1]^2 \times \mathfrak{Op}\, K) \subset N.
$$

Proof. Applying Complement 2.6, one obtains a C-structure (h_0, H_0) over B with respect to f, β, and which satisfies (ii) above. Applying Theorem 2.12 to this initial C-structure (h_0, H_0), one obtains a C-structure (h, H) over B which satisfies both of the properties (i), (ii). □

Complement 2.14 (C^∞-structures). *Suppose $p\colon E \to B$ is a smooth affine $\mathbf{R}^q$-bundle, and that $f \in \Gamma(E)$, $\beta \in \Gamma(\mathcal{R})$ are smooth sections. There is a C-structure (h, H) over B with respect to f, β such that H (and hence h) is a smooth map and such that for all $b \in B$, $f(b) = \int_0^1 h(t, b)\, dt$.*

Proof. Applying Complement 2.7, there is a smooth C-structure (h_0, H_0) over B with respect to f, β. The complement follows from Theorem 2.12 applied to this initial C-structure (h_0, H_0), subject to the following modifications to ensure a smooth reparametrization. One may assume that the measures $d\mu_i$ on the interval $[0, 1]$, $1 \leq i \leq q + 1$, are defined by positive smooth density functions. Hence the local barycentric coordinate functions p_i, of the smooth section f in (2.12), are smooth functions, $1 \leq i \leq q + 1$. Consequently, with respect to a smooth partition of unity $\{q_j\}_{j \geq 1}$, the measures $d\mu(b)$ defined by (2.13) are smooth, from which it follows that the change of coordinates map, $\lambda(t, b) = \int_0^1 d\mu(b)$, is smooth. One concludes that the reparametrized C-structure (h, H), $H(t, s, b) = G(\lambda^{-1}(t, b), s, b)$, satisfies the property that H (and hence h) is a smooth map. □

Complement 2.15 (Parameters). *Let P be a compact Hausdorff space (a parameter space). Let $f\colon P \to \Gamma(E)$, $\beta\colon P \to \Gamma(\mathcal{R})$ be continuous maps such that for all $(p, b) \in P \times B$, the following convex hull property obtains:*

$$f(p, b) \in \mathrm{Conv}(\mathcal{R}_b, \beta(p, b)).$$

There is a P-parameter family of C-structures (h, H) over B with respect to f, β such that for all $(p, b) \in P \times B$, $f(p, b) = \int_0^1 h(p, t, b)\, dt$.

Proof. Let $\mathrm{id} \times p\colon P \times E \to P \times B$ be the pullback of the affine bundle $p\colon E \to B$ along the projection map onto the second factor, $\pi\colon P \times B \to B$. Applying Complement 2.9, one obtains a C-structure (h_0, H_0) over $P \times B$, with respect to f, β. The complement follows from Theorem 2.12 applied to this initial C-structure (h_0, H_0) over $P \times B$. □

Corollary 2.16 (Ample Relations). *Let $\mathcal{R}$ be open and ample in E and suppose $\mathcal{R}$ admits a section $\beta \in \Gamma(\mathcal{R})$. For each section $f \in \Gamma(E)$ there is a C-structure (h, H) over B with respect to f, β such that, for all $b \in B$, $f(b) = \int_0^1 h(t, b)\, dt$.*

Proof. Applying Complement 2.10, it follows that for each $f \in \Gamma(E)$, there is a C-structure (h_0, H_0) over B with respect to f, β. The complement follows from Theorem 2.12 applied to this initial C-structure (h_0, H_0). □

CHAPTER 3

ANALYTIC THEORY

§1. The One-Dimensional Theorem

Let B be a compact Hausdorff space which is split: $B = C \times [0,1]$. For the purposes of the one-dimensional theorem below, the compact space C serves as a space of parameters and plays no essential role. Let $\pi \colon E = B \times \mathbf{R}^q \to B$, be the product $\mathbf{R}^q$-bundle over the base space B. The space of continuous sections $\Gamma(E)$ is identified naturally with $C^0(B, \mathbf{R}^q)$. Let $f \in \Gamma(E)$. Employing the splitting of B, one defines the derivative map $\partial_t f \colon B \to \mathbf{R}^q$ where $t \in [0,1]$ and $\partial_t = \partial/\partial t$. A section $f \in \Gamma(E)$ is C^1 in t if $\partial_t f \in \Gamma(E)$. Let $\|\ \|$ be the sup-norm on $C^0(B, \mathbf{R}^q)$.

In terms of the above data the basic approximation problem underlying Convex Integration theory may be stated as follows. Let $\mathcal{R} \subset E$ be a relation. Suppose $\beta \in \Gamma(\mathcal{R})$, $f_0 \in \Gamma(E)$ are sections such that f_0 is C^1 in t and for all $b \in B$,

$$\partial_t f_0(b) \in \operatorname{IntConv}(\mathcal{R}_b, \beta(b)).$$

The basic approximation problem is to construct, for each $\epsilon > 0$, a section $f = f_\epsilon \in \Gamma(E)$ which is C^1 in t and for which the following properties obtain:

(i) $\partial_t f \in \Gamma(\mathcal{R})$. More precisely, one requires that for each $b \in B$, $\partial_t f(b)$ lies in the arc-component of $\mathcal{R}_b$ to which $\beta(b)$ belongs.

(ii) $\|f - f_0\| \leq \epsilon$.

In addition to the basic properties (i), (ii), the full development of the theory requires two additional properties: a homotopy property and a relative property (properties (iii), (iv) below). Both of these additional properties are required in subsequent chapters for the proof of the h-principle. Briefly, the h-principle is proved inductively over a locally finite covering of the base manifold by charts. Thus, relative properties are essential in order to preserve what was established in previous steps of the induction.

(iii)   There is a homotopy $F \colon [0,1] \to \Gamma(\mathcal{R})$ which connects β to $\partial_t f \colon F_0 = \beta$, $F_1 = \partial_t f$.

(iv)   Let $K \subset B$ be closed such that $\partial_t f_0 = \beta$ on $\mathfrak{Op}\, K$. One can further arrange that $f = f_0$ over (a smaller) $\mathfrak{Op}\, K$.

D. Spring, *Convex Integration Theory: Solutions to the h-principle in geometry and topology*, Modern Birkhäuser classics, DOI 10.1007/978-3-0348-0060-0_3, © Springer Basel AG 2010

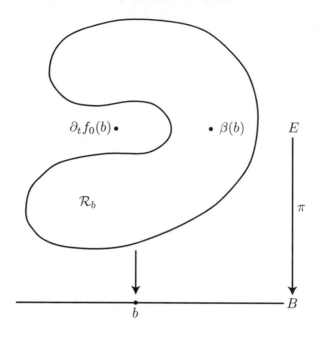

$$\text{FIGURE } 3.1$$

From (i) it follows that, in general, the displacement from $\partial_t f_0$ to $\partial_t f$ is large i.e. in general f is not a C^1-approximation to f_0. The strategy is to solve the above approximation problem first in the case that $\mathcal{R}$ is open in E (note that, fiberwise, the interior of the convex hull coincides with the convex hull, in the case of open relations $\mathcal{R}$ in E). In case $\mathcal{R}$ is closed in E, one solves the approximation problem for an open neighbourhood of $\mathcal{R}$; passing to the limit of a sequence of solutions obtained from a corresponding sequence of open neighbourhoods whose intersection is $\mathcal{R}$, one solves the approximation problem for $\mathcal{R}$. Additional hypotheses on $\mathcal{R}$ are required to ensure that the limit converges. Only open relations are treated in this chapter; a more general treatment of the approximation problem, which includes the case of closed relations, is studied in Part II.

In case $\mathcal{R} \subset E$ is open this approximation problem was first formulated and solved in Gromov [17], reproduced with more details in Adachi [1], and with some analytic refinements in Spring [37]. A more complete solution in the C^r case appears in Gromov [18]. In case $B = [0,1]$ i.e. the parameter space C is a point, this problem, with a different analytic formulation, occurs in Optimal Control theory and is solved by the Relaxation Theorem (cf. A. F. Filippov [13], and Chapter X below for a discussion of the Relaxation Theorem). The One-Dimensional Theorem 3.4 may be viewed as a C^1-Relaxation Theorem with parameters, and applies to relations $\mathcal{R}$ which are open in E. On the other hand, in case $B = [0,1]$ and under Lipschitz hypotheses on a closed relation $\mathcal{R}$, the Relaxation Theorem of Filippov [13] provides a section $f \in \Gamma(E)$ for which the C^0-approximation (ii) holds and such that the section $\partial_t f\colon B \to \mathcal{R}$ is absolutely continuous in t (not C^1 in t in general). Further research is required to investigate the connections which recently have been observed between Convex Integration theory and the Relaxation Theorem aspects of Optimal Control theory.

The solution of the above approximation problem, following Gromov, employs in an essential way the Integral Representation Theorem, II Theorem 2.12. Suppose $\mathcal{R}$ is open. Applying the Integral Representation Theorem, there is a C-structure (h, H) over B with respect to $\partial_t f_0, \beta$ such that for all $b \in B$,

$$\partial_t f_0(b) = \int_0^1 h(s,b)\,ds. \tag{3.1}$$

Employing (3.1), a natural approach to solving the approximation problem is as follows. Let $f \in \Gamma(E)$ be the section $((c,t) \in C \times [0,1])$,

$$f(c,t) = f_0(c,0) + \int_0^t h(s,(c,s))\,ds. \tag{3.2}$$

Evidently, $\partial_t f(c,t) = h(t,(c,t)) \in \mathcal{R}$. Thus $\partial_t f \in \Gamma(\mathcal{R})$ and, since $h(0,b) = \beta(b)$, it follows that for all $b \in B$, $\partial_t f(b)$, $\beta(b)$ lie in the same path component of $\mathcal{R}_b$ i.e. property (i) is satisfied. Employing the homotopy H, one checks that property (iii) is satisfied. In general however, the C^0-approximation property (ii) is far from satisfied. The remedy, due to Gromov, is to modify (3.2) by introducing an auxiliary continuous function $\theta\colon [0,1] \to [0,1]$ into the integrand as follows.

$$f(c,t) = f_0(c,0) + \int_0^t h(\theta(s),(c,s))\,ds. \tag{3.3}$$

Again $\partial_t f \in \Gamma(\mathcal{R})$ and properties (i), (iii) are satisfied. Following Gromov, if θ oscillates sufficiently rapidly (the analytic role of θ, explained below, exploits the integral representation (3.1)), then the C^0-approximation property (ii) is satisfied, thus solving the approximation problem. The following Proposition (which does not employ the relation $\mathcal{R}$) constructs θ in a more precise form than in Gromov [18], and in a suitably general setting that is appropriate for the refined approximation theorems which follow.

Proposition 3.1. *Let $\pi\colon E = B \times \mathbf{R}^q \to B$ be the product $\mathbf{R}^q$-bundle over a compact Hausdorff space $B = C \times [0,1]$. Let $g_0 \in \Gamma(E)$ be C^1 in t and let $\psi\colon [0,1] \to \Gamma(E)$ be a continuous map which is an integral representation of the derivative map $\partial_t g_0$:*

$$\partial_t g_0(b) = \int_0^1 \psi(s,b)\,ds, \quad \text{for all } b \in B. \tag{3.4}$$

For each $\epsilon > 0$, there is a continuous function $\theta_\epsilon \colon [0,1] \to [0,1]$ such that if $g_\epsilon \in \Gamma(E)$ is the section $((c,t) \in C \times [0,1])$,

$$g_\epsilon(c,t) = g_0(c,0) + \int_0^t \psi(\theta_\epsilon(s),(c,s))\,ds, \tag{3.5}$$

then g_ϵ is C^1 in t and $\lim_{\epsilon \to 0} \|g_\epsilon - g_0\| = 0$.

Proof. Let $\epsilon \in (0,1)$ and let $I_j = [t_j, t_j + \epsilon]$, $1 \le j \le m$, be a sequence of disjoint subintervals in [0,1], each of length ϵ and such that the total length of these subintervals is $\ge 1 - \epsilon$. Let $\theta \colon [0,1] \to [0,1]$ be a continuous function; we specify below the construction of a suitable $\theta = \theta_\epsilon$. Let $\ell, k \colon B \to \mathbf{R}^q$ be the auxiliary step functions (in what follows, $b = (c,t) \in C \times [0,1] = B$),

$$\ell(c,t) = g_0(c,0) + \epsilon \sum_{j=1}^{[t]} \partial_t g_0(c,t_j), \tag{3.6}$$

$$k(c,t) = g_0(c,0) + \sum_{j=1}^{[t]} \int_{I_j} \psi(\theta(s),(c,s))\,ds. \tag{3.7}$$

Here and for the rest of the proof, one employs $[t]$ as an index of summation to indicate that the sum is taken over all j such that $I_j \subset [0,t]$ (thus $k(c,t) = \ell(c,t) = g_0(c,0)$ in case the sum is empty). Note the estimate $[t] \le 1/\epsilon$ for all $t \in [0,1]$. As in (3.5), let $g \in \Gamma(E)$ be the section defined with respect to the function θ,

$$g(c,t) = g_0(c,0) + \int_0^t \psi(\theta(s),(c,s))\,ds. \tag{3.8}$$

Lemma 3.2. $\lim_{\epsilon \to 0} \|\ell - g_0\| = 0$; $\lim_{\epsilon \to 0} \|k - g\| = 0$ (*independent of the auxiliary function θ*).

Before proving Lemma 3.2, we sketch the proof of the Proposition. Employing (3.8), The Proposition states that $\lim_{\epsilon \to 0} \|g - g_0\| = 0$, where $g = g_\epsilon$ is defined in (3.5) with respect to a suitably chosen function $\theta = \theta_\epsilon$. By Lemma 1.2, this is equivalent to proving $\lim_{\epsilon \to 0} \|\ell - k\| = 0$, for a suitable function $\theta = \theta_\epsilon$. To better understand this latter limit, employing the integral representation (3.4), the step function ℓ in (3.6) is rewritten as follows.

$$\ell(c,t) = g_0(c,0) + \epsilon \sum_{j=1}^{[t]} \int_0^1 \psi(s,(c,t_j))\,ds. \tag{3.9}$$

Briefly, the function θ_ϵ is chosen so that the restriction of θ_ϵ to each subinterval I_j is a (linear) homeomorphism onto the interval [0,1]. The effect of θ_ϵ is to transform (by changing variables) each integral summand $\int_{I_j} \psi(\theta_\epsilon(s), (c,s))\, ds$ in (3.7), associated to the map k, into approximately the integral $\epsilon \int_0^1 \psi(t, (c, t_j))\, dt$ which is the corresponding summand in (3.9), associated to the map ℓ:

$$\int_{I_j} \psi(\theta_\epsilon(s), (c,s))\, ds \approx \epsilon \int_0^1 \psi(t, (c, t_j))\, dt, \quad 1 \le j \le m. \tag{3.10}$$

Thus θ_ϵ is a highly oscillatory piecewise linear function which magnifies ("blows up") the subintervals I_j to all of $[0,1]$, $1 \le j \le m$, such that (3.10) holds. Simple estimates on the errors are employed to show that $\lim_{\epsilon \to 0} \|\ell - k\| = 0$, which proves Proposition 1.1. The details are as follows.

Proof of Lemma 3.2. Since $\psi \colon [0,1] \times B \to \mathcal{R}$ is continuous, hence bounded, and since the total length of the subintervals I_j is $\ge 1 - \epsilon$, it is clear from (3.7), (3.8) that $\lim_{\epsilon \to 0} \|k - g\| = 0$, independent of θ.

Let $\ell_0 \colon B \to \mathbf{R}^q$ be the auxiliary step function,

$$\ell_0(c,t) = g_0(c,0) + \sum_{j=1}^{[t]} g_0(c, t_j + \epsilon) - g_0(c, t_j). \tag{3.11}$$

Since the map g_0 is C^1 in t, and since the total length of the subintervals I_j is $\ge 1 - \epsilon$, it follows from the telescoping sum (3.11) that,

$$\lim_{\epsilon \to 0} \|\ell_0 - g_0\| = 0. \tag{3.12}$$

On the other hand, employing (3.6), (3.11), one computes,

$$\ell_0(c,t) - \ell(c,t) = \epsilon \sum_{j=1}^{[t]} \frac{g_0(c, t_j + \epsilon) - g_0(c, t_j)}{\epsilon} - \partial_t g_0(c, t_j). \tag{3.13}$$

Since g_0 is C^1 in t it follows that, given $\delta > 0$, the summand for each j of (3.13) is bounded above in the sup-norm by δ, provided $\epsilon > 0$ is sufficiently small. Since $[t] \le 1/\epsilon$, employing (3.13), $\lim_{\epsilon \to 0} \|\ell - \ell_0\| = 0$; hence from (3.12), $\lim_{\epsilon \to 0} \|\ell - g_0\| = 0$. $\square$

Returning to the proof of Proposition 3.1, let $\theta \equiv \theta_\epsilon \colon [0,1] \to [0,1]$ be the piecewise C^1-function such that on each subinterval $I_j = [t_j, t_j + \epsilon]$, $\theta \colon I_j \to [0,1]$ is the linear homeomorphism,

$$\theta(s) = \frac{s - t_j}{\epsilon} \quad 1 \le j \le m. \tag{3.14}$$

Employing the change of variables $t = \theta(s)$ on the subinterval I_j, $1 \leq j \leq m$, one computes,

$$\int_{I_j} \psi(\theta(s), (c, s)) \, ds = \epsilon \int_0^1 \psi(t, (c, t_j + \delta_\epsilon)) \, dt, \quad \text{where } \delta_\epsilon = \epsilon t. \qquad (3.15)$$

Replacing each integral summand of the step function k in (3.7) by the corresponding summand (3.15), $1 \leq j \leq [t]$, and employing the integral representation (3.4), one obtains the following sum for the function k:

$$\begin{aligned} k(c, t) &= g_0(c, 0) + \epsilon \sum_{j=1}^{[t]} \int_0^1 \psi(t, (c, t_j + \delta_\epsilon)) \, dt \\ &= g_0(c, 0) + \epsilon \sum_{j=1}^{[t]} \partial_t g_0(c, t_j + \delta_\epsilon). \end{aligned} \qquad (3.16)$$

Consequently, from (3.6), (3.16), it follows that,

$$k(c, t) - \ell(c, t) = \epsilon \sum_{j=1}^{[t]} \partial_t g_0(c, t_j + \delta_\epsilon) - \partial_t g_0(c, t_j). \qquad (3.17)$$

Since g_0 is C^1 in t and $[t] \leq 1/\epsilon$, $\lim_{\epsilon \to 0} \delta_\epsilon = 0$ uniformly on $[0,1]$, one concludes from (3.17) that $\lim_{\epsilon \to 0} \|k - \ell\| = 0$. Let $g = g_\epsilon \in \Gamma(E)$ be the section in (3.8), defined with respect to $\theta = \theta_\epsilon$. Employing Lemma 3.1 and the above $\lim_{\epsilon \to 0} \|k - \ell\| = 0$, it follows that $\lim_{\epsilon \to 0} \|g_\epsilon - g_0\| = 0$. Also from (3.8), $\partial_t g_\epsilon = \psi(\theta_\epsilon(t), (c, t))$ i.e. g_ϵ is C^1 in t, which completes the proof of the Proposition. $\square$

Remark 3.3. The auxiliary function θ_ϵ of Proposition 3.1 is piecewise C^1. For technical reasons, we arrange, as in Gromov [18], for θ_ϵ to be C^∞. Briefly, assume $\epsilon \in (0, 1/2)$ and let $\theta_\epsilon \colon [0, 1] \to [0, 1]$ be a C^∞-function such that on each subinterval $I_j = [t_j, t_j + \epsilon]$, θ_ϵ is a linear homeomorphism onto the subinterval $[\epsilon, 1 - \epsilon]$. Specifically, on each interval I_j,

$$\theta_\epsilon(s) = as + b, \quad a = \frac{1 - 2\epsilon}{\epsilon}, \, b = \epsilon - at_j, \quad 1 \leq j \leq m. \qquad (3.18)$$

Employing the change of variables $t = \theta_\epsilon(s)$ on each subinterval I_j, $1 \leq j \leq m$, on computes (cf. (3.15)):

$$\int_{I_j} \psi(\theta_\epsilon, (c, s)) \, ds = \frac{\epsilon}{1 - 2\epsilon} \int_\epsilon^{1 - \epsilon} \psi(t, (c, t_j + \delta_\epsilon)) \, dt,$$

$$\text{where } \delta_\epsilon = \frac{\epsilon(t - \epsilon)}{1 - 2\epsilon}. \qquad (3.19)$$

Replacing each integral summand of the step function k in (3.7) (associated to $\theta = \theta_\epsilon$) by the corresponding summand (3.19), $1 \leq j \leq [t]$, one obtains an analogous sum for (3.17):

$$k(c,t) - \ell(c,t) = \epsilon \sum_{j=1}^{[t]} \left(\frac{1}{1-2\epsilon} \int_\epsilon^{1-\epsilon} \psi(t,(c,t_j+\delta_\epsilon))\,dt - \int_0^1 \psi(s,(c,t_j))\,ds \right). \tag{3.20}$$

Again, since $[t] \leq 1/\epsilon$, and $\lim_{\epsilon \to 0} \delta_\epsilon = 0$ uniformly on [0,1], one concludes from (3.20) that $\lim_{\epsilon \to 0} \|k - \ell\| = 0$, from which it follows, employing Lemma 1.2, that $\lim_{\epsilon \to 0} \|g_\epsilon - g_0\| = 0$, where $g = g_\epsilon \in \Gamma(E)$ is the section in (3.8), defined with respect to $\theta = \theta_\epsilon$ above.

One-Dimensional Theorem 3.4. *Let $\pi \colon E \to B$ be the product $\mathbf{R}^q$-bundle over a compact Hausdorff space $B = C \times [0,1]$. Let $\mathcal{R} \subset E$ be open and suppose there are sections $\beta \in \Gamma(\mathcal{R})$, $f_0 \in \Gamma(E)$ such that f_0 is C^1 in t and, for all $b \in B$,*

$$\partial_t f_0(b) \in \mathrm{Conv}(\mathcal{R}_b, \beta(b)).$$

For each $\epsilon > 0$, there is a section $f_\epsilon \in \Gamma(E)$ which is C^1 in t, and a continuous map $F = F_\epsilon \colon [0,1] \to \Gamma(\mathcal{R})$ such that the following properties obtain:

(i) $\lim_{\epsilon \to 0} \|f_\epsilon - f_0\| = 0$.

(ii) $\partial_t f_\epsilon \in \Gamma(\mathcal{R})$; $F_0 = \beta$, $F_1 = \partial_t f_\epsilon$, *and the image of F is contained in the image of a C-structure with respect to $\partial_t f, \beta$.*

Suppose furthermore that B is a compact C^1-manifold and that $K \subset B$ is closed such that $\partial_t f_0 = \beta$ on $\mathfrak{Op}\,K$. Let N be a neighbourhood of $\beta(B)$ in $\mathcal{R}$. For $\epsilon > 0$ sufficiently small, then up to a small perturbation in N over $\mathfrak{Op}\,K$, one can assume that the section $f_\epsilon \in \Gamma(E)$ and the homotopy $F = F_\epsilon \colon [0,1] \to \Gamma(\mathcal{R})$ satisfy the additional property (relative theorem) :

(iii) $f_\epsilon = f_0$ *on (a smaller)* $\mathfrak{Op}_1 K$; *the homotopy F is constant on $\mathfrak{Op}_1 K$: for all $t \in [0,1]$, $F_t = \beta$ on $\mathfrak{Op}_1 K$.*

Proof. Let $N(\delta) \subset N$ be a δ-ball neighbourhood of $\beta(B)$ i.e. $N(\delta)$ meets each fiber E_b in an open ball of radius $\delta > 0$, center $\beta(b)$, $b \in B$. Applying the Relative Integral Representation Theorem, II Complement 2.13, there is a C-structure (h, H) with respect to $\partial_t f_0, \beta$, such that for all $b \in B$,

$$\partial_t f_0(b) = \int_0^1 h(s,b)\,ds, \quad H([0,1]^2 \times \mathfrak{Op}\,K) \subset N(\delta/2). \tag{3.21}$$

Since the map $h \colon [0,1] \to \Gamma(E)$ is an integral representation of the derivative map $\partial_t f_0$, it follows from Proposition 3.1 applied to the data $f_0 \in \Gamma(E)$, $h \colon [0,1] \to$

$\Gamma(\mathcal{R})$, that for each $\epsilon > 0$, there is a continuous function $\theta_\epsilon \colon [0,1] \to [0,1]$ such that the corresponding section $f_\epsilon \in \Gamma(E)$,

$$f_\epsilon(c,t) = f_0(c,0) + \int_0^t h(\theta_\epsilon(s),(c,s))\,ds, \qquad (3.22)$$

satisfies conclusion (i) above. From (3.22), $\partial_t f_\epsilon(c,t) = h(\theta_\epsilon(t),(c,t))$ i.e. f_ϵ is C^1 in t and $\partial_t f_\epsilon \in \Gamma(\mathcal{R})$. Furthermore, employing the C-structure (h,H), let $F\colon [0,1] \to \Gamma(\mathcal{R})$ be the homotopy $((c,t) \in B = C \times [0,1])$,

$$F(u,(c,t)) = H(\theta_\epsilon(t),u,(c,t)), \quad 0 \le u \le 1. \qquad (3.23)$$

Since $H(t,0,b) = \beta(b)$; $H(t,1,b) = h(t,b)$, it follows from (3.23) that $F_0 = \beta$, $F_1 = \partial_t f_\epsilon$, which proves conclusion (ii).

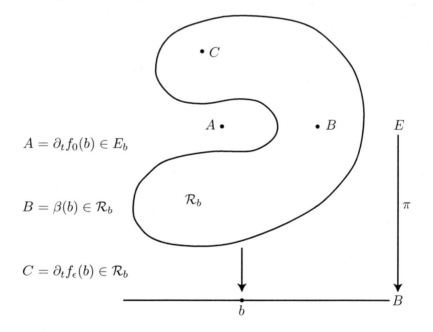

$A = \partial_t f_0(b) \in E_b$

$B = \beta(b) \in \mathcal{R}_b$

$C = \partial_t f_\epsilon(b) \in \mathcal{R}_b$

FIGURE 3.2

Turning now to property (iii), from (3.23) the image of the homotopy F is a subset of the image of H. Employing (3.21) it follows that for all $\epsilon > 0$, over $\mathfrak{Op}\,K$, the image of F, hence also the image of $\partial_t f_\epsilon$, lies in $N(\delta/2)$ In particular, over $\mathfrak{Op}\,K$,

$$\|\partial_t f_\epsilon - \partial_t f_0\| = \|\partial_t f_\epsilon - \beta\| < \delta/2. \qquad (3.24)$$

Since B is a C^1-manifold there is a function $\lambda \in C^1(B, [0, 1])$, compactly supported in $\mathfrak{Op}\, K$, such that $\lambda = 1$ in a neighbourhood W of K. Let $f'_\epsilon = f_\epsilon + \lambda(f_0 - f_\epsilon)$. Thus f'_ϵ is C^1 in t, $f'_\epsilon = f_0$ on W and $f'_\epsilon = f_\epsilon$ on $B \setminus \mathfrak{Op}\, K$. Evidently, since $\lim_{\epsilon \to 0} \| f_\epsilon - f_0 \| = 0$, it follows also that $\lim_{\epsilon \to 0} \| f'_\epsilon - f_0 \| = 0$ i.e. f'_ϵ satisfies the approximation property (i).

Let $F' = F'_\epsilon \colon [0, 1] \to \Gamma(E)$ be the homotopy,

$$F'(s, b) = F(s, b) + s\, \partial_t(\lambda(f_0 - f_\epsilon))(b). \tag{3.25}$$

Thus $F' = F$ on $[0, 1] \times (B \setminus \mathfrak{Op}\, K)$; $F'_0 = \beta$, and since $F_1 = \partial_t f_\epsilon$, it follows that $F'_1 = \partial_t f'_\epsilon$. For sufficiently small $\epsilon > 0$, we prove that in fact $F' \colon [0, 1] \to \Gamma(\mathcal{R})$. Indeed, computing the derivative ∂_t in (3.25) ($F_s(b) = F(s, b)$; $F'_s(b) = F'(s, b)$):

$$F'_s = F_s + s\lambda\, \partial_t(f_0 - f_\epsilon) + s(\partial_t \lambda)(f_0 - f_\epsilon). \tag{3.26}$$

The function $\partial_t \lambda$ is bounded since λ is compactly supported in $\mathfrak{Op}\, K$. Employing $\lim_{\epsilon \to 0} \| f_0 - f_\epsilon \| = 0$, it follows from (3.24), (3.26) that, over $\mathfrak{Op}\, K$, for all $s \in [0, 1]$,

$$\| F'_s - F_s \| < \delta/2, \tag{3.27}$$

provided $\epsilon > 0$ is sufficiently small. Recall (3.21), over $\mathfrak{Op}\, K$, the image of H, hence from (3.23) the image of F, lies in $N(\delta/2)$. Consequently, from (3.27), the image of the homotopy F' over $\mathfrak{Op}\, K$ lies in $N(\delta) \subset N$, and hence $F' \colon [0, 1] \to \Gamma(\mathcal{R})$, provided $\epsilon > 0$ is sufficiently small. Evidently, the pair (f'_ϵ, F') satisfies conclusion (iii) of the theorem, except possibly the requirement that the homotopy F' be constant on $\mathfrak{Op}\, K$. To satisfy this additional property, note that, over $\mathfrak{Op}\, K$, $F'_0 = \beta$, $F'_1 = \partial_t f'_\epsilon = \beta$ and that, fiberwise over $b \in \mathfrak{Op}\, K$, the image of F' lies in a δ-ball in $\mathcal{R}_b$, centered at $\beta(b)$. Consequently, over $\mathfrak{Op}\, K$, F' is homotopic in $N(\delta)$, $\mathrm{rel}\{0, 1\}$, to the constant homotopy β. Employing standard arguments with bump functions to construct a homotopy of homotopies in $N(\delta)$, F' is homotopic to $F'' = F''_\epsilon \colon [0, 1] \to \Gamma(\mathcal{R})$, such that $F''_0 = \beta$, $F''_1 = \partial_t f'_\epsilon$, and such that F'' is the constant homotopy β over (a smaller) $\mathfrak{Op}_1 K$, and $F'' = F'$ over $[0, 1] \times (B \setminus \mathfrak{Op}\, K)$. Details are left to the reader. The pair $(f'_\epsilon, F''_\epsilon)$ satisfies conclusion (iii) of the theorem. $\qquad\square$

We remark here that the relative version (iii) is an essential feature that is required in the proofs of the h-principle by induction over a system of charts: constructions of holonomic sections and homotopies in a chart are relative to constructions already carried out inductively in previous charts. From (3.23), note that the image of the homotopy F' is contained in the image of a C-structure, except for a small perturbation that is needed in order to obtain conclusion (iii). This unavoidable exception is a technical inconvenience that requires additional attention in the proofs of the h-principle (Chapter IV) and of the h-Stability

Theorem (Chapter VIII). The One-Dimensional Theorem admits several refinements which depend on the corresponding smooth and parametric Complements 2.14, 2.15 of the Integral Representation Theorem, II Theorem 2.12. The above remarks on the relative version (iii) apply also to these Complements.

A continuous map $g\colon P \to \Gamma(E)$ induces a section (same notation) $g \in \Gamma(P \times E)$ of the pullback bundle $\mathrm{id} \times \pi\colon P \times E \to P \times B$; the pullback relation, $P \times \mathcal{R}$, is open in $P \times E$. If P, B are C^r-manifolds, $r \geq 1$, the map $g\colon P \to \Gamma(E)$ is defined to be of class C^r if the corresponding section $g \in \Gamma(P \times E)$ is of class C^r.

Complement 3.5 (Parameters). *Let P be a compact Hausdorff space and suppose $f_0\colon P \to \Gamma(E)$, $\beta\colon P \to \Gamma(\mathcal{R})$ are continuous maps such that for all $(p, b) \in P \times B$, $f_0(p)$ is C^1 in t and,*

$$\partial_t f_0(p, b) \in \mathrm{Conv}(\mathcal{R}_b, \beta(p, b)).$$

For each $\epsilon > 0$ there is a continuous map $f_\epsilon\colon P \to \Gamma(E)$ and a homotopy $F = F_\epsilon\colon [0, 1] \times P \to \Gamma(\mathcal{R})$, such that for all $p \in P$ $f_\epsilon(p)$ is C^1 in t, and such that,

(i) $\lim_{\epsilon \to 0} \|f_\epsilon(p) - f_0(p)\| = 0$ *uniformly on P.*

(ii) $\partial_t f_\epsilon(p) \in \Gamma(\mathcal{R})$; $F_0(p) = \beta(p)$, $F_1(p) = \partial_t f_\epsilon(p)$, *and the image of F is contained in the image of a P-parameter family of C-structures.*

Suppose furthermore that P, B are compact C^1-manifolds and that $K \subset P \times B$ is closed such that $\partial_t f_0 = \beta$ on $\mathfrak{Op}\, K$. Let N be a neighbourhood of $\beta(P \times B)$ in $P \times \mathcal{R}$. For $\epsilon > 0$ sufficiently small, then up to a small perturbation in N over $\mathfrak{Op}\, K$, one can assume that the maps f_ϵ, F satisfy the additional property (relative theorem) :

(iii) $f_\epsilon = f_0$ *on (a smaller) $\mathfrak{Op}_1 K$; for all $t \in [0, 1]$, $F_t = \beta$ over $\mathfrak{Op}_1 K$.*

Proof. As in the relative One-Dimensional Theorem, let $N(\delta)$ be a δ-ball neighbourhood of $\beta(P \times B)$ in $P \times E$ such that $N(\delta) \subset N$. Employing II Complements 2.13, 2.15 of the Integral Representation Theorem, there is a P-parameter family of C-structures (h, H) over B with respect to $\partial_t f_0$, β, such that for all $(p, b) \in P \times B$,

$$\partial_t f_0(p, b) = \int_0^1 h(p, s, b)\, ds; \quad H([0, 1]^2 \times \mathfrak{Op}\, K) \subset N(\delta/2).$$

Employing (h, H) above, the Complement follows from the Theorem 3.4 applied to the open relation $P \times \mathcal{R}$ in the pullback bundle $P \times E \to P \times B$ over the split compact Hausdorff space $P \times B = (P \times C) \times [0, 1]$. In this way, the parameter space P is "absorbed" into the auxiliary space C. In particular, in the non-relative case, the required map $f_\epsilon\colon P \to \Gamma(E)$ is of the form $(b = (c, t) \in C \times [0, 1])$:

$$f_\epsilon(p, c, t) = f_0(p, c, 0) + \int_0^t h(p, \theta_\epsilon(s), (c, s))\, ds,$$

where $\theta_\epsilon\colon [0, 1] \to [0, 1]$ is a suitable auxiliary C^∞-function. $\square$

We prove now a smooth refinement of the one-dimensional theorem over an n-cube in $\mathcal{R}^n$. Suppose $B = I^n$, the unit cube $[0,1]^n \subset \mathbf{R}^n$, $n \geq 1$. Thus B is the split manifold $B = I^{n-1} \times [0,1]$. Let $(u_1, u_2, \ldots, u_{n-1}, t)$ be coordinates on $\mathbf{R}^n$. Employing the notation $\partial^p_{u_j}$ to denote the derivative of order p in the variable u_j, $1 \leq j \leq n - 1$, let ∂^α_U denote the U-derivatives in the variables u_j, $1 \leq j \leq n - 1$ $(n \geq 2)$:

$$\partial^\alpha_U = \partial^{p_1}_{u_1} \circ \partial^{p_2}_{u_2} \circ \cdots \circ \partial^{p_{n-1}}_{u_{n-1}}; \quad |\alpha| = p_1 + p_2 + \cdots + p_{n-1}. \tag{3.28}$$

Thus the U-derivatives exclude the derivative ∂_t.

The Space $C^{s,r}(B, \mathbf{R}^q)$. Fix integers $s \geq r \geq 0$ and let $\| \ \|^{s,r} \equiv \| \ \|^{s,r}_U$ be the following sup-norm on $C^s(B, \mathbf{R}^q)$:

$$\|f\|^{s,r} = \sup\{\|\partial^k_t \circ \partial^\alpha_U f\| : k + |\alpha| \leq s; \ k \leq r\}. \tag{3.29}$$

It is convenient to suppress the superscript r, in case $r = 0$:

$$\|f\|^s = \|f\|^{s,0} = \sup\{\|\partial^\alpha_U f\| : |\alpha| \leq s\}, \quad f \in C^s(B, \mathbf{R}^q). \tag{3.30}$$

$C^{s,r}(B, \mathbf{R}^q)$ is the space $C^s(B, \mathbf{R}^q)$, topologized with the above norm $\| \ \|^{s,r}$. The purpose of the space $C^{s,r}(B, \mathbf{R}^q)$ is to allow for C^s-approximations, s large, in the U-coordinates of split manifold $B = I^{n-1} \times [0,1]$.

Example 3.6. For all $r \geq 0$, $C^{r,r}(B, \mathbf{R}^q) = C^r(B, \mathbf{R}^q)$ i.e. the norm $\| \ \|^{r,r}$ coincides with the usual sup-norm for the C^r-topology on the function space $C^r(B, \mathbf{R}^q)$. For all $k \geq s$, the inclusion $i : C^k(B, \mathbf{R}^q) \to C^{s,r}(B, \mathbf{R}^q)$ is continuous. An example which occurs in the $C^\perp$-Approximation Theorem 3.8 below is as follows: Let $T : C^s(B, \mathbf{R}^q) \to C^s(B, \mathbf{R}^q)$ be the map,

$$T(f)(u,t) = \int_0^t f(u,s)\, ds.$$

Then, $T : C^{s,r}(B, \mathbf{R}^q) \to C^{s,r+1}(B, \mathbf{R}^q)$ is continuous $(s \geq r + 1)$.

Employing these norms, one obtains the following refinement of the One-Dimensional Theorem which provides C^s-approximations on the corresponding U-derivatives in case $B = I^n$.

Complement 3.7 (Smooth Parameters). *Let $\pi : E \times \mathbf{R}^q \to B$ be the product $\mathbf{R}^q$-bundle over the n-cube I^n in $\mathbf{R}^n$, $n \geq 1$, and let $\mathcal{R} \subset E$ be open. Let also P be a compact C^∞-manifold and suppose $f_0 : P \to \Gamma(E)$, $\beta : P \to \Gamma(\mathcal{R})$ are smooth maps such that for all $(p,b) \in P \times B$,*

$$\partial_t f_0(p,b) \in \mathrm{Conv}(\mathcal{R}_b, \beta(p,b)).$$

Fix an integer $s \geq 1$. For each $\epsilon > 0$ there is a C^∞-map $f_\epsilon \colon P \to \Gamma(E)$ and a C^∞-homotopy $F = F_\epsilon \colon [0,1] \times P \to \Gamma(\mathcal{R})$ such that, for all $p \in P$, the following properties obtain:

(i) $\lim_{\epsilon \to 0} \| f_\epsilon(p) - f_0(p) \|^s = 0$.

(ii) $\partial_t f_\epsilon(p) \in \Gamma(\mathcal{R})$; $F_0(p) = \beta(p)$, $F_1(p) = \partial_t f_\epsilon(p)$, *and the image of F is contained in the image of a P-parameter family of C-structures.*

Suppose furthermore $K \subset P \times B$ is closed such that $\partial_t f_0 = \beta$ on $\mathfrak{Op}\, K$. Let N be a neighbourhood of $\beta(P \times B)$ in $P \times \mathcal{R}$. If $\epsilon > 0$ is sufficiently small, then up to a small perturbation in N over $\mathfrak{Op}\, K$, one may assume that $f_\epsilon \colon P \to \Gamma(E)$, $F \colon [0,1] \times P \to \Gamma(\mathcal{R})$ satisfy the additional properties (relative theorem),

(iii) $f_\epsilon = f_0$ *on (a smaller)* $\mathfrak{Op}_1 K$; *for all $t \in [0,1]$, $F_t = \beta$ on* $\mathfrak{Op}_1 K$.

Proof. Applying Complement 3.5, for each $\epsilon > 0$, there is a continuous map $f_\epsilon \colon P \to \Gamma(E)$ and a homotopy $F_\epsilon \colon [0,1] \times P \to \Gamma(\mathcal{R})$ such that conclusions (i), (ii), (iii) are satisfied with respect to the norm $\| \ \|$ on $C^0(B, \mathbf{R}^q)$. To obtain further smoothness of the map f_ϵ and to obtain conclusion (i) with respect to the norm $\| \ \|^s$, one proceeds as follows. Since the sections f, β are smooth, applying II Complements 2.14, 2.15, one may assume that the C-structure (h, H) employed in the proof of the Complement 3.5 of the One-Dimensional theorem is smooth. Explicitly, for all $(p, b) \in P \times B$, $(b = (u, t) \in I^{n-1} \times [0,1])$:

$$\partial_t f_0(p, b) = \int_0^1 h(p, s, b)\, ds. \tag{3.31}$$

$$f_\epsilon(p, u, t) = f_0(u, 0) + \int_0^t h(p, \theta_\epsilon(s), (u, s))\, ds. \tag{3.32}$$

where $\theta_\epsilon \colon [0,1] \to [0,1]$ is a suitable auxiliary C^∞-function. Since h is smooth and since the continuous function θ does not depend on the U-coordinates in I^{n-1}, it follows from (3.31) (3.32), that the U-derivatives of $\partial_t f_0$, f_ϵ, are computed under the integral sign:

$$\partial_U^\alpha \partial_t f_0(p, u, t) = \int_0^1 \partial_U^\alpha h(p, s, (u, t))\, ds; \tag{3.33}$$

$$\partial_U^\alpha f_\epsilon(p, u, t) = \partial_U^\alpha f_0(p, u, 0) + \int_0^t \partial_U^\alpha h(p, \theta_\epsilon(s), (u, s))\, ds. \tag{3.34}$$

Employing the pullback bundle $\mathrm{id} \times \pi \colon P \times E \to P \times B$, for each derivative ∂_U^α, (3.33) states that the continuous map $\partial_U^\alpha h \colon [0,1] \to \Gamma(P \times E)$ is an integral representation of the t-derivative map $\partial_t \partial_U^\alpha f_0$, associated to the section $\partial_U^\alpha f_0 \in$

$\Gamma(P \times E)$. Applying Proposition 3.1 to the data $g_0 = \partial_U^\alpha f_0 \in \Gamma(P \times E)$, $\psi = \partial_U^\alpha h \colon [0,1] \to \Gamma(P \times E)$ it follows that, for all $p \in P$,

$$\lim_{\epsilon \to 0} \|\partial_U^\alpha (f_\epsilon(p) - f_0(p))\| = 0, \tag{3.35}$$

Indeed, as explained above, the U-derivative $\partial_U^\alpha f_\epsilon$ satisfies (3.34), to which the conclusion of the Proposition applies (i.e. $g_\epsilon = \partial_U^\alpha f_\epsilon$). Since h is smooth note also from (3.32) that, for each $\epsilon > 0$, $f_\epsilon \colon P \times B \to \mathbf{R}^q$ is a C^∞-map.

Since (3.35) holds for all the (finitely many) derivatives ∂_U^α for which $|\alpha| \leq s$, one concludes that, for all $p \in P$, $\lim_{\epsilon \to 0} \|f_\epsilon(p) - f_0(p)\|_U^s = 0$, which proves conclusion (i).

Applying Complement 3.5 (iii), the conclusion of Complement 3.7 (iii) is satisfied for a small perturbation $(f_\epsilon', F_\epsilon')$ of (f_ϵ, F_ϵ) in N over $\mathfrak{Op}\, K$, provided $\epsilon > 0$ is sufficiently small. These perturbations also satisfy conclusion (i) for the norm $\| \ \|^s$. Indeed, employing $f_\epsilon' = f_\epsilon + \lambda(f_0 - f_\epsilon)$, since $\lim_{\epsilon \to 0} \|f_0 - f_\epsilon\|_U^s = 0$, it is clear that $\lim_{\epsilon \to 0} \|f_\epsilon' - f_0\|_U^s = 0$ i.e. the perturbed section f_ϵ' also satisfies conclusion (i). $\qquad \square$

§2. The $C^\perp$-Approximation Theorem

In what follows, Complement 3.7 to the One-Dimensional Theorem is generalized to the $C^\perp$-approximation theorem, which is the main technical result that is employed in subsequent chapters for proving the h-principle. In case B is the n-cube $I^n = I^{n-1} \times [0,1]$, Complement 3.7 proves C^s-approximations in the corresponding U-derivatives with respect to coordinates in I^{n-1}. The generalization consists of extending these approximations to include the pure derivatives ∂_t^j, $0 \leq j \leq r - 1$.

The $C^\perp$-Approximation Theorem 3.8. *Let $\pi \colon E = B \times \mathbf{R}^q \to B$ be the product $\mathbf{R}^q$-bundle over the n-cube $B = [0,1]^n$ in $\mathbf{R}^n$, $n \geq 1$, and let $\mathcal{R} \subset E$ be open. Let also P be a compact C^∞-manifold (a parameter space) and let $s \geq r \geq 1$ be integers. Suppose $g_0 \colon P \to \Gamma(E)$, $\beta \colon P \to \Gamma(\mathcal{R})$ are smooth maps such that for all $(p, b) \in P \times B$,*
$$\partial_t^r g_0(p, b) \in \mathrm{Conv}(\mathcal{R}_b, \beta(p, b)).$$
For each $\epsilon > 0$ there is a C^∞-map $g_\epsilon \colon P \to \Gamma(E)$ and a C^∞-homotopy $F = F_\epsilon \colon [0,1] \times P \to \Gamma(\mathcal{R})$ such that, for all $p \in P$, the following properties obtain:

(i) $\lim_{\epsilon \to 0} \|g_\epsilon(p) - g_0(p)\|^{s, r-1} = 0$ *uniformly in $p \in P$.*

(ii) $\partial_t^r g_\epsilon(p) \in \Gamma(\mathcal{R})$; $F_0(p) = \beta(p)$, $F_1(p) = \partial_t^r g_\epsilon(p)$, *and the image of F is contained in the image of a P-parameter family of C-structures.*

Suppose furthermore $K \subset P \times B$ is closed such that $\partial_t^r g_0 = \beta$ on $\mathfrak{Op}\, K$. Let N be a neighbourhood of $\beta(P \times B)$ in $P \times \mathcal{R}$. If $\epsilon > 0$ is sufficiently small, then up *to a small perturbation in N over $\mathfrak{Op}\,K$, one may assume that $g_\epsilon\colon P\to\Gamma(E)$, $F\colon[0,1]\times P\to\Gamma(\mathcal{R})$ satisfy the additional properties (relative theorem),*

(iii) $g_\epsilon = g_0$ *on (a smaller) $\mathfrak{Op}_1 K$; for all $t\in[0,1]$, $F_t=\beta$ on $\mathfrak{Op}_1 K$.*

Proof. Note that the approximation (i) excludes the pure rth order derivative ∂_t^r. Indeed, from (ii), $\partial_t^r g_\epsilon \in \Gamma(\mathcal{R})$; in general, therefore, the displacement from $\partial^r g_0$ to $\partial^r g_\epsilon$ is large i.e. in general, g_ϵ is not a C^r-approximation of g_0. However, with respect to the U-derivatives, the approximation (i) is C^s-close and is C^0-close with respect to the derivatives ∂_t^j, $0\le j\le r-1$.

Evidently, in case $r=1$, the $C^\perp$-approximation theorem reduces to Complement 3.7 of the One-Dimensional Theorem. In general, one proceeds as follows. Applying Complement 3.7 to the data $f_0 = \partial_t^{r-1} g_0\colon P\to\Gamma(E)$, $\beta\colon P\to\Gamma(\mathcal{R})$, for each $\epsilon>0$ there is a C^∞-map $\ell_\epsilon\colon P\to\Gamma(E)$, and a homotopy $F=F_\epsilon\colon[0,1]\times P\to\Gamma(\mathcal{R})$ such that, for all $p\in P$, the following properties obtain:

(iv) $\lim_{\epsilon\to 0}\|\ell_\epsilon(p)-\partial_t^{r-1}g_0(p)\|^s = 0$.

(v) $\partial_t\ell_\epsilon(p)\in\Gamma(\mathcal{R})$; $F_0(p)=\beta(p)$, $F_1(p)=\partial_t\ell_\epsilon(p)$.

More precisely, according to the proof of Complement 3.7, the map ℓ_ϵ is of the form $((u,t)\in B = I^{n-1}\times[0,1])$:

$$\ell_\epsilon(p,u,t) = \partial_t^{r-1} g_0(p,u,0) + \int_0^t h(p,\theta_\epsilon(s),(u,s))\,ds, \qquad (3.36)$$

where (h,H) is a smooth P-parameter family of C-structures over B with respect to the maps $\partial_t^r g_0$, β, and $\theta_\epsilon\colon[0,1]\to[0,1]$ is a suitably chosen C^∞-function. Note that the theorem is proved in the case $r=1$ by choosing $g_\epsilon=\ell_\epsilon$. Indeed, in case $r=1$, conclusions (i), (ii), of the theorem are equivalent, respectively, to properties (iv), (v) of the section ℓ_ϵ. Suppose now $r\ge 2$. The map g_ϵ is obtained by "$(r-1)$-fold" integration of the map ℓ_ϵ. In detail, let $\ell_{r-1}=\ell_\epsilon\colon P\to C^\infty(B,\mathbf{R}^q)$. By downward recursion on k define,

$$\ell_{k-1}(p,u,t) = \partial_t^{k-1} g_0(p,u,0) + \int_0^t \ell_k(p,u,s)\,ds, \qquad 1\le k\le r-1. \qquad (3.37)$$

In what follows we show that the map $g_\epsilon=\ell_0$ (i.e. $k=1$ in (3.37)) satisfies conclusions (i), (ii), of the theorem. Note that ℓ_0 is a C^∞-map and that furthermore, applying ∂_t successively in (3.37) and employing (v), it follows that $\partial_t^r\ell_0 = \partial_t\ell_\epsilon\colon P\to\Gamma(\mathcal{R})$. Again from (v), $F_0=\beta$, $F_1=\partial_t^r\ell_0=\partial_t^r g_\epsilon$, which proves conclusion (ii) of the theorem for $g_\epsilon=\ell_0$.

Turning now to the approximation property (i), note that the smooth map $g_0 \colon P \times B \to \mathbf{R}^q$ satisfies the following properties, analogous to (3.37) for the sequence of maps ℓ_k,

$$\partial_t^{k-1} g_0(p, u, t) = \partial_t^{k-1} g_0(p, u, 0) + \int_0^t \partial_t^k g_0(p, u, s)\, ds,$$

$$1 \le k \le r - 1. \quad (3.38)$$

Consequently, employing (3.37), (3.38), it follows that,

$$\ell_{k-1}(p, u, t) - \partial_t^{k-1} g_0(p, u, t) =$$

$$\int_0^t \ell_k(p, u, s) - \partial_t^k g_0(p, u, s)\, ds, \quad 1 \le k \le r - 1. \quad (3.39)$$

Evidently, the derivative ∂_t^m, $m \ge 1$, operating on the left side of (3.39) is equivalent to the derivative ∂_t^{m-1} operating on the integrand of the right side of (3.39). It therefore follows from (3.39) that one obtains the following increasing sup-norm estimates for each $p \in P$:

$$\|\ell_{k-1}(p) - \partial_t^{k-1} g_0(p)\|^{s, r-k} \le \|\ell_k(p) - \partial_t^k g_0(p)\|^{s, r-k-1},$$

$$1 \le k \le r - 1. \quad (3.40)$$

In particular, $\|\ell_0(p) - g_0(p)\|^{s, r-1} \le \|\ell_{r-1}(p) - \partial_t^{r-1} g_0(p)\|^{s, 0}$, from which it follows, employing (iv) ($\ell_{r-1} = \ell_\epsilon$), that $\lim_{\epsilon \to 0} \|\ell_0(p) - g_0(p)\|_U^{s, r-1} = 0$, which proves conclusion (i) for $g_\epsilon = \ell_0$.

It remains to prove conclusion (iii). Let $N(\delta) \subset N$ be a δ-ball neighbourhood of $\beta(P \times B)$. As in the proof of (relative) Complement 3.5 (iii) one may assume that the above P-parameter family of smooth C-structures (h, H) over B with respect to $\partial_t^r g_0, \beta$ satisfies the property that,

$$H([0, 1]^2 \times \mathfrak{Op}\, K) \subset N(\delta/2).$$

Since the image of F is contained in the image of H (cf. (3.23)), it follows that, over $\mathfrak{Op}\, K$, for all $\epsilon > 0$, the image of the homotopy F lies in $N(\delta/2)$. In particular, since $F_1 = \partial_t^r g_\epsilon$, it follows that over $\mathfrak{Op}\, K$, for all $p \in P$,

$$\|\partial_t^r g_\epsilon(p) - \partial_t^r g_0(p)\| = \|\partial_t^r g_\epsilon(p) - \beta(p)\| < \delta/2. \quad (3.41)$$

As in the proof of Theorem 3.4, let $g_\epsilon' = g_\epsilon + \lambda\, (g_0 - g_\epsilon)$, where $\lambda \colon P \times B \to [0, 1]$ is a C^∞ function, compactly supported in $\mathfrak{Op}\, K$, such that $\lambda = 1$ in a neighbourhood of K. Thus $g_\epsilon' = g_0$ in (a smaller) $\mathfrak{Op}_1 K$ and $g_\epsilon' = g_\epsilon$ on $B \setminus$

$\mathfrak{Op}\, K$. Since, for all $p \in P$, $\lim_{\epsilon \to 0} \|g_\epsilon(p) - g_0(p)\|^{s,r-1} = 0$, it follows also that $\lim_{\epsilon \to 0} \|g'_\epsilon(p) - g_0(p)\|^{s,r-1} = 0$; hence g'_ϵ is a C^r-map which also satisfies conclusion (i).

Let $F' = F'_\epsilon \colon [0,1] \times P \to \Gamma(E)$ be the homotopy $((v, p, b) \in [0,1] \times P \times B)$:

$$F'(v, p, b) = F_\epsilon(v, p, b) + v \partial^r_t(\lambda\,(g_0 - g_\epsilon))(p, b).$$

Evidently, $F' = F$ on $B \setminus \mathfrak{Op}\, K$, $F'_0 = \beta$; employing $F_1 = \partial^r_t g_\epsilon$, it is clear that $F'_1 = \partial^r_t g'_\epsilon$. We show that, for $\epsilon > 0$ sufficiently small, the homotopy $F' \colon [0,1] \times P \to \Gamma(\mathcal{R})$. Indeed, computing derivatives ∂^r_t,

$$F'(v, p, b) = F(v, p, b) + v \lambda(p, b)\,(\partial^r_t(g_0 - g_\epsilon)(p, b)) + E, \qquad (3.42)$$

where E is a linear combination of derivatives of the form, $\partial^j_t \lambda \cdot \partial^k_t(g_0 - g_\epsilon)$, $k + j = r$; $k \leq r-1$. Since $\lim_{\epsilon \to 0} \|g_\epsilon - g_0\|^{s,r-1} = 0$ it follows that $\lim_{\epsilon \to 0} \|E\| = 0$. Since λ is compactly supported in $\mathfrak{Op}\, K$, it follows from (3.41), (3.42) that, over $\mathfrak{Op}\, K$, for all $v \in [0,1]$ $(F_v(p, b) = F(v, p, b); F'_v(p, b) = F'(v, p, b))$:

$$\|F'_v - F_v\| < \delta/2,$$

provided $\epsilon > 0$ is sufficiently small. It follows that, over $\mathfrak{Op}\, K$, the image of the homotopy F' lies in $N(\delta) \subset N$ and hence $F' \colon [0,1] \times P \to \Gamma(\mathcal{R})$, provided $\epsilon > 0$ is sufficiently small. Note that over $\mathfrak{Op}_1 K$, $F'_0 = \beta$, $F'_1 = \partial^r_t g_0 = \partial^r_t g'_\epsilon = \beta$. Consequently over $\mathfrak{Op}_1 K$ F' is homotopic in $N(\delta)$, rel$\{0,1\}$, to the constant homotopy β. Employing standard arguments with bump functions to construct a homotopy of homotopies in $N(\delta)$ (as in the proof of Theorem 3.4 (iii)), F' is homotopic to a homotopy $F'' \colon [0,1] \times P \to \Gamma(\mathcal{R})$ such that $F'' = F'$ on $[0,1] \times (P \times B) \setminus \mathfrak{Op}\, K$, F'' is the constant homotopy β in (a smaller) $\mathfrak{Op}_1 K$. Hence the pair (g'_ϵ, F'') satisfies conclusion (iii) of the theorem. $\qquad \square$

Corollary 3.9 (Ample Relations). *Suppose in addition $\mathcal{R} \subset E$ is open and ample. Let $\beta \colon P \to \Gamma(\mathcal{R})$ be a smooth map, where P is compact C^∞ manifold (a parameter space) and let $s \geq r \geq 1$ be integers. For each smooth map $g_0 \colon P \to \Gamma(E)$, the following properties obtain:*

For each $\epsilon > 0$ there is a C^∞-map $g_\epsilon \colon P \to \Gamma(E)$ and a homotopy $F = F_\epsilon \colon [0,1] \times P \to \Gamma(\mathcal{R})$ such that, for all $p \in P$, conclusions (i), (ii), (iii) *are satisfied.*

Proof. Since $\mathcal{R}$ is ample it follows that for all $(p, b) \in P \times B$,

$$\partial^r_t g_0(p, b) \in \mathrm{Conv}(\mathcal{R}_b, \beta(p, b)).$$

Thus the main hypothesis of the $C^\perp$-Approximation Theorem is satisfied, from which the Corollary follows. $\qquad \square$

CHAPTER 4

OPEN AMPLE RELATIONS IN 1-JET SPACES

In this chapter we apply the analytic theory of Chapter III to prove the h-principle, Theorem 4.2, for differential relations in spaces of 1-jets which are open and ample. Differential relations in spaces of higher order jets and also non-ample relations are treated in subsequent chapters. There are good reasons for treating separately the cases of open, ample differential relations that occur in the context of spaces of 1-jets:

1. Included in these cases are many of the historically important results in immersion-theoretic topology such as the h-principle for immersions and for maps of rank $\geq k$. Included also are the h-principle for some more special applications to symplectic topology and to proving the existence of linearly independent divergence-free vector fields on manifolds. However, the h-principle for submersions of open manifolds, Phillips [31], does not follow directly from Theorem 4.2. Indeed, the submersion relation is not ample. Phillips' Theorem is proved in VIII Theorem 8.28 as an application of the theory of relations over jet spaces.

2. In these cases also the analytic theory developed in Chapter III applies directly to prove the h-principle. By contrast, in the cases of ample and non-ample relations in spaces of higher order jets, and of over-relations $\rho\colon \mathcal{R} \to X^{(r)}$, $r \geq 1$, more topological technique, specifically the theory of iterated convex hull extensions, needs to be introduced. These additional techniques are developed in Part II. Furthermore, the strong C^0-dense h-principle is trivially equivalent to the C^0-dense h-principle for relations in spaces of 1-jets (cf. I §3). In spaces of higher order jets the proof of the strong C^i-dense h-principle, $i \geq 1$, is technically more difficult and was an open question in Gromov [18] in the case of general non-ample relations. In Part II we prove strong C^i-dense h-principles (Theorems 8.4, 8.12), based on our strong version of the h-Stability Theorem 7.2.

3. The inductive proof procedure employed in the proof of Theorem 4.2 is the model for the proof procedure of the h-principle for relations $\rho\colon \mathcal{R} \to X^{(r)}$, $r \geq 1$, over spaces of higher order jets. This proof procedure is much easier to visualize and also is decidedly easier to implement in the case of spaces of 1-jets. In Gromov [17], where only spaces of 1-jets are treated, a somewhat different

D. Spring, *Convex Integration Theory: Solutions to the h-principle in geometry and topology*, Modern Birkhäuser classics, DOI 10.1007/978-3-0348-0060-0_4, © Springer Basel AG 2010

inductive proof procedure is employed, based on a cubical triangulation of the base manifold V (Adachi [1] also follows the proof procedure of Gromov [17]). This proof procedure works well for spaces of 1-jets since the corresponding $C^{\perp}$-Approximation Theorem in Gromov [17] is easily proved relative to the boundary of the cubes. In the case of higher order derivatives the proof of the relative version of the $C^{\perp}$-Approximation Theorem, III Theorem 3.8, is more delicate and there is no special advantage pertaining to cubical triangulations. Our proof procedure (adapted from Spring [37]) constructs a holonomic section inductively over a countable, locally finite covering by charts of the base manifold, at the nth stage of which one constructs a section which is holonomic in the nth chart and which coincides suitably on the overlap with the sections constructed during previous steps of the induction.

Analytically, the distinguishing feature of spaces of 1-jets is that locally, in coordinates $(u_1, u_2, \ldots, u_n)$ on the base manifold, the 1-jet of a function is expressed in terms of the "pure derivatives" $\partial/\partial u_i$, $1 \le i \le n$. Since the analytic approximation theory of Chapter III is also developed in terms of pure rth order derivatives, this theory applies directly to prove the main results in the 1-jet case.

The appearance of mixed derivatives in spaces of higher order jets requires the theory of convex hull extensions, developed in Part II, in order to reduce to the pure derivative case. However, the case of 2-jets is rather special. Although in this book we have not attempted to carry out the details, the approximation results of Chapter III can be applied to mixed derivatives of the form $\partial^2/\partial u_i \partial u_j$, $i \ne j$, by introducing suitable linear changes of coordinates which reduce to the pure derivative case. By treating these mixed derivatives of order 2 first, and then the remaining pure derivatives of order 2, the theorems in this chapter for 1-jet spaces carry over to open, ample relations in spaces of 2-jets.

§1. C^0-Dense h-Principle

Let $p\colon X \to V$ be a smooth fiber bundle, fiber dimension q, over a smooth base manifold V, $\dim V = n \ge 1$. Let $\mathcal{R} \subset X^{(1)}$ i.e., $\mathcal{R}$ is a differential relation in the space of 1-jets $X^{(1)}$. Locally near a point in X, $p\colon X \to V$ is a product bundle, $X_U = U \times \mathbf{R}^q \to U$, where $U \subset V$ is a chart. We identify $\Gamma^r(X_U) \equiv C^r(U, \mathbf{R}^q)$, $0 \le r \le \infty$. The manifold $X^{(1)}$ is represented locally as,

$$X_U^{(1)} = U \times W \times \prod_1^n \mathbf{R}^q = J^1(U, W). \tag{4.1}$$

where $W \subset \mathbf{R}^q$ is open. With respect to local coordinates $(u_1, u_2, \ldots, u_n)$ in U, a section $f \in \Gamma^1(X_U) = C^1(U, \mathbf{R}^q)$ induces the section $j^0 f(x) = (x, f(x))$, and the section $j^1 f \in \Gamma(X_U^{(1)})$, the 1-jet extension of f: $(\partial_i \equiv \partial/\partial u_i, \ 1 \leq i \leq n)$: for all $x \in U$,

$$j^1 f(x) = (x, f(x), \partial_1 f(x), \partial_2 f(x), \ldots, \partial_n f(x)) \in X_U^{(1)}. \qquad (4.2)$$

Let $X_U^{(1)} = X_U^\perp \times \mathbf{R}^q$, where the $\mathbf{R}^q$-factor corresponds to the derivative ∂_n. Thus for all $x \in U$,

$$j^1 f(x) = (j^\perp f(x), \partial_n f(x)),$$

where $j^\perp f(x) \in X_U^\perp$ is the "perp"-jet of f i.e., all derivatives except ∂_n are included. Evidently, the projection $p_\perp^1 : X_U^{(1)} \to X_U^\perp$ is a product $\mathbf{R}^q$-bundle. $X_z^{(1)} = z \times \mathbf{R}^q$ denotes the fiber over $z \in X_U^\perp$ of this bundle.

Convex Integration theory is concerned principally with the properties of the projection map (same notation) $p_\perp^1 : \mathcal{R}_U \to X_U^\perp$, where $\mathcal{R}_U = \mathcal{R} \cap X_U^{(1)}$. Analogously, let $\mathcal{R}_z$ denote the fiber of $\mathcal{R}$ over $z \in X_U^\perp$ with respect to the projection map $p_\perp^1$.

$$\mathcal{R} \xrightarrow{\ i\ } X^{(1)}$$
$$\downarrow{\scriptstyle p_\perp^1}$$
$$X^\perp$$

Ample Relations in $\mathbf{X}^{(1)}$. $\mathcal{R} \subset X^{(1)}$ is *ample* if with respect to all local coordinate representations $X_U^{(1)}$ above, the following property obtains for the $\mathbf{R}^q$-bundle $p_\perp^1 : X_U^{(1)} \to X_U^\perp$: for all $(z, w) \in X_U^\perp \times \mathcal{R}_z$, $\mathrm{Conv}(\mathcal{R}_z, w) = X_z^{(1)} \equiv \mathbf{R}^q$ i.e., in the fiber $X_z^{(1)}$, the convex hull of each path component of $\mathcal{R}_z$ is all of $\mathbf{R}^q$. The vacuous case $\mathcal{R}_z = \emptyset$ is included.

Example 4.1. The immersion relation in the extra dimensional case is open and ample. Specifically, let V, W be smooth manifolds $\dim V = n$, $\dim W = q$, $n \leq q$. A C^1-map $f : V \to W$ is an immersion if the tangent bundle map $df : T(V) \to T(W)$ has maximal rank $= n$. Let $p : X = V \times W \to V$ be the product bundle, fiber W. The immersion relation $\mathcal{R} \subset X^{(1)}$ is the subspace defined by germs of sections $f \in \Gamma^1(X)$ such that, with respect to the local representation (4.2) of $j^1 f$, the derivatives $\partial_i f(x) \in \mathbf{R}^q$, $1 \leq i \leq n$, are linearly independent. Thus, with respect to the local coordinates (4.1) on $X_U^{(1)}$ of the manifold $X^{(1)}$, $\mathcal{R}_U \subset X_U^{(1)}$ is the subspace of points $(x, y, (v_i)_{1 \leq i \leq n})$, such that the n vectors $v_i \in \mathbf{R}^q$, $1 \leq i \leq n$, are linearly independent. Clearly, $\mathcal{R}_U$ is open in $X_U^{(1)}$ and hence $\mathcal{R}$ is open in $X^{(1)}$.

The projection $p_\perp^1 : X_U^{(1)} \to X_U^\perp$ deletes the last vector $v_n \in \mathbf{R}^q$ in the above coordinates on $X_U^{(1)}$, i.e.,

$$p_\perp^1(x, y, v_1, v_2, \ldots, v_n) = (x, y, v_1, v_2, \ldots, v_{n-1}) \in X_U^\perp. \qquad (4.3)$$

Let $w = (x, y, (v_i)_{1 \leq i \leq n}) \in \mathcal{R}_U$ and let L be the $(n-1)$-dimensional subspace of $\mathbf{R}^q$ generated by the linearly independent vectors $v_1, v_2, \ldots, v_{n-1}$. Evidently, from (4.3), the fiber $\mathcal{R}_z$, $z = p_\perp^1(w) \in X_U^\perp$, consists of all points $(z, v) \in z \times \mathbf{R}^q \equiv \mathbf{R}^q$ such that the n vectors $v_1, v_2, \ldots v_{n-1}, v$ are linearly independent i.e., $v \notin L$ and $\mathcal{R}_z$ is the complement of the subspace L in $\mathbf{R}^q$. Consequently, in case $q = n$ (the equidimensional case), L has codimension 1, and $\mathcal{R}_z$ consists of two path connected $1/2$-spaces, the convex hull of each of which is itself i.e., these convex hulls are not $\mathbf{R}^q$. Hence the immersion relation is not ample in the equidimensional case. In case $q \geq n+1$, then L has codimension ≥ 2; hence $\mathcal{R}_z$ is path connected and its convex hull is $\mathbf{R}^q$ i.e., $\mathcal{R}$ is open and ample in the extra dimensional case.

Let $s \colon X^{(1)} \to V$ be the source map, $s(j^1 f(x)) = x$, $x \in V$. $\Gamma^r(\mathcal{R})$ denotes the space of C^r-sections, in the compact-open C^r-topology, with respect to the source map (same notation) $s \colon \mathcal{R} \to V$, $0 \leq r \leq \infty$. Recall the projection map $p_0^1 \colon X^{(1)} \to X^{(0)}$, $p_0^1(j^1 f(x)) = f(x) \in X$, $x \in V$, the 0-jet component of $j^1 f(x)$.

h-Principle. Recall I §3, a section $\alpha \in \Gamma(\mathcal{R})$ is *holonomic* if there is a C^1-section $h \in \Gamma^1(X)$ such that $\alpha = j^1 h$. The relation $\mathcal{R} \subset X^{(1)}$ satisfies the *h-principle* if for each $\alpha \in \Gamma(\mathcal{R})$ there is a homotopy $H \colon [0,1] \to \Gamma(\mathcal{R})$, $H_0 = \alpha$, such that H_1 is holonomic. The principle result of this chapter is the following theorem.

Theorem 4.2 (C^0-dense h-principle). *Let $p \colon X \to V$ be a smooth fiber bundle, fiber dimension q, over a smooth manifold V, $\dim V = n \geq 1$, and let $\mathcal{R} \subset X^{(1)}$ be open and ample. Let $h \in \Gamma(X)$ be a continuous section and let $\varphi \in \Gamma(\mathcal{R})$ be a continuous extension of h: $p_0^1 \circ \varphi = h \in \Gamma(X)$. Let K_0 be closed in V and suppose h is C^1 on $\mathfrak{Op}\, K_0$ and satisfies $j^1 h = \varphi$ on K_0. Furthermore, let $\mathcal{N}$ be a neighbourhood of the image $h(V)$ in X.*

There is a section $f \in \Gamma^1(X)$ and a homotopy $F \colon [0,1] \to \Gamma(\mathcal{R})$ such that the following properties are satisfied.

  (i) *$j^1 f \in \Gamma(\mathcal{R})$ i.e. f solves the relation $\mathcal{R}$.*
  (ii) *$f(V) \subset \mathcal{N}$ i.e. f is C^0-close to h in the fine C^0-topology.*
  (iii) *$F_0 = \varphi$, $F_1 = j^1 f$; for all $t \in [0,1]$, the image $p_0^1 \circ F_t(V) \subset \mathcal{N}$.*
  (iv) *(Relative theorem) For all $t \in [0,1]$, $F_t = \varphi$ on K_0.*

Remark 4.3. Conclusion (iii) of the theorem proves the h-principle for open, ample relations $\mathcal{R} \subset X^{(1)}$. Indeed, let $\varphi \in \Gamma(\mathcal{R})$. Let $p_0^1 \circ \varphi = h \in \Gamma(X)$, the induced 0-jet component of the section φ. Conclusion (iii) of the theorem, applied to the data φ, h ($K_0 = \emptyset$) states that φ is homotopic in $\mathcal{R}$ to a holonomic section, which proves that $\mathcal{R}$ satisfies the h-principle. Furthermore, in case no relative theorem is required i.e. $K_0 = \emptyset$, one may choose f to be smooth.

Proof of Theorem. In order to apply the results of Chapter III, we first arrange for the data h, φ to be smooth on $V \setminus K_0$. Let N be a neighbourhood of the image $\varphi(V)$ in $\mathcal{R}$. One identifies a small neigbourhood U of $h(V)$ in X with a vertical disk bundle (= $\ker Ds$) with respect to the smooth source map $s\colon X \to V$. Since $p_0^1\colon X^{(1)} \to X$ is a Serre fibration, there is a continuous lift $\nu\colon U \to N$ such that: (i) $p_0^1 \circ \nu = \mathrm{id}_U$; (ii) $\nu \circ h = \varphi \in \Gamma(X^{(1)})$ (cf. Bredon [5], p. 451 and V Theorem 5.1).

Employing the smooth approximation theorem, I Theorem 1.1 (h is C^1 in $\mathfrak{Op}\, K_0$), up to a small homotopy rel K_0 in U one may assume that h is smooth on $V \setminus K_0$. Employing the lift ν, one lifts the above homotopy into N. Thus, up to a small homotopy of φ in $\mathcal{R}$, $p_0^1 \circ \varphi = h$ is smooth on $V \setminus K_0$. Again employing the bundle $p_0^1\colon X^{(1)} \to X$, up to a further small homotopy on φ in $\mathcal{R}$ over the base $h(V) \subset X$ one may assume that φ is smooth on $V \setminus K_0$ and that $\varphi = j^1 h$ on $\mathfrak{Op}\, K_0$.

Let $(W_i)_{i \geq 1}$, $(U_i)_{i \geq 1}$ be locally finite coverings by closed charts of $\overline{V \setminus K_1}$ where K_1 is a closed neighbourhood of K_0 in $\mathfrak{Op}\, K_0$ such that for all $i \geq 1$: (i) $W_i \subset \mathrm{int}\, U_i$; (ii) U_i is diffeomorphic to the cube $[0,1]^n \subset \mathbf{R}^n$; (iii) $U_i \subset V \setminus K_0$. In particular $\varphi = j^1 h$ on $\mathfrak{Op}\, K_1$ and h, φ are smooth on each chart of the cover $(U_i)_{i \geq 1}$.

The proof proceeds by induction, which we briefly describe as follows. Let $\mathcal{N}_1$ be a neighbourhood of the image $h(V)$ in X such that $\overline{\mathcal{N}}_1 \subset \mathcal{N}$. Beginning with the section h, one constructs a sequence of sections h_i, $i = 0, 1, 2, \ldots$, $h_0 = h$, such that for all $i \geq 1$, $h_i = h_{i-1}$ on the complement of the chart U_i in V, and $j^1 h_i \in \Gamma(\mathcal{R})$ on $\mathfrak{Op}\, W_i$ i.e., the section h_i "improves" the section h_{i-1} by solving the relation $\mathcal{R}$ over $W_i \subset U_i$. This improvement requires the approximation theory of Chapter III which ensures also that $h_i(V) \in \mathcal{N}_1$, and that for all $i \geq 1$ ($W_0 = \emptyset$),

$$j^1 h_i = \varphi \text{ on } \mathfrak{Op}\, K_1; \quad h_i = h_{i-1} \text{ on } \bigcup_0^{i-1} \mathfrak{Op}\, W_i.$$

Then, $f = \lim_{i \to \infty} h_i \in \Gamma^1(X)$ satisfies all the conclusions of the theorem. The modification of the section h_{i-1} in the chart U_i to obtain the section h_i, $i \geq 1$, is itself an inductive process, the main step of which is as follows.

The Main Inductive Step. Let U be closed chart on the base manifold $V \setminus K_0$ such that U is diffeomorphic to the cube $I^n = [0,1]^n \subset \mathbf{R}^n$, with local coordinates $(u_1, u_2, \ldots, u_n)$. We identity U with I^n.

We are given the following data: a local section $g \in \Gamma^\infty(X_U)$ such that $g(U) \cap \partial X = \emptyset$ i.e., $g \in C^\infty(I^n, \mathbf{R}^q)$; an integer $k \in \{0, 1 \ldots, n-1\}$ and a section $\rho \equiv \rho_k \in \Gamma^\infty(\mathcal{R}_U)$; a closed set $L \subset V$; a closed set W in the interior of U.

This data is assumed to satisfy the following properties with respect to the local coordinate representation (4.1) of $X_U^{(1)}$.

(a_k) $\rho = j^1 g$ on $U \cap \mathfrak{Op}\, L$.

(b_k) $p_0^1 \circ \rho = g$ i.e. g is the 0-jet component of ρ. Furthermore, for all $x \in W$,

$$\rho(x) = (x, g(x), \partial_1 g(x), \dots, \partial_k g(x), \varphi_{k+1}(x), \dots, \varphi_n(x)) \in \mathcal{R}_U,$$

where $\varphi_j \in C^\infty(I^n, \mathbf{R}^q)$, $k + 1 \le j \le n$.

Note that in case $k = 0$, for all $x \in W$, the map ρ includes none of the derivatives of g: $\rho(x) = (x, g(x), \varphi_1(x), \varphi_2(x), \dots, \varphi_n(x)) \in \mathcal{R}_U$.

Lemma 4.4. *Let $\delta > 0$. Let $k \in \{0, \dots, n-1\}$ and let the pair (g, ρ), $\rho \equiv \rho_k$, satisfy (a_k), (b_k) above. Let also $\mathcal{M}$ be a neighbourhood of the image $g(U)$ in X_U. There is a map $f \in C^\infty(U, \mathbf{R}^q)$ and a C^∞-homotopy $F \colon [0,1] \to \Gamma(\mathcal{R}_U)$ such that the following properties are satisfied ($\| \; \|$ is the sup-norm on $C^0(I^n, \mathbf{R}^q)$):*

(v) $\|f - g\| < \delta$. *For all $t \in [0,1]$ the image $p_0^1 \circ F_t(U) \subset \mathcal{M}$.*

(vi) $F_0 = \rho_k$; $p_0^1 \circ F_1 = f$, *and for all $x \in W$,*

$$F_1(x) = (x, f(x), \partial_1 f(x), \dots, \partial_{k+1} f(x), \varphi_{k+2}(x), \dots, \varphi_n(x)) \in \mathcal{R}_U.$$

(vii) $f = g$ on $U \cap \mathfrak{Op}\, L$; *the homotopy F is constant on $\mathfrak{Op}\, L$: for all $t \in [0,1]$,*
    $F_t = \rho = j^1 g$ on $U \cap \mathfrak{Op}\, L$.

In particular, the pair (f, F_1) satisfies the properties (a_{k+1}), (b_{k+1}).

Proof. The point of the lemma is that on W the section ρ, which includes the first k derivatives of the section g, is improved through a homotopy rel L in $\Gamma(\mathcal{R}_U)$ to a section F_1 which includes the first $(k+1)$-derivatives of the section f. In particular, in case $k = n - 1$, $F_1 = j^1 f \in \Gamma_W(\mathcal{R})$. Note that this improvement occurs in the $\mathbf{R}^q$-factor of $X_U^{(1)}$ which corresponds to the derivative $\partial_{k+1} = \partial/\partial u_{k+1}$.

For the purposes of the lemma, notationally it is convenient to shift the ∂_{k+1}-coordinates in $X_U^{(1)}$ to the last factor i.e., let $X_U^{(1)} = X_U^\perp \times \mathbf{R}^q$, where the $\mathbf{R}^q$-factor corresponds to the derivative ∂_{k+1}. Thus, with respect to this product structure, for all $x \in U$,

$$\rho(x) = (\rho^\perp(x), \varphi_{k+1}(x)) \in \mathcal{R}_U. \tag{4.4}$$

where $\rho^\perp(x) = (x, g(x), \varphi_1(x), \dots, \widehat{\varphi}_{k+1}(x), \dots, \varphi_n(x)) \in X_U^\perp$ i.e., $\rho^\perp(x)$ deletes the component $\varphi_{k+1}(x)$ of $\rho(x)$.

Let $\pi\colon E = I^n \times \mathbf{R}^q$ be the product $\mathbf{R}^q$-bundle over I^n obtained by pulling back the product bundle $p^\perp\colon X_U^{(1)} \to X_U^\perp$ (associated to the above product decomposition with respect to the derivative ∂_{k+1}) along the embedding $\rho^\perp\colon I^n \to X_U^\perp$:

$$
\begin{array}{ccc}
E & \longrightarrow & X_U^{(1)} \\
\pi \downarrow & & \downarrow p^\perp \\
I^n & \overset{\rho^\perp}{\longrightarrow} & X_U^\perp
\end{array}
$$

Let $\mathcal{S} \subset E$ be the pull-back of the relation $\mathcal{R}$. Since $\mathcal{R}$ is open in $X^{(1)}$ and ample with respect to all coordinate directions in the chart U, it follows that $\mathcal{S}$ is open and ample in E. One now reformulates the data of the lemma in terms of the relation $\mathcal{S}$ in E. Since $\rho \in \Gamma^\infty(\mathcal{R}_U)$, it follows from (4.4) that $\varphi_{k+1} \in \Gamma^\infty(\mathcal{S})$. Since $\mathcal{S}$ is ample in E, for all $x \in I^n$,

$$\partial_{k+1}g(x) \in \mathrm{Conv}(\mathcal{S}_x, \varphi_{k+1}(x)). \tag{4.5}$$

Employing property (a_k) of the data, it follows that on $\mathfrak{Op}\, L$, $\varphi_{k+1} = \partial_{k+1}g$. Fix an integer $s \geq 1$. Applying III Complement 3.7 (without parameters) to the above data, and for which the main hypothesis is satisfied by (4.5), one concludes that, for each $\epsilon > 0$ sufficiently small, there is a section $g_\epsilon \in \Gamma^\infty(E)$ and a C^∞-homotopy $G = G_\epsilon\colon [0,1] \to \Gamma(\mathcal{S})$, such that the following properties obtain (for convenience of notation, let $t = u_{k+1}$; $\partial_t = \partial_{k+1}$):

(viii) $\lim_{\epsilon \to 0} \|g_\epsilon - g\|^s = 0$.

(ix) $\partial_t g_\epsilon \in \Gamma(\mathcal{S})$; $G_0 = \varphi_{k+1}$; $G_1 = \partial_t g_\epsilon$.

(x) $g_\epsilon = g$ on $U \cap \mathfrak{Op}\, L$; for all $r \in [0,1]$, $G_r = \varphi_{k+1} = \partial_{k+1}g$ on $U \cap \mathfrak{Op}\, L$.

To prove the lemma we suitably "push forward" the pair of maps (g_ϵ, G_ϵ) back to $\mathcal{R}_U$. With respect to the source map $s\colon X_U^{(1)} \to U$, let $F = F_\epsilon\colon [0,1] \to \Gamma(X_U^{(1)})$ be the homotopy $((t,x) \in [0,1] \times U)$,

$$
\begin{aligned}
F_\epsilon(t,x) = {} & (x, g(x) + t(g_\epsilon - g)(x), \varphi_1(x) + t\partial_1(g_\epsilon - g)(x), \\
& \ldots, \varphi_k + t\partial_k(g_\epsilon - g)(x), \varphi_{k+2}(x), \ldots, \varphi_n(x), G_\epsilon(t,x)) \\
= {} & (y_\epsilon(t,x), G_\epsilon(t,x)) \in X_U^\perp \times \mathbf{R}^q = X_U^{(1)}, \tag{4.6}
\end{aligned}
$$

where $y_\epsilon(t, \cdot) \in \Gamma(X_U^\perp)$ is the $\perp$-component of the homotopy F_ϵ. In particular, at $t = 1$, $p_0^1 \circ F_\epsilon(1, \cdot) = g_\epsilon \in \Gamma(X_U)$. Since $G_0 = \varphi_{k+1}$, it follows that $F_0 = \rho$. Furthermore, from (x), $g_\epsilon = g$ on $U \cap \mathfrak{Op}\, L$, and the homotopy G is constant $= \varphi_{k+1}$ on $U \cap \mathfrak{Op}\, L$; hence, employing (a_k), it follows that the homotopy F_ϵ is constant $= \rho = j^1 g$ on $\mathfrak{Op}\, L$. Setting $t = 1$, and employing (ix), $G_1 = \partial_t g_\epsilon = \partial_{k+1} g_\epsilon$; hence, from (4.6), (b_k), for all $x \in W$,

$$F_\epsilon(1, x) = (x, g_\epsilon(x), \partial_1 g_\epsilon(x), \dots, \partial_k g_\epsilon(x),$$
$$\varphi_{k+2}(x), \dots, \varphi_n(x), \partial_{k+1} g_\epsilon(x)). \quad (4.7)$$

Note also from (viii) that in particular, since $s > 0$, $\lim_{\epsilon \to 0} \|g_\epsilon - g\| = 0$. Setting $f = g_\epsilon$, for $\epsilon > 0$ sufficiently small, $\|f - g\| < \delta$ i.e. the first part of conclusion (v) is satisfied. From (4.6), the 0-jet component of the homotopy F_ϵ is explicitly:

$$p_0^1 \circ F_\epsilon(t, x) = (x, g(x) + t(g_\epsilon(x) - g(x))) \in X_U.$$

Since $\lim_{\epsilon \to 0} \|g_\epsilon - g\| = 0$ it follows that, for $\epsilon > 0$ sufficiently small, for all $(t, x) \in [0, 1] \times U$, $p_0^1 \circ F_\epsilon(t, x) \in \mathcal{M}$, which completes the proof of (v).

The above properties established for the homotopy F_ϵ prove that the pair of maps (f, F_ϵ) satisfies all the conclusions of Lemma 4.4 except possibly the property that F_ϵ is a homotopy in $\Gamma(\mathcal{R}_U)$.

Remark 4.5. The push forward of the bundle $\pi \colon E \to I^n$ by $\rho^\perp$ lies in $\mathcal{R}_U$ over the base $\rho^\perp(U)$ in $X_U^\perp$. Employing (4.6), the $\perp$-jet component of the homotopy $F_\epsilon(t, \cdot)$ is $y_\epsilon(t, \cdot)$, a small perturbation (which depends on ϵ) of $\rho^\perp$. However the homotopy G_ϵ (the ∂_t-component of the homotopy F_ϵ) is constructed as in III Complement 3.7 in terms of the rapidly oscillation function θ_ϵ (cf. (3.23)). Hence the homotopy $G_\epsilon(t, \cdot)$ oscillates rapidly and does not satisfy uniform approximations as $\epsilon \to 0$. Further arguments, presented below, are therefore required to prove that F_ϵ is a homotopy in $\Gamma(\mathcal{R}_U)$.

The proof that F_ϵ is a homotopy in $\Gamma(\mathcal{R}_U)$ is somewhat delicate and depends in an essential way on the C-structure (h, H) employed in the proof of Complement 3.7, as it applies to the relation $\mathcal{S} \subset E$ above. Specifically, employing Complement 3.7 (ii) (cf. also the remarks following the proof of Theorem 3.4), except for a small perturbation near L to arrange that the homotopy G_ϵ is constant $= \varphi_{k+1}$ on $U \cap \mathfrak{Op}\, L$, for all $\epsilon > 0$ the image of the homotopy G_ϵ is contained in the image of H, where $H \colon [0, 1]^2 \to \Gamma(\mathcal{S})$. Consequently, for all $\epsilon > 0$, the image of G_ϵ is contained in a compact set in $\mathcal{S}$, thus providing the control required to prove that, for $\epsilon > 0$ sufficiently small, F_ϵ, defined by (4.6), is a homotopy in $\Gamma(\mathcal{R}_U)$. The details are as follows.

Let $H' \colon [0, 1]^2 \to \Gamma(\mathcal{R}_U)$ be the push forward of the map H along the embedding $\rho^\perp \colon U \to X_U^\perp$: for all $(t, s, x) \in [0, 1]^2 \times U$,

$$H'(t, s, x) = (\rho^\perp(x), H(t, s, x)) \in \mathcal{R}_U. \quad (4.8)$$

Let V be a neighbourhood of the compact set im H' in the open set $\mathcal{R}_U$. Since I^n is compact, there is a $\delta > 0$, such that, with respect to a metric d on the manifold $X^\perp$, the following property obtains: for all $(t, s, x) \in [0, 1]^2 \times U$,

$$(y, H(t, s, x)) \in V \subset \mathcal{R}_U \quad \text{for all } y \in X_U^\perp \text{ such that } d(\rho^\perp(x), y) < \delta. \quad (4.9)$$

Employing (viii), since $s \geq 1$, it follows in particular that,

$$\lim_{\epsilon \to 0} \|\partial_j(g_\epsilon - g)\| = 0 \quad 1 \leq j \leq n - 1. \quad (4.10)$$

Consequently, for $\epsilon > 0$ sufficiently small, one concludes from (4.6), (4.10) that for all $(t, x) \in [0, 1] \times U$,

$$d(y_\epsilon(t, x), \rho^\perp(x)) < \delta. \quad (4.11)$$

Employing (4.9), (4.11), for $\epsilon > 0$ sufficiently small, it follows that for all $(t, s, u, x) \in [0, 1]^3 \times U$,

$$(y_\epsilon(t, x), H(u, s, x)) \in \mathcal{R}_U. \quad (4.12)$$

As explained above, on the complement of a neighbourhood $N(L)$ of $L \cap U$ in U, the image of the homotopy G_ϵ is contained in im H. Consequently, for sufficiently small $\epsilon > 0$, employing (4.6), (4.12),

$$F_\epsilon \colon [0, 1] \times (U \setminus N(L)) \to \mathcal{R}_U. \quad (4.13)$$

It remains to prove that the image of the homotopy F_ϵ on $N(L)$ is also contained in $\mathcal{R}_U$. To this end, let N_ρ be a neighbourhood of $\rho(U)$ in $\mathcal{R}_U$. Employing (4.4), $\varphi_{k+1} \in \Gamma(\mathcal{S})$ is the pullback of $\rho \in \Gamma(\mathcal{R}_U)$; hence there is a compact neighbourhood N_{k+1} of $\varphi_{k+1}(U)$ in $\mathcal{S}$ such that N_{k+1} is contained in the pullback of N_ρ. Applying Complement 3.7 (iii) to the neighbourhood N_{k+1}, one can assume that, for $\epsilon > 0$ sufficiently small, the small perturbation to obtain the homotopy G_ϵ in $N(L)$ satisfies the property,

$$G_\epsilon \colon [0, 1] \times N(L) \to N_{k+1} \subset \mathcal{S}. \quad (4.14)$$

Since the push forward of the compact neighbourhood N_{k+1} is contained in N_ρ, it follows from (4.11), (4.14) that, for $\epsilon > 0$ sufficiently small, the homotopy F_ϵ in (4.6) satisfies the property,

$$F_\epsilon \colon [0, 1] \times N(L) \to N_\rho \subset \mathcal{R}_U. \quad (4.15)$$

Employing (4.13), (4.15), it follows that $F_\epsilon \colon [0, 1] \to \Gamma(\mathcal{R}_U)$, which completes the proof of the lemma. $\qquad \square$

Corollary 4.6. *Suppose the data* (g, ρ) *satisfies properties* (a_0), (b_0) *(i.e.* $k = 0$*), with respect to a closed set* $L \subset U$*. Let* $\delta > 0$*. Let also* $\mathcal{M}$ *be a neighbourhood of the image* $g(U)$ *in* X_U*. There is a map* $f \in C^\infty(U, \mathbf{R}^q)$ *and a* C^∞*-homotopy* $F[0,1] \to \Gamma(\mathcal{R}_U)$ *such that the following properties are satisfied.*

(i) $\|f - g\| < \delta$*. For all* $t \in [0,1]$*, the image* $p_0^1 \circ F_t(U) \subset \mathcal{M}$*.*

(ii) $F_0 = \rho$*;* $p_0^1 \circ F_1 = f \in \Gamma(X_U)$*;* $F_1 = j^1 f$ *on* $\mathfrak{Op}\, W$ *i.e.* $j^1 f \in \Gamma_W(\mathcal{R})$*.*

(iii) $f = g$ *on* $U \cap \mathfrak{Op}\, L$*; the homotopy* F *is constant on* $\mathfrak{Op}\, L$*: for all* $t \in [0,1]$*,* $F_t = \rho = j^1 g$ *on* $U \cap \mathfrak{Op}\, L$*.*

Proof. Let $\mu \in (0, \delta/n)$ be an auxiliary constant. Inductively on k, one employs Lemma 4.4 n-times, starting with the pair (g, ρ) at $k = 0$ $(f^0 = g)$, to obtain a sequence of pairs (f^k, F^k) which satisfy the properties (a_k), (b_k) of the data, $1 \le k \le n$, where $f^k \in C^\infty(U, \mathbf{R}^q)$, $F^k : [0,1] \to \Gamma^\infty(\mathcal{R}_U)$, and such that the following properties obtain for all k, $1 \le k \le n$.

(iv) $\|f^k - f^{k-1}\| < \mu$*. For all* $t \in [0,1]$ the image $p_0^1 \circ F_t^k(U) \subset \mathcal{M}$.

(v) $F_0^1 = \rho$; $F_1^k = F_0^{k+1}$; $p_0^1 \circ F^k(1, \cdot) = f^k \in \Gamma(X_U)$.

(vi) $f^k = g$ on $U \cap \mathfrak{Op}\, L$; the homotopy F^k is constant on $\mathfrak{Op}\, L$: for all $t \in [0,1]$, $F_t^k = \rho = j^1 g$ on $U \cap \mathfrak{Op}\, L$.

From (v), the time-1 map of the homotopy F^k is equal to the time-0 map of the homotopy F^{k+1}. One may therefore concatenate the homotopies F^k, $1 \le k \le n$, to obtain a homotopy $F : [0,1] \to \Gamma(\mathcal{R}_U)$ such that $F_0 = \rho$, $F_1 = j^1 f^n \in \Gamma_W(\mathcal{R}_U)$. Employing (iv), for all $t \in [0,1]$ the 0-jet component $p_0^1 \circ F_t(U) \subset \mathcal{M}$. Let $f = f^n$. Employing (iv), $\|f - g\| < n\mu < \delta$, which completes the proof of (i), and the corollary is proved. $\square$

Returning to the proof of Theorem 4.2 suppose, inductively, there are sections $h_i \in \Gamma^1(X)$, smooth on $V \setminus K_0$, $0 \le i \le m$, such that $h_0 = h$, and such that the following properties are satisfied with respect to the covers $(W_i)_{i \ge 1}$, $(U_i)_{i \ge 1}$ of the base manifold $\overline{V \setminus K_1}$. For all i, $1 \le i \le m$ $(W_0 = \emptyset)$:

(p_1) $h_i = h_{i-1}$ on $\mathfrak{Op}\,(V \setminus U_i)$.

(p_2) $h_i = h$ on $\mathfrak{Op}\, K_1$.

(p_3) $j^1 h_i(\mathfrak{Op}\, W_i) \subset \mathcal{R}$.

(p_4) $h_i = h_{i-1}$ on $\bigcup_0^{i-1} \mathfrak{Op}\, W_j$.

(p_5) The image $h_i(V) \in \mathcal{N}_1$.

Furthermore, we suppose inductively that there are homotopies $H_i : [0,1] \to \Gamma(\mathcal{R})$, smooth on $[0,1] \times (V \setminus K_0)$, $0 \le i \le m$, where H_0 is the constant homotopy $= \varphi$, and such that the following properties are satisfied. For all $(t, x) \in [0,1] \times V$, and for all i, $1 \le i \le m$:

(q_1) $H_1(0, x) = \varphi(x)$; for all $i \geq 1$, $H_i(0, x) = H_{i-1}(1, x)$.

(q_2) $H_i(t, x) = H_{i-1}(1, x)$ for all $x \in \mathfrak{Op}\,(V \setminus U_i)$ i.e., the homotopy H_i is constant on $\mathfrak{Op}\,(V \setminus U_i)$.

(q_3) The homotopy H_i is constant $= \varphi$ on $\mathfrak{Op}\,K_1$.

(q_4) $H_i(1, x) = j^1 h_i(x)$ on $\bigcup_0^i \mathfrak{Op}\,W_j$.

(q_5) At $t = 1$, $p_0^1 \circ H_i(1, \cdot) = h_i \in \Gamma(X)$; for all $(t, x) \in [0, 1] \times V$, $p_0^1 \circ H_i(t, x) \in \mathcal{N}_1$.

We construct below $h_{m+1} \in \Gamma^1(X)$ and a homotopy $H_{m+1} \colon [0, 1] \to \Gamma(\mathcal{R})$, smooth on $V \setminus K_0$, such that the pair (h_{m+1}, H_{m+1}) satisfies the above properties for $i = m + 1$. Note that the induction begins with $m = 0$ i.e., with the section $h_0 = h \in \Gamma^\infty(X)$ and the constant homotopy $H_0 = \varphi(x)$.

For notational convenience in what follows, we set $U = U_{m+1}$, $W = W_{m+1}$, $g = h_m$ on U and let $\rho \in \Gamma(\mathcal{R}_U)$ be the restriction of $H_m(1, \cdot)$ to U. In case $m = 0$, ρ is the restriction of φ to U. Thus the pair of maps (g, ρ) satisfies the following properties:

(r_1) In case $m \geq 1$, from (q_5), $p_0^1 \circ \rho = g \in \Gamma(X_U)$; in case $m = 0$, $g = h_U$ is the 0-jet component of φ over U by hypothesis.

(r_2) From (q_4), $\rho = j^1 g$ on $U \cap \bigcup_0^m W_j$; from (q_3), (p_2), $\rho = \varphi = j^1 g$ on $\mathfrak{Op}\,(K_1 \cap U)$.

Employing (r_2), it follows that,

(r_3) $\rho = j^1 g$ on $U \cap \mathfrak{Op}\,L$, where $L = K_1 \cup \bigcup_0^m W_j$.

Consequently, employing (r_1), (r_3), the pair of maps (g, ρ) satisfies properties (a_0), (b_0) (i.e. $k = 0$) of the data for Lemma 4.4, with respect to the closed set $L \subset V$ defined in (r_3). Let $\delta > 0$. Applying Corollary 4.6 to the data (g, ρ) and the closed set L, it follows that there is a map $f \in C^\infty(U, \mathbf{R}^q)$ and a C^∞-homotopy $F \colon [0, 1] \to \Gamma(\mathcal{R}_U)$ such that the following properties are satisfied.

(i) $\|f - g\| < \delta$. For all $s \in [0, 1]$ $p_0^1 \circ F_s(U) \subset \mathcal{N}_1$.

(ii) $F_0 = \rho$; $F_1 = j^1 f$ on $\mathfrak{Op}\,W$; at $t = 1$, $p_0^1 \circ F_1(1, \cdot) = f \in \Gamma(X_U)$.

(iii) $f = g = h_m$ on $U \cap \mathfrak{Op}\,L$; the homotopy F is constant $= \rho = j^1 g$ on $U \cap \mathfrak{Op}\,L$.

Let $\mu \colon V \to [0, 1]$ be a C^∞-function such that $\mu = 0$ on $\mathfrak{Op}\,(V \setminus U)$ and $\mu = 1$ on $\mathfrak{Op}\,W$. Employing the cut-off function μ, the homotopy $H_{m+1} \colon [0, 1] \to \Gamma(\mathcal{R})$ is constructed as follows. For all $(t, x) \in [0, 1] \times V$,

$$H_{m+1}(t, x) = \begin{cases} H_m(1, x) & \text{if } x \in \mathfrak{Op}\,(V \setminus U) \\ F(\mu(x)t, x) & \text{if } x \in U. \end{cases} \tag{4.16}$$

Since $\mu = 0$ on $\mathfrak{Op}(\partial U)$ and $F_0 = \rho$, it follows that H_{m+1} is well defined, continuous, and smooth on $V \setminus K_0$. Evidently, $H_{m+1}(0, \cdot) = H_m(1, \cdot)$. Employing (iii), the homotopy H_{m+1} is constant $= j^1 h_m$ on $\mathfrak{Op}\, L$. In particular, from (p_2), the homotopy H_{m+1} is constant $= j^1 h = \varphi$ on $\mathfrak{Op}\, K_1$. Hence properties (q_1), (q_2), (q_3) are satisfied. Also from (iii), (q_4) (for $i = m$), the homotopy H_{m+1} is constant $= j^1 h_m$ on $\bigcup_0^m \mathfrak{Op}\, W_j$.

Let $h_{m+1} \in \Gamma^1(X)$, smooth on $V \setminus K_0$, be the section such that $h_{m+1} = p_0^1 \circ H_{m+1}(1, \cdot) \in \Gamma(X)$ i.e. the 0-jet component of the homotopy H_{m+1} at $t = 1$. Thus the first part of (q_5) is satisfied by definition. From the properties established above for the homotopy H_{m+1}, it follows that the section h_{m+1} satisfies (p_1), (p_2) and (p_4) (for $i = m+1$). Furthermore, since $\mu = 1$ on $\mathfrak{Op}\, W$, employing (ii), it follows that property (p_3) is satisfied for $i = m + 1$, from which one concludes also that the homotopy H_{m+1} satisfies (q_4).

It remains to show the approximation properties (p_5), (q_5) for $i = m + 1$. Applying (q_5) to the homotopy H_m, and property (i) above for the 0-jet component of the homotopy F, it follows from (4.16) that for all $(t, x) \in [0, 1] \times V$,

$$p_0^1 \circ H_{m+1}(t, x) \in \mathcal{N}_1 \tag{4.17}$$

which completes the proof of (q_5). In particular, the image $h_{m+1}(V) \subset \mathcal{N}_1$, which proves (p_5) and the inductive construction of the sequences $(h_i)_{i \geq 0}$, $(H_i)_{i \geq 0}$ is complete.

To complete the proof of Theorem 4.2, let $f \in \Gamma(X)$ be the section $f = \lim_{i \to \infty} h_i$ (pointwise). Since the cover $(W_i)_{i \geq 1}$ of $V \setminus K_1$ is locally finite, it follows that for all compact sets K in V, there is an integer $n \geq 1$ such that $K \subset \mathfrak{Op}\, K_1 \bigcup_0^n W_j$. Employing property (p_4) of the sequence $(h_i)_{i \geq 1}$, it follows that $f = h_m$ on K for all $m \geq n$. Since the manifold V is a countable union of compact subsets, one concludes that f is well-defined and $f \in \Gamma^1(X)$. In particular, employing (p_5) the image $f(V) \subset \overline{\mathcal{N}_1} \subset \mathcal{N}$. Employing property (q_4) of the sequence of homotopies $(H_i)_{i \geq 1}$, it follows also that $j^1 f \in \Gamma(\mathcal{R})$ thus proving properties (i), (ii) of the theorem. Furthermore, employing property (p_2), $f = h$ on $\mathfrak{Op}\, K_1$.

It remains to show that φ is homotopic in $\Gamma(\mathcal{R})$ rel K_0, to $j^1 f$. Employing (q_1), for each $i \geq 1$, the time-0 map $H_i(0, \cdot)$ is equal to the time-1 map $H_{i-1}(1, \cdot)$. The homotopy F is obtained by setting $F_1 = j^1 f$; $F_t = \varphi$ (constant homotopy) on $[0, 1/2]$; on the interval $[1/2, 1]$, F is the concatenation of the homotopies H_i', where for all $i \geq 1$, H_i' is obtained from H_i by transforming $[0, 1]$ into $[1 - 2^{-i}, 1 - 2^{-(i+1)}]$. Since the cover $(W_i)_{i \geq 1}$ is locally finite, employing (q_2), (q_4), it follows that $F \colon [0, 1] \to \Gamma(\mathcal{R})$ is well-defined, continuous, and $F_0 = \varphi$, $F_1 = j^1 f$. Employing (q_3), (q_5), the homotopy F is constant $= \varphi$ on $\mathfrak{Op}\, K_1$ and for all $t \in [0, 1]$, the projection $p_0^1 \circ F_t(V) \subset \mathcal{N}$, which proves conclusions (iii), (iv) and the proof of the theorem is complete. $\qquad \square$

The main corollary of theorem 4.2 is the weak homotopy equivalence theorem below (cf. I §1.3.2). In particular, in the case of the immersion relation, Example 4.1, one recovers the classical weak homotopy equivalence results of Hirsch [21], Smale [36].

Let $\mathcal{R} \subset X^{(1)}$ be a relation and let $\Gamma_{\mathcal{R}}(X)$ denote the subspace of $\Gamma^1(X)$ consisting of all sections h such that $j^1 h \in \Gamma(\mathcal{R})$. The induced map $J\colon \Gamma_{\mathcal{R}}(X) \to \Gamma(\mathcal{R})$, $J(h) = j^1 h$, is continuous.

Weak Homotopy Equivalence Theorem 4.7. *Let $\mathcal{R} \subset X^{(1)}$ be open and ample. The map $J\colon \Gamma_{\mathcal{R}}(X) \to \Gamma(\mathcal{R})$ is a weak homotopy equivalence.*

Theorem 4.7 is a consequence of the following general parametric h-principle. Let Z be an auxiliary compact m-dimensional smooth manifold (a parameter space). Associated to Z are the bundles,

$$\mathrm{id} \times p\colon Z \times X \to Z \times V; \quad \mathrm{id} \times p_0^1\colon Z \times X^{(1)} \to Z \times X.$$

In particular $Z \times \mathcal{R}$ is an open relation in $Z \times X^{(1)}$. For the purposes of the proof of Theorem 4.7, $Z = S^i$ in the case of the surjectivity of the ith homotopy group; $Z = S^i \times [0,1]$ in the case of the injectivity of the ith homotopy group. The following theorem generalizes Theorem 4.2 to the case of an auxiliary compact space of parameters Z.

Theorem 4.8 Parametric h-Principle. *Let $\alpha \in C^0(Z, \Gamma(\mathcal{R}))$ and let $h \in C^0(Z, \Gamma(X))$ be the induced map such that for all $z \in Z$, $p_0^1 \circ \alpha(z) = h_z \in \Gamma(X)$. Suppose $Z_0 \subset Z$ is closed, $\partial Z \subset Z_0$, such that $h \in C^1(\mathfrak{Op}\, Z_0, X)$ and for all $z \in \mathfrak{Op}\, Z_0$, $\alpha(z) = j^1 h_z \in \Gamma(\mathcal{R})$.*
There is a homotopy rel Z_0, $H \in C^0(Z \times [0,1], \Gamma(\mathcal{R}))$, $H_0 = \alpha$, such that for all $z \in Z$, $H(z,1) \in \Gamma(\mathcal{R})$ is holonomic: $H(z,1) = j^1 f_z \in \Gamma(\mathcal{R})$ where $f \in C^1(Z, \Gamma(X))$.

Proof. As in the proof of Theorem 4.2, we first arrange for the data α, h to be smooth on $Z \setminus Z_0$. Let N be a neighbourhood of $\mathrm{ev}\,\alpha(V)$ in $Z \times \mathcal{R}$. Let $\mathrm{ev}\, h \in \Gamma(Z \times X)$ denote the section $(z, x) \mapsto (z, h_z(x))$. Employing the bundle $\mathrm{id} \times p_0^1\colon Z \times X^{(1)} \to Z \times X$, there is a neighbourhood U of the image $\mathrm{ev}\, h(Z \times V)$ in $Z \times X$ and a continuous lift $\nu\colon U \to N$ such that $(\mathrm{id} \times p_0^1) \circ \nu = \mathrm{id}_U$; $\nu \circ \mathrm{ev}\, h = \mathrm{ev}\,\alpha \in \Gamma(Z \times \mathcal{R})$. Employing the approximation theorem I 1.1, up to a small homotopy rel $Z_0 \times V$ of $\mathrm{ev}\, h$ in U, one may assume that $h \in C^\infty(Z \setminus Z_0, \Gamma(X))$. Lifting this homotopy into N, up to a small homotopy of $\mathrm{ev}\,\alpha$ in N, one may assume that $p_0^1 \circ \alpha = h \in C^\infty(Z \setminus Z_0, X)$. Again employing the bundle $\mathrm{id} \times p_0^1\colon Z \times X^{(1)} \to Z \times X$, up to a further small homotopy of $\mathrm{ev}\,\alpha$ in N one may assume that $\alpha \in C^\infty(Z \setminus Z_0, \Gamma(\mathcal{R}))$ and that for all $z \in \mathfrak{Op}\, Z_0$, $\alpha(z) = j^1(h_z) \in \Gamma(X^{(1)})$.

Let $Z_1 \subset Z \setminus Z_0$ be a smooth compact submanifold with boundary, $\dim Z_1 = m$, such that $\partial Z_1 \subset \mathfrak{Op}\, Z_0$. Thus $\alpha \in C^\infty(Z_1, \Gamma(\mathcal{R}))$, $h \in C^\infty(Z_1, \Gamma(X))$, and for all $z \in \mathfrak{Op}\,(\partial Z_1)$, $\alpha(z) = j^1(h_z) \in \Gamma(\mathcal{R})$. Following the proof of Theorem 4.2 $(K_0 = \emptyset)$, let $(W_i)_{i \geq 1}$, $(U_i)_{i \geq 1}$, be locally finite coverings of the base manifold V by closed charts such that $W_i \subset \mathrm{int}\,U_i$ for all $i \geq 1$. Applying III Complement 3.7 (with parameters), the main inductive step Lemma 4.4 in the proof of Theorem 4.2 carries over to the case of parametrized smooth maps with respect to the smooth space of parameters Z_1. Thus inductively one constructs a homotopy rel $\mathfrak{Op}\,Z_0$ of α which is holonomic on $Z_1 \times \bigcup_1^i W_j$, for all $i \geq 1$. Taking the limit as $i \to \infty$ the theorem is proved. Details are left to the reader. □

Proof of Theorem 4.7. The parametric h-principle above is employed to prove the theorem as follows. Fix an integer $i \geq 0$, and let $\ell \in \Gamma_{\mathcal{R}}(X)$ be a base point. We show that $J_* \colon \pi_i(\Gamma_{\mathcal{R}}(X), \ell) \to \pi_i(\Gamma(\mathcal{R}), j^1\ell)$ is onto. Let $x_0 \in S^i$ be a base point and let $\varphi \colon (S^i, x_0) \to (\Gamma(\mathcal{R}), j^1\ell)$ be continuous. Up to a small homotopy, one may suppose that φ is constant $(= j^1\ell)$ on $\mathfrak{Op}\,x_0$ in S^i. Surjectivity now follows from Theorem 4.8 applied to the case $Z = S^i$ $(\partial Z = \emptyset)$ and $Z_0 = x_0 \in S^i$.

We now show that $J_* \colon \pi_i(\Gamma_{\mathcal{R}}(X), \ell) \to \pi_i(\Gamma(\mathcal{R}), j^1\ell)$ is injective. To this end let $h_0, h_1 \colon (S^i, x_0) \to (\Gamma_{\mathcal{R}}(X), \ell)$ such that $J \circ h_0, J \circ h_1 \colon (S^i, x_0) \to (\Gamma(\mathcal{R}), j^1\ell)$ are homotopic: there is a homotopy $H \colon [0,1] \times S^i \to \Gamma(\mathcal{R})$, rel x_0, such that $H_i = J \circ h_i$, $i = 0, 1$. In particular H is constant $(= j^1\ell)$ on the line segment L, where $L = [0,1] \times x_0$ in $[0,1] \times S^i$. Up to a small homotopy, one can assume that H is constant $(= j^1\ell)$ on $\mathfrak{Op}\,L$ in $[0,1] \times S^i$, and that for some $\epsilon \in (0, 1/2)$, $H_t = j^1h_0$; $H_{1-t} = j^1h_1$ for all $t \in [0, \epsilon]$.

Let $Z = [0,1] \times S^i$, and let $Z_0 = (\{0,1\} \times S^i) \cup L$. Thus $\partial Z \subset Z_0$ and, employing the above small homotopies, $H \colon Z \to \Gamma(\mathcal{R})$ satisfies the hypotheses of Theorem 4.8 on $\mathfrak{Op}\,Z_0$. Injectivity now follows from Theorem 4.8. Note that Theorem 4.8 yields additional information. The homotopy $H \colon Z \to \Gamma(\mathcal{R})$ itself is homotopic to a parametrized family of holonomic maps. □

§2. Examples

Let V, W be smooth manifolds, $\dim V = n$, $\dim W = q$. Let $X = V \times W \to V$ be the product bundle, fiber W. In particular $\Gamma^r(X) \equiv C^r(V, W)$, $r \in \{0, 1, 2, \ldots, \infty\}$.

In case $q \geq n$, let $\mathcal{I} \subset X^{(1)}$ denote the immersion relation, as in Example 4.1. Thus $\mathcal{I}$ is open and, in case $q \geq n + 1$, $\mathcal{I}$ is ample. Applying Theorem 4.7, one recovers the classical immersion theorem of Hirsch [21] in the extra dimensional case.

Theorem 4.9 (Hirsch). *Let $\mathcal{I} \subset X^{(1)}$ be the immersion relation and suppose $q \geq n + 1$. The map $J \colon \Gamma_{\mathcal{I}}(X) \to \Gamma(\mathcal{I})$, $h \mapsto j^1h$, induces a weak homotopy equivalence.*

With respect to the above bundle $X = V \times W \to V$, let $\mathcal{R}^k \subset X^{(1)}$ be the k-mersion relation: $\mathcal{R}^k$ is the subspace defined by germs of sections $f \in \Gamma^1(X)$

of rank $\geq k$ $(k \leq \inf\{n,q\})$. Thus, analogous to Example 4.1, with respect to the local coordinates (4.1) on $X_U^{(1)}$, $\mathcal{R}_U^k \subset X_U^{(1)}$ is the subspace of points $w = (x, y, (v_i)_{1 \leq i \leq n})$ such that the vectors $v_i \in \mathbf{R}^q$, $1 \leq i \leq n$, span a subspace of dimension $\geq k$. $\mathcal{R}^k$ is open in $X^{(1)}$. We prove that in case $k \leq q - 1$ then $\mathcal{R}^k$ is also ample. Indeed, with respect to the projection $p_\perp^1 \colon X_U^{(1)} \to X_U^\perp$ which deletes the last vector $v_n \in \mathbf{R}^q$ (cf. (4.3)), Let $L \subset \mathbf{R}^q$ denote the subspace spanned by the remaining vectors $v_1, \ldots, v_{n-1}$. Thus $\dim L \geq k - 1$. Let $\mathcal{R}_z^k$ be the fiber over $z = p_\perp^1(w) \in X_U^\perp$. Thus $\mathcal{R}_z^k$ consists of all points $(z, v) \in \{z\} \times \mathbf{R}^q \equiv \mathbf{R}^q$ such that the subspace of $\mathbf{R}^q$ spanned by L, v is of dimension $\geq k$. If $\dim L \geq k$, then $\mathcal{R}_z^k = \mathbf{R}^q$. In case $\dim L = k - 1$, then $\mathcal{R}_z^k$ is the complement of the subspace L in $\mathbf{R}^q$. Since $k \leq q - 1$, it follows that L has codimension ≥ 2 from which it follows that $\mathcal{R}_z^k$ is path connected and its convex hull is $\mathbf{R}^q$ i.e. $\mathcal{R}^k$ is ample in the extra dimensional case $k \leq q - 1$. Applying Theorem 4.8, one recovers the classical theorem of S. Feit [12]. Note that the case $k = n \leq q - 1$ is the Hirsch Theorem above.

Theorem 4.10 (Feit). *Let $\mathcal{R}^k \subset X^{(1)}$ be the k-mersion relation above, and suppose $k \leq q - 1$. The map $J \colon \Gamma_{\mathcal{R}^k}(X) \to \Gamma(\mathcal{R}^k)$, $h \mapsto j^1 h$, induces a weak homotopy equivalence.*

We develop some formal properties of 1-jet spaces of differential forms that are employed in both of the examples below on the existence of divergence free vector fields and the existence of non-degenerate 2-forms.

Let $Y = T^*(V)$ the cotangent bundle of a smooth n-dimensional manifold V. Thus $\Gamma^r(Y)$ is the space of C^r 1-forms on V; $s \colon Y^{(r)} \to V$ is the vector bundle of r-jets of germs of C^r-sections of Y; $p_0^1 \colon Y^{(1)} \to Y$ is a vector bundle, fiber $\mathcal{L}(\mathbf{R}^n, \mathbf{R}^n)$. In local coordinates $(u_1, \ldots, u_n)$ in a chart U on V, $Y_U^{(1)} = Y_U \times \mathcal{L}(\mathbf{R}^n, \mathbf{R}^n)$; if $\alpha = \sum_i a_i(u)\, du_i$ is a 1-form on U of class C^1 then for all $u \in U$,

$$j^1 \alpha(u) = (\alpha(u), A(u)) \in Y_U^{(1)}, \qquad (4.18)$$

where $A(u)$ is the $n \times n$ matrix $(\partial a_i / \partial u_j(u)) \in \mathcal{L}(\mathbf{R}^n, \mathbf{R}^n)$. Let $Z = \Lambda^2(V)$, the 2nd exterior power of the cotangent bundle of V. Thus $\Gamma(Z)$ is the space of continuous 2-forms on V. The exterior derivative $d \colon \Gamma^1(Y) \to \Gamma(Z)$, $\alpha \mapsto d\alpha$, induces a vector bundle homomorphism $\Delta \colon Y^{(1)} \to Z$ that covers the bundle projection $p \colon Y \to V$ such that on the level of 1-forms, for all $x \in V$, $\Delta(j^1 \alpha(x)) = d\alpha(x)$, where $\alpha \in \Gamma^1(Y)$.

$$
\begin{array}{ccc}
Y^{(1)} & \xrightarrow{\ \Delta\ } & Z \\
\downarrow{\scriptstyle p_0^1} & & \downarrow{\scriptstyle p} \\
Y & \xrightarrow[\ p\ ]{} & V
\end{array}
$$

In the above local coordinates $\Delta(y, B) = \sum_{j<k}(b_{jk} - b_{kj})\,du_j \wedge du_k$, where $y \in Y$ and $B = (b_{ij}) \in \mathcal{L}(\mathbf{R}^n, \mathbf{R}^n)$. (Employing the coordinates (4.18) note that for all $u \in U$, $\Delta(j^1\alpha(u)) = d\alpha(u) \in Z_u$.) Evidently Δ is a linear epimorphism on each fiber and, in the above local coordinates, $(y, B) \in \ker \Delta$ if and only if B is a symmetric matrix.

Let $p_0^1 \colon E \to V$ be the restriction of the vector bundle $p_0^1 \colon Y^{(1)} \to Y$ to the zero section ($\equiv V$). Since the zero section is a strong deformation retract of the total space $Y^{(1)}$ it follows that the bundle $p_0^1 \colon Y^{(1)} \to Y$ is equivalent to the pullback of $p_0^1 \colon E \to V$ along the projection map $p \colon Y \to V$. Furthermore, in the above local coordinates for Δ, the image of Δ is independent of $y \in Y$. Thus, employing the pullback equivalence, Δ factors through the restriction epimorphism (same notation) $\Delta \colon E \to Z$ over the base space V:

$$
\begin{array}{ccccc}
Y^{(1)} & \longrightarrow & E & \overset{\Delta}{\longrightarrow} & Z \\[2pt]
{\scriptstyle p_0^1}\big\downarrow & & {\scriptstyle p_0^1}\big\downarrow & & \big\downarrow{\scriptstyle p} \\[2pt]
Y & \underset{p}{\longrightarrow} & V & \underset{\mathrm{id}}{\longrightarrow} & V
\end{array}
$$

Since a vector bundle epimorphism over a manifold always splits (Hirsch [22]) it follows that there is a vector bundle monomorphism $g \colon Z \to E \subset Y^{(1)}$ such that $\Delta \circ g = \mathrm{id} \colon Z \to Z$. In particular any 2-form $\lambda \in \Gamma(Z)$ lifts to a section $\mu \in \Gamma(E)$ such that $\Delta \circ \mu = \lambda$. In fact one can lift λ to a section of the restriction of the bundle $Y^{(1)}$ to the subbundle over the image of any 1-form in Y. With these bundle preliminaries, we consider the following applications of Theorem 4.2 to problems involving differential forms.

Divergence Free Vector Fields. We apply Convex Integration theory to prove the existence of a global basis of divergence free vector fields on a smooth orientable 3-manifold. Let V be a smooth orientable 3-manifold, and let Ω be a non-zero volume form on V. A 2-form ω on V defines a unique vector field F on V such that the contraction $i_F(\Omega) = \omega$: for all vector fields Y, Z on V,

$$
\omega(Y, Z) = \Omega(F, Y, Z).
$$

In local coordinates, if $\Omega = f\,dx \wedge dy \wedge dz$; $\omega = A\,dx \wedge dy + B\,dy \wedge dz + C\,dx \wedge dz$, then $F = \dfrac{1}{f}(B\partial/\partial x - C\partial/\partial y + A\partial/\partial z)$. This correspondence preserves linear independence. A vector field F on V is divergence free with respect to Ω if the Lie derivative $L_F(\Omega) = 0$. Let $\omega = i_F(\Omega)$. Employing the classical relation $L_F = d\,i_F + i_F d$, it follows that $L_F(\Omega) = d\omega$. Thus F is divergence free if and only if $d\omega = 0$ i.e. the corresponding 2-form ω is closed. In particular if $\omega = d\alpha$

is an exact 2-form then the corresponding vector field F is divergence free. Thus to construct divergence free vector fields which are linearly independent at each point of V it is sufficient to construct three linearly independent exact 2-forms on V. This is accomplished in the following theorem, adapted from Gromov [18] who proves the corresponding existence theorem for divergence free vector fields on manifolds V, $\dim V = n \geq 3$.

Theorem 4.11 (Gromov). *Let F_1, F_2, F_3 be a basis of vector fields on a smooth orientable 3-manifold V i.e. linearly independent at each point $v \in V$. There is a homotopy F_i^t, $t \in [0,1]$, of linear independent vector fields on V, $F_i^0 = F_i$, $1 \leq i \leq 3$, such that the vector fields F_i^1, $1 \leq i \leq 3$, are divergence free.*

Proof. Let $Y = T^*(V)$, the cotangent bundle of V. $\Gamma^r(Y)$ is the space of C^r 1-forms on V; $s\colon Y^{(r)} \to V$ is the vector bundle of germs of r-jets of sections of Y. In local coordinates (x, y, t) in a chart U on V, $Y_U^{(1)} = Y_U^{(0)} \times \mathbf{R}^9$ ($\mathbf{R}^9 \equiv \mathcal{L}(\mathbf{R}^3, \mathbf{R}^3)$). Thus if $\alpha = a\,dx + b\,dy + c\,dt$ is a C^1 1-form on U then,

$$j^1\alpha(p) = (j^0\alpha(p), (a_x, b_x, c_x), (a_y, b_y, c_y), (a_t, b_t, c_t)) \in Y \times \mathbf{R}^9.$$

Employing the decomposition (4.1), let $Y_U^{(1)} = Y_U^{\perp} \times \mathbf{R}^3$, where the $\mathbf{R}^3$-factor corresponds to the derivative $\partial/\partial t$ (coordinates $(a_t, b_t, c_t) \in \mathbf{R}^3$).

Let also $Z = \Lambda^2 Y$, the 2nd exterior power of the cotangent bundle of V. Thus $\Gamma(Z)$ is the space of continuous 2-forms on V. As explained above, the exterior derivative $d\colon \Gamma^1(Y) \to \Gamma(Z)$, $\alpha \mapsto d\alpha$, induces a vector bundle homomorphism, $\Delta\colon Y^{(1)} \to Z$ such that on the level of 1-forms, $\Delta(j^1\alpha) = d\alpha$, where $\alpha \in \Gamma^1(Y)$. Explicitly, in the above local coordinates for a 1-form $\alpha \in \Gamma^1(Y)$,

$$\Delta(j^1\alpha) = (b_y - a_x)dx \wedge dy + (c_x - a_t)dx \wedge dt + (c_y - b_t)dy \wedge dt.$$

Let $R = \{z\} \times \mathbf{R}^3 \equiv \mathbf{R}^3$ be the fiber in $Y_z^{(1)}$ over the base point $z \in Y_U^{\perp}$. Then in local coordinates, Δ induces a map (same notation) $\Delta\colon R \to \mathbf{R}^3$, $(u, v, w) \mapsto (A, B - u, C - v)$ where A, B, C are constants. Hence the induced map Δ is affine, whose image is an affine 2-plane in $\mathbf{R}^3$ with normal vector $(1, 0, 0)$. Furthermore, as explained above there is a bundle monomorphism $g\colon Z \to Y^{(1)}$ such that $\Delta \circ g = \mathrm{id}\colon Z \to Z$.

Let $X = \oplus_1^3 Y$, the 3-fold direct sum of the cotangent bundle of V. Thus $X^{(1)} = \oplus_1^3 Y^{(1)}$, and the map on 1-jets $\Delta\colon Y^{(1)} \to Z$ induces a map (same notation) $\Delta\colon X^{(1)} \to \oplus_1^3 Z$, such that on the level of 1-forms:

$$\Delta(j^1\alpha_1, j^1\alpha_2, j^1\alpha_3) = (d\alpha_1, d\alpha_2, d\alpha_3).$$

Again employing (4.1), $X_U^{(1)} = X_U^\perp \times \mathbf{R}^9$, where the $\mathbf{R}^9 = \oplus_1^3 \mathbf{R}^3$ factor corresponds to the derivative $\partial/\partial t$.

Define now the relation $\mathcal{R} \subset X^{(1)}$ to be the open subset consisting of all triples of 1-jets $(z_1, z_2, z_3) \in X_v^{(1)}$, for all $v \in V$, such that the 2-forms $\Delta(z_1), \Delta(z_2), \Delta(z_3) \in Z_v$ are linearly independent. In what follows we show that $\mathcal{R}$ is ample.

Note that in the above local coordinates Δ maps the three $\mathbf{R}^3$-factors (corresponding to the derivative $\partial/\partial t$) to parallel 2-planes in $\mathbf{R}^3 = Z_v$ with normal vector $(1, 0, 0)$. The dependence of the 2-forms $\Delta(z_i)$, $1 \le i \le 3$, in these parallel 2-planes is analyzed in the following lemma. In what follows i, j, k are the standard basis vectors in $\mathbf{R}^3$.

Lemma 4.12. *Let P_i, $1 \le i \le 3$, be parallel 2-planes (hence affine subspaces) in $\mathbf{R}^3$. Let $P = \prod_1^3 P_i$, and let $\Sigma \subset P$ be the singular set,*

$$\Sigma = \{(w_1, w_2, w_3) \in P \mid w_1, w_2, w_3 \text{ are dependent in } \mathbf{R}^3\}.$$

The convex hull of each path component of the complement of Σ in P is either empty or all of P.

Proof. The parallel 2-planes are normalized to have normal vector k. Let $w_r = x_r i + y_r j + L_r k \in P_r$, $1 \le r \le 3$. One computes,

$$w_1 \wedge w_2 \wedge w_3 = Q(x_1, y_1, x_2, y_2, x_3, y_3) i \wedge j \wedge k = (L_3(x_1 y_2 - x_2 y_1) +$$
$$L_2(x_3 y_1 - x_1 y_3) + L_1(x_2 y_3 - x_3 y_2)) i \wedge j \wedge k.$$

Thus $\Sigma \subset P \equiv \mathbf{R}^6$ is the locus $Q = 0$. In the degenerate case when all $L_r = 0$, $1 \le r \le 3$, then $\Sigma = P$ (all three planes coincide with the plane $z = 0$). Otherwise, one may assume $L_3 \ne 0$. Note that the complement of the quadric $xy - z = 0$ in $\mathbf{R}^3$ has 2 path components and the convex hull of each component is $\mathbf{R}^3$. Similarly, by considering levels $x_2 = c$, the convex hull of each path component of the complement of the quadric $x_1 y_2 - x_2 y_1 = 0$ in $\mathbf{R}^4$ is $\mathbf{R}^4$. Again considering levels $Q(x_1, y_1, x_2, y_2, c, d) = 0$ ($x_3 = c, y_3 = d$), the lemma follows. $\square$

The ampleness of $\mathcal{R}$ follows from Lemma 4.12. Indeed, recall that in the above local product structure, $X_U^{(1)} = X_U^\perp \times \mathbf{R}^9$, and the map $\Delta : \mathbf{R}^9 \to \oplus_1^3 Z_v$ sends (affinely) the three $\mathbf{R}^3$-factors onto parallel 2-planes in Z_v. In particular, if p is a point in one of the 2-planes, then $\Delta^{-1}(p) = p' \times \mathbf{R}$, p' a point in the corresponding $\mathbf{R}^3$-factor. Since paths in these 2-planes lift back to paths in the corresponding $\mathbf{R}^3$-factors, employing Lemma 4.12, it follows from elementary geometrical arguments that, for each $z \in X^\perp$, the convex hull of each path component of $\mathcal{R}_z \subset \{z\} \times \mathbf{R}^9$ is all of $\mathbf{R}^9$ (if not empty). Consequently $\mathcal{R} \subset X^{(1)}$ is open and ample.

Returning to the proof of Theorem 4.11, the independent vector fields F_i, $1 \leq i \leq 3$ on V induce linear independent 2-forms ω_i, $1 \leq i \leq 3$ on V. Lifting via the vector bundle monomorphism $g\colon Z \to Y^{(1)}$, there is a continuous section $\beta \in \Gamma(\mathcal{R})$ such that $\Delta \circ \beta = (\omega_1, \omega_2, \omega_3) \in \oplus_1^3 \Gamma(Z)$. Applying Theorem 4.2 to the open, ample relation $\mathcal{R} \subset X^{(1)}$, it follows that there is a homotopy $\beta_t \in \Gamma(\mathcal{R})$, $\beta_0 = \beta$, $t \in [0,1]$, such that β_1 is holonomic: $\beta_1 = j^1 h \in \Gamma(\mathcal{R})$, where $h \in \Gamma^1(X)$ (since no relative theorem is involved, one may assume h is smooth). Thus $h = (\alpha_1, \alpha_2, \alpha_3) \in \oplus_1^3 \Gamma^1(Y)$ such that:

$$\Delta \circ \beta_1 = (d\alpha_1, d\alpha_2, d\alpha_3) \in \oplus_1^3 \Gamma(Z).$$

In particular, the exact 2-forms $d\alpha_i$, $1 \leq i \leq 3$, are linearly independent at each point $v \in V$. It follows that the vector fields L_i on V corresponding to $d\alpha_i$, $1 \leq i \leq 3$, are linearly independent and divergence free. Furthermore the homotopy of linearly independent vector fields F_i^t, $1 \leq i \leq 3$, induced from the homotopy of linearly independent 2-forms $\Delta \circ \beta_t \in \oplus_1^3 \Gamma(Z)$, $t \in [0,1]$, connect F_i to L_i, $1 \leq i \leq 3$, which completes the proof of the theorem. $\square$

Applications to Symplectic Topology. A symplectic structure on a smooth manifold V, $\dim V = 2m$, $m \geq 1$, consists of a symplectic 2-form σ on V i.e. σ is a closed 2-form on V which is non-degenerate: σ^m is a nowhere zero $2m$-form (volume form) on V. Symplectic structures (V_1, σ_1), (V_2, σ_2) are smoothly equivalent if there is a diffeomorphism $f\colon V_1 \to V_2$ such that $f^* \sigma_1 = \sigma_2$. The existence and classification of symplectic structures is a topological problem of current research interest. In case V is an open manifold which admits a non-degenerate 2-form (perhaps not closed) then the covering homotopy method, as generalized in Gromov [16], proves the existence of an exact symplectic structure σ on V: $\sigma = d\alpha$, where α is a 1-form on V, and $(d\alpha)^m$ is nowhere zero. Following McDuff [28], we prove the following theorem for odd-dimensional smooth manifolds (possibly compact and without boundary). Let V be a smooth manifold, $\dim V = 2m + 1$. A 2-form σ on V is *non-degenerate* if the $2m$-form σ^m is nowhere zero on V.

Theorem 4.13 (McDuff). *Let V be a smooth manifold, $\dim V = 2m + 1$, $m \geq 1$, and let $a \in H^2(V) \equiv H^2(V; \mathbf{R})$. Each non-degenerate 2-form μ on V is homotopic through non-degenerate 2-forms μ_t, $t \in [0,1]$, $\mu_0 = \mu$, to a symplectic 2-form μ_1 such that $[\mu_1] = a \in H^2(V)$.*

(Relative case) Furthermore suppose $K \subset V$ is a closed submanifold such that $d\mu = 0$ on $\mathfrak{Op}\, K$ and $[\mu|_K] = a|_K \in H^2(K)$ The homotopy μ_t can be chosen to be constant on K: $\mu_t = \mu$ rel K, $0 \leq t \leq 1$.

Proof. We follow the proof of McDuff [28] where this theorem is applied to prove concordance results for symplectic forms ([28] Theorem 4.2). Let $Y = T^*(V)$ the cotangent bundle of V. Thus $\Gamma^r(Y)$ is the space of C^r 1-forms on V, $s\colon Y^{(r)} \to V$ is the vector bundle of r-jets of germs of C^r-sections of Y, and $p_0^1\colon Y^{(1)} \to Y$ is a vector bundle, fiber $\mathcal{L}(\mathbf{R}^n, \mathbf{R}^n)$. In local coordinates $(u_1, \ldots, u_n)$ in a chart U on V $(n = 2m+1)$, $Y_U^{(1)} = Y_U \times \mathcal{L}(\mathbf{R}^n, \mathbf{R}^n)$; if $\alpha = \sum_i a_i(u)\, du_i$ is a 1-form on U then for all $u \in U$,

$$j^1\alpha(u) = (\alpha(u), A(u)) \in Y_U^{(1)},$$

where $A(u)$ is the $n \times n$ matrix $(\partial a_i / \partial u_j(u)) \in \mathcal{L}(\mathbf{R}^n, \mathbf{R}^n)$. Let $Z = \Lambda^2(V)$, the 2nd exterior power of the cotangent bundle of V. Thus $\Gamma(Z)$ is the space of continuous 2-forms on V. The exterior derivative $d\colon \Gamma^1(Y) \to \Gamma(Z)$, $\alpha \mapsto d\alpha$, induces a vector bundle homomorphism $\Delta\colon Y^{(1)} \to Z$ such that on the level of 1-forms, for all $x \in V$, $\Delta(j^1\alpha(x)) = d\alpha(x)$, where $\alpha \in \Gamma^1(Y)$. In the above local coordinates,

$$\Delta(y, B) = \sum_{j<k}(b_{jk} - b_{kj})\, du_j \wedge du_k,$$

where $y \in Y$ and $B = (b_{ij}) \in \mathcal{L}(\mathbf{R}^n, \mathbf{R}^n)$. Fix $a \in H^2(V)$ and let $\lambda \in \Gamma^1(Z)$ be a closed 2-form on V such that $[\lambda] = a$. Let $\mathcal{R} \subset Y^{(1)}$ be the relation defined as follows $(s\colon Y^{(1)} \to V$ is the source map):

$$\mathcal{R} = \{z \in Y^{(1)} \mid \lambda(x) + \Delta(z) \in Z_x,\ x = s(z) \in V,\ \text{is non-degenerate}\}.$$

In the above local coordinates let $\lambda(x) = \sum_{j<k} \lambda_{jk}(x)\, du_j \wedge du_k$. Then $z = (y, B) \in \mathcal{R}$ $(p(y) = x \in V)$ if and only if the 2-form,

$$\sigma(z) = \sum_{j<k}(\lambda_{jk}(x) + b_{jk} - b_{kj})\, du_j \wedge du_k \in Z_x,$$

is non-degenerate, where $B = (b_{ij}) \in \mathcal{L}(\mathbf{R}^n, \mathbf{R}^n)$ i.e. $\sigma^m(z) \in \Lambda^{2m}(V_x)$ is nowhere zero. Since non-degeneracy is an open condition it follows that $\mathcal{R} \subset Y^{(1)}$ is open.

Lemma 4.14. *The relation $\mathcal{R} \subset Y^{(1)}$ is ample.*

Proof. In the above local coordinates on a chart U we employ the product decomposition, $Y_U^{(1)} = Y_U^{\perp} \times \mathbf{R}^n$, where the $\mathbf{R}^n$-factor corresponds to the ith row of an $n \times n$ matrix $B = (b_{ij}) \in \mathcal{L}(\mathbf{R}^n, \mathbf{R}^n)$. To prove $\mathcal{R}$ is ample it is sufficient to prove ampleness in the above product decomposition in the case of the first row $i = 1$. Setting $t = (t_1, \ldots, t_n) \in \mathbf{R}^n$, let

$$\sigma_t(z) = \sum_{i=2}^{n} t_i\, du_1 \wedge du_i + \sum_{2 \leq j < k}(b_{jk} - b_{kj} + \lambda_{jk}(x))\, du_j \wedge du_k.$$

To prove the lemma we show that for all $B \in \mathcal{L}(\mathbf{R}^n, \mathbf{R}^n)$ the set,

$$Q = \{(t_1, \ldots, t_n) \in \mathbf{R}^n \mid \sigma_t(z)^m \in \Lambda^{2m}(V_x) \neq 0\}$$

is ample in $\mathbf{R}^n$. To this end $(n = 2m + 1)$ one calculates,

$$\sigma_t(z)^m = \left(\sum_{i,k=2}^{n} \epsilon_{ik} B_{ik} t_i \right) du_1 \cdots \wedge \widehat{du_k} \cdots \wedge du_n + C \, du_2 \wedge \cdots \wedge du_n,$$

where $C \in \mathbf{R}$ is constant (independent of the variables t_i); $B_{ik} = B_{ki}$ is constant and $\epsilon_{ik} B_{ik} t_i$ is the coefficient of the term,

$$(du_1 \wedge du_i) \wedge du_2 \cdots \wedge \widehat{du_i} \cdots \wedge \widehat{du_k} \cdots \wedge du_n,$$

in the expansion of $\sigma_t(z)^m$; $\epsilon_{ik} = \pm 1$ $(B_{ii} = 0,\ 2 \leq i \leq n)$. Evidently $\sigma_t(z)^m \neq 0$ if $C \neq 0$ i.e. $Q = \mathbf{R}^n$ in this case. Suppose now $C = 0$. Thus Q is the complement in $\mathbf{R}^n$ of the subspace L of solutions to the system of $(n-1)$ linear equations,

$$\sum_{i=2}^{n} \epsilon_{ik} B_{ik} t_i = 0, \quad 2 \leq k \leq n.$$

In case all $B_{ik} = 0$, then $L = \mathbf{R}^n$ and $Q = \emptyset$. If some term $B_{ik} \neq 0$ then the ith and kth equations are linearly independent since $B_{ii} = B_{kk} = 0$. Consequently the above system of equations has rank ≥ 2 (for example, in case $m = 1$, there are two equations, $B_{23}t_2 = 0$, $B_{32}t_3 = 0$). Since $\operatorname{codim} L \geq 2$ it follows that the complement Q is open, path connected and the convex hull of Q is all of $\mathbf{R}^n$. Thus Q is ample in $\mathbf{R}^n$ which completes the proof of the lemma. $\qquad\square$

Returning to the proof of Theorem 4.13, let $\mu \in \Gamma(Z)$ be a non-degenerate 2-form on V. Employing the continuous lift property, explained above, there is a section $\varphi \in \Gamma(Y^{(1)})$ such that $\Delta \circ \varphi = \mu - \lambda$. Consequently $\varphi \in \Gamma(\mathcal{R})$. In the relative case, since $[\mu] = [\lambda]$ on $\mathfrak{Op}\, K$, then up to addition of an exact 2-form to λ one may assume that $\varphi = 0$ on $\mathfrak{Op}\, K$. Applying Theorem 4.2 there is a homotopy $\varphi_t \in \Gamma(\mathcal{R})$, $t \in [0,1]$, $\varphi_0 = \varphi$, such that φ_1 is holonomic i.e. there is a 1-form $\alpha \in \Gamma^1(Y)$ such that $\varphi_1 = j^1\alpha \in \Gamma(\mathcal{R})$. Furthermore $\varphi_t = 0$ on $\mathfrak{Op}\, K$, $0 \leq t \leq 1$. In particular,

$$\mu_1 = \lambda + \Delta \circ \varphi_1 = \lambda + d\alpha \in \Gamma(Z),$$

is a non-degenerate 2-form on V. Thus μ_1 is closed, non-degenerate and $[\mu_1] = [\lambda] \in H^2(V)$. The homotopy of non-degenerate 2-forms on V, $\mu_t = \lambda + \Delta \circ \varphi_t$, $0 \leq t \leq 1$, proves the theorem. $\qquad\square$

CHAPTER 5

MICROFIBRATIONS

§1. Introduction

5.1.1. The theory in Part I proves the h-principle for open ample relations $\mathcal{R} \subset X^{(1)}$. In general however, the theory in Part I does not extend naturally to prove the h-principle for open, ample relations $\mathcal{R} \subset X^{(r)}$ in case $r \geq 2$. In effect, the analytic theory in Chapter III allows for controlled "large" moves in the pure derivatives $\partial^r / \partial t^r$ while maintaining small perturbations in all the complementary $\perp$-derivatives. This analytic technique works well in spaces of 1-jets $X^{(1)}$ since in local coordinates first order derivatives are all pure. As mentioned in the introduction to Chapter IV, by suitable local changes of coordinates it is possible to apply this technique also in the case of open, ample relations in 2-jet spaces $X^{(2)}$, although we have not attempted to develop the details in this book.

However, in the case of rth order derivatives, $r \geq 3$, some additional technique is required to prove the h-principle. Indeed, the above method of applying convex integration sequentially to each of the rth order partial derivatives in local coordinates simply does not work. The difficulty here is illustrated by considering for example, the mixed partial derivatives $\partial^3 / \partial x^2 \partial y$, $\partial^3 / \partial x \partial y^2$. It is not possible to change coordinates locally so that each of these derivatives lies in the $\perp$-space of the other in the new local coordinates. Thus large changes in one of these derivatives, produced by convex integration (applied to pure 3rd order derivatives in new local coordinates) inevitably involves large changes in the other of these derivatives. Thus the proof procedure in Part I which consists of applying the analytic theory of Chapter III in local coordinates sequentially to all of the (pure) first order derivatives will not work in case $r \geq 3$.

The way around this difficulty, first developed by Gromov [18], is to introduce a sequence of convex hull extensions of a relation $\mathcal{R} \subset X^{(r)}$, $r \geq 1$. The solution of the h-principle for $\mathcal{R}$ reduces to solving the h-principle successively for each of the convex hull extensions in the sequence. The point of this procedure is that the solution of the h-principle for each of these convex hull extensions reduces locally to convex integration in a pure derivative $\partial^r / \partial t^r$ with respect to some local coordinate system in the base manifold V i.e. the analytic theory of Chapter III applies.

D. Spring, *Convex Integration Theory: Solutions to the h-principle in geometry and topology*, Modern Birkhäuser classics, DOI 10.1007/978-3-0348-0060-0_5, © Springer Basel AG 2010

5.1.2. Microfibrations. A continuous map $\rho\colon X \to Y$ is a *microfibration* if the following "micro" homotopy lifting property is satisfied. If $f\colon P \times [0,1] \to Y$, $g\colon P \to X$ are continuous maps, P a compact polyhedron, such that $\rho \circ g = f_0$, then there is an $\epsilon \in (0,1]$ (which in general depends on the data) and a continuous map $G\colon P \times [0,\epsilon] \to X$ such that $\rho \circ G = f\colon P \times [0,\epsilon] \to Y$ and $G_0 = g$.

$$
\begin{array}{ccc}
P \times \{0\} & \xrightarrow{\;g\;} & X \\
{\scriptstyle i}\downarrow & & \downarrow{\scriptstyle \rho} \\
P \times [0,1] & \xrightarrow[f]{} & Y
\end{array}
$$

Classically, ρ is a (Serre) fibration if in addition $\epsilon = 1$ in the above definition. Examples of microfibrations that occur in the general theory are as follows:

(i) If $X \subset Y$ is open then the inclusion map $i\colon X \to Y$ is a microfibration.

(ii) The composition of microfibrations is a microfibration.

(iii) Let $\rho\colon X \to Y$ be a microfibration. Then $\mathrm{id} \times \rho\colon Z \times X \to Z \times Y$ is a microfibration for each auxiliary topological space Z. More generally, if $g\colon W \to Y$ is continuous then the induced pullback map (same notation) $\rho\colon g^*X \to Y$ is a microfibration.

(iv) If M, N are smooth manifolds then a submersion $\rho\colon M \to N$ is a microfibration. Indeed, since ρ has maximal rank, it follows that at each $x \in M$, $\rho(x) \in N$, there are local coordinates with respect to which M admits a local product structure such that the map ρ locally is a projection map; the microfibration property easily follows. In addition, if M is compact, then ρ is a fibration.

In case Y is a manifold then a microfibration $\rho\colon X \to Y$ is an open map, though not conversely. There are microfibrations $\rho\colon X \to S^1$, $X \subset \mathbf{R}^3$ is compact (not a polyhedron), that are not Serre fibrations. Let $\rho\colon X \to Y$ be a microfibration where X, Y are compact polyhedra and ρ is piecewise linear. It seems reasonable to conjecture that ρ is a Serre fibration. The literature does not seem to address this question.

5.1.3. Relations Over Jet Spaces. A relation *over* $X^{(r)}$ is a continuous map $\rho\colon \mathcal{S} \to X^{(r)}$. Following Gromov [18], the relation $\rho\colon \mathcal{S} \to X^{(r)}$ is *open* if ρ is a microfibration. The microfibration property of ρ is required to ensure that, under appropriate conditions, embeddings into the jet space $X^{(r)}$ lift back to $\mathcal{S}$. Note that if $\rho\colon Z \to X^{(r)}$ is a Serre fibration and $\mathcal{S} \subset Z$ is open, then the restriction map (same notation) $\rho\colon \mathcal{S} \to X^{(r)}$ is a microfibration, hence an open relation over $X^{(r)}$. In this chapter open relations over $X^{(r)}$ are studied in their own right, with applications in subsequent chapters to convex hull extension relations over $X^{(r)}$, referred to above, as well as to other relations over $X^{(r)}$ of topological interest. A primary example of an open relation over $X^{(r)}$ is constructed as follows. Let $\mathcal{R} \subset X^{(s)}$ be open, $s \geq r$, and let $\rho\colon \mathcal{R} \to X^{(r)}$ be the natural projection map (the restriction of the projection map $p_r^s\colon X^{(s)} \to X^{(r)}$). Since $\mathcal{R}$ is open and the projection map p_r^s is a Serre fibration it follows that $\rho\colon \mathcal{R} \to X^{(r)}$ is a microfibration.

$$
\begin{array}{ccc}
\mathcal{R} & \overset{i}{\longrightarrow} & X^{(s)} \\
\rho \downarrow & & \downarrow p_r^s \\
X^{(r)} & =\!=\!= & X^{(r)}
\end{array}
$$

Note that in case $s = r$, the relation $\rho\colon \mathcal{R} \to X^{(r)}$ discussed above reduces to the inclusion $i\colon \mathcal{R} \to X^{(r)}$ i.e. an open set $\mathcal{R} \subset X^{(r)}$ (conveniently described as a relation $\mathcal{R}$ in the jet space $X^{(r)}$). Thus the theory of open relations over $X^{(r)}$, developed in this chapter, includes as a special case the theory developed in previous chapters for open relations in spaces of r-jets $X^{(1)}$. In subsequent chapters open relations over jet spaces will also be employed to prove the h-principle in cases where the base manifold V is open and the relation $\mathcal{R} \subset X^{(r)}$ may not be ample, i.e. the theory developed in previous chapters does not apply. For example the submersion relation associated to smooth maps $f\colon V \to W$ is not ample. Thus convex integration theory reproves many of the classical theorems in immersion theoretic topology such as Phillips' Theorem [31] for open manifolds.

5.1.4. Neighbourhood Lifting Properties. Let $\pi\colon E \to K$ be a Euclidean vector bundle, fiber dimension $q \geq 1$, over a locally finite (hence locally compact) CW-complex K. We identify $K \subset E$ via the inclusion by the zero section $i\colon K \to E$. If $M \subset K$ is a subcomplex and $\epsilon\colon M \to (0, \infty)$ is a continuous function then $N_M(\epsilon)$ denotes the ϵ-disk neighbourhood of M in E: the fiber of $\pi\colon N_M(\epsilon) \to M$ over $x \in M$ is the disk of radius $\epsilon(x) > 0$, centre x, in $E_x = \mathbf{R}^q$. In what follows we employ the abridged notation $\mathfrak{Op}_E(M) = N_M(\epsilon)$ to denote a disk bundle neighbourhood of M in E, where the radius function $\epsilon\colon M \to (0, \infty)$ is not specified explicitly and may be replaced by a smaller radius function as required by the mathematical context. Microfibrations are useful in the theory of relations over $X^{(r)}$ developed in this chapter principally because of the following neighbourhood lifting property.

Theorem 5.1. Let $\rho\colon X \to Y$ be a microfibration and let $\pi\colon E \to K$ be a Euclidean vector bundle, fiber dimension $q \geq 1$, over a locally finite CW-complex K. Suppose $L \subset K$ is a subcomplex and $g\colon \mathfrak{Op}_E(K) \to Y$, $\mu\colon \mathfrak{Op}_E(L) \cup K \to X$

are continuous maps such that $\rho \circ \mu = g \colon \mathfrak{Op}_E(L) \cup K \to Y$.

$$\mathfrak{Op}_E(L) \cup K \xrightarrow{\ \mu\ } X$$

$$\left. i \right\downarrow \qquad\qquad \left\downarrow \rho\right.$$

$$\mathfrak{Op}_E(K) \xrightarrow[\ g\ ]{} Y$$

There is a (perhaps smaller) neighbourhood $\mathfrak{Op}_E(K)$ *of* K *in* E *and a continuous lift* $\nu \colon \mathfrak{Op}_E(K) \to X$ *such that* $\rho \circ \nu = g \colon \mathfrak{Op}_E(K) \to Y$ *and* $\nu = \mu \colon \mathfrak{Op}_E(L) \cup K \to X$.

Proof. Note that in case $\rho \colon X \to Y$ is a Serre fibration (not just a microfibration) and $\mathfrak{Op}_E(L) = N_L(1)$, $\mathfrak{Op}_E(K) = N_K(1)$ (unit disk bundles) the theorem is classical and follows from the homotopy lifting property for Serre fibrations (cf. Bredon [5], p. 451). In case K is compact and $L = \emptyset$ then the theorem is trivial and follows form the microfibration property applied globally to a strong deformation retract of $\mathfrak{Op}_E K$ to the zero section ($\equiv K$). In general however one has to proceed inductively over closed cells in K, relative to lifts constructed at previous inductive steps. This requires some care because $\rho \colon X \to Y$ is assumed only to be a microfibration. We provide explicit details based on the following lemma. Let $E = D^n \times \mathbf{R}^q \to D^n$ be the product bundle over D^n.

Lemma 5.2. *Let* $g \colon \mathfrak{Op}_E(D^n) \to Y$, $\mu \colon D^n \times \{0\} \cup \mathfrak{Op}_E(\partial D^n) \to X$ *be continuous maps such that* $\rho \circ \mu = g \colon D^n \times \{0\} \cup \mathfrak{Op}_E(\partial D^n) \to Y$. *There is continuous lift* $\nu \colon \mathfrak{Op}_E(D^n) \to X$ *such that* $\rho \circ \nu = g \colon \mathfrak{Op}_E(D^n) \to Y$, *and* $\nu = \mu \colon D^n \times \{0\} \cup \mathfrak{Op}_E(\partial D^n) \to X$.

Proof. With respect to the maps g, μ, $\mathfrak{Op}_E(D^n) = D^n \times [-\kappa, \kappa]^q$, respectively $\mathfrak{Op}_E(\partial D^n) = \partial D^n \times [-\kappa, \kappa]^q$ for $\kappa > 0$. One may assume $\kappa = 1$. It is standard that there is a homeomorphism $h \colon D^n \times [0,1] \to D^n \times [0,1]$ such that $h(D^n \times \{0\}) = D^n \times \{0\} \cup \partial D^n \times [0,1]$ (cf. Bredon [5], p. 451). There is a similar homeomorphism in case $[0,1]$ is replaced throughout by $[-1, 0]$. Since $\rho \colon X \to Y$ is a microfibration, employing this pair of homeomorphisms and a strong deformation retract from $D^n \times [-1,1]^q$ to $D^n \times \{0\}$, the lemma is proved. $\square$

$$D^n \times \{0\} \cup \mathfrak{Op}_E(\partial D^n) \xrightarrow{\ \mu\ } X$$

$$\left. i \right\downarrow \qquad\qquad \left\downarrow \rho\right.$$

$$\mathfrak{Op}_E(D^n) \xrightarrow[\ g\ ]{} Y$$

Returning to the proof of the theorem, we proceed by induction on the dimension of the skeleta $K^{(j)}$, $j \geq 0$, of the CW-complex K i.e., we follow the corresponding proof of the classical theorem in the case of Serre fibrations $\rho \colon X \to Y$, with appropriate modifications based on the above lemma to prove the theorem for microfibrations. Let $v \in K^{(0)}$ be a vertex in $K \setminus L$. Since the unit disk $N_v(1)$ in the fiber $E_v = \mathbf{R}^q$ deformation retracts to v, and since $\rho\colon X \to Y$ is a microfibration, it follows that there is an $\epsilon \in (0,1]$ and a continuous lift $\overline{\nu}\colon N_v(\epsilon) \to X$ such that $\rho \circ \overline{\nu} = g\colon N_v(\epsilon) \to Y$. One concludes that there is a continuous lift $\nu_0\colon \mathfrak{Op}_E(K^{(0)}) \cup K \to X$ such that $\rho \circ \nu_0 = g\colon \mathfrak{Op}_E(K^{(0)}) \to Y$, and the restriction $\nu_0 = \mu\colon \mathfrak{Op}_E(L^{(0)}) \cup K \to X$. Inductively, suppose that there is a continuous lift $\nu_j\colon \mathfrak{Op}_E(K^{(j)}) \cup K \to X$, $j \geq 0$, such that:

(i) $\rho \circ \nu_j = g\colon \mathfrak{Op}_E(K^{(j)}) \to Y$.

(ii) $\nu_j = \mu\colon \mathfrak{Op}_E(L^{(j)}) \cup K \to X$.

(iii) In case $j \geq 1$, $\nu_j = \nu_{j-1}\colon \mathfrak{Op}_E(K^{(j-1)}) \to X$.

To construct a continuous lift $\nu_{j+1}\colon \mathfrak{Op}_E(K^{(j+1)}) \cup K \to X$, we proceed by induction over the closed $(j+1)$-cells in $K^{(j+1)}$. Let $f_\sigma\colon D^{j+1} \to K$ represent a $(j+1)$-cell σ; $f_\sigma(\partial D^{j+1}) \subset K^{(j)}$. Since D^{j+1} is contractible, the pullback bundle $f_\sigma^* E$ over D^{j+1} is trivial which, for convenience, we assume to be the product bundle, $\pi\colon D^{j+1} \times \mathbf{R}^q \to D^{j+1}$. Let $h\colon f_\sigma^* E \to E$ be the induced bundle map (a linear isomorphism on each fiber) over $f_\sigma\colon D^{j+1} \to K$. Thus, with respect to this product bundle, we have the following data:

$$
\begin{array}{ccc}
D^{j+1} \times \{0\} \cup \mathfrak{Op}_E(\partial D^{j+1}) & \xrightarrow{(\mu \cup \nu_j) \circ h} & X \\
\ \downarrow{\scriptstyle i} & & \ \downarrow{\scriptstyle \rho} \\
\mathfrak{Op}_E(D^{j+1}) & \xrightarrow[g \circ h]{} & Y
\end{array}
$$

Applying the lemma, there is a continuous lift $\nu_\sigma\colon \mathfrak{Op}_E(D^{j+1}) \to X$ such that,

(iv) $\rho \circ \nu_\sigma = g \circ h\colon \mathfrak{Op}_E(D^{j+1}) \to Y$.

(v) $\nu_\sigma = (\mu \cup \nu_j) \circ h\colon D^{j+1} \times \{0\} \cup \mathfrak{Op}_E(\partial D^{j+1}) \to X$.

Pushing the map ν_σ forward to $K_\sigma = f_\sigma(D^{j+1}) \subset K^{(j+1)}$, one obtains a lift (same notation) $\nu_\sigma\colon N_{K_\sigma}(\epsilon_\sigma) \to X$, compatible with $\nu_j\colon \mathfrak{Op}_E(K^{(j)}) \to X$. Thus inductively one constructs a family of continuous lifts, $\nu_\sigma\colon N_{K_\sigma}(\epsilon_\sigma) \to X$, compatible with $\nu_j\colon \mathfrak{Op}_E(K^{(j)}) \to X$, where σ runs over all closed $(j+1)$-cells in $K^{(j+1)}$.

Since K is locally finite, hence locally compact, a simple partition of unity argument constructs a continuous map $\epsilon\colon K^{(j+1)} \to (0, \infty)$ and a continuous lift $\nu_{j+1}\colon N_{K^{(j+1)}}(\epsilon) \cup K \to X$ such that $\epsilon \leq \epsilon_\sigma$ and $\nu_{j+1} = \nu_\sigma\colon N_{K_\sigma}(\epsilon) \to X$ for each $(j+1)$-cell σ.

In particular, $\nu_{j+1} = \nu_j\colon \mathfrak{Op}_E(K^{(j)}) \to X$, $\rho \circ \nu_{j+1} = g\colon N_{K^{(j+1)}}(\epsilon) \to Y$, and $\nu_{j+1} = \mu\colon \mathfrak{Op}_E(L^{(j+1)}) \cup K \to X$. This inductive construction of the lift

$\nu_{j+1}\colon \mathfrak{Op}_E(K^{(j+1)}) \cup K \to X$ completes the construction of the sequence of lifts $(\nu_j)_{j\geq 0}$.

Since K is locally compact, employing a partition of unity one constructs a continuous map $\nu\colon \mathfrak{Op}_E(K) \to X$ (on each compact set $M \subset K$, $\nu = \nu_j\colon \mathfrak{Op}_E(M) \to X$ for some $j = j(M)$) such that, $\rho\circ\nu = g\colon \mathfrak{Op}_E(K) \to Y$, $\nu = \mu\colon \mathfrak{Op}_E(L)\cup K \to X$, which completes the proof of the theorem. $\qquad\square$

Theorem 5.3 (Approximations). *Let $\rho\colon X \to Y$ be a microfibration, Y a smooth manifold, and let $p\colon Y \to V$ be a smooth fiber bundle with q-dimensional fibers over a base manifold V. Let $\alpha \in \Gamma(X)$ be a continuous section $(p \circ \rho \circ \alpha = \mathrm{id}_V)$. For each neighbourhood N of the image $\alpha(V)$ in X there is a neighbourhood U of the image $\rho \circ \alpha(V)$ in Y and a continuous lift $\nu\colon U \to X$, $\rho \circ \nu = \mathrm{id}_U$, such that $\nu(U) \subset N$ and $\nu \circ \rho \circ \alpha = \alpha\colon V \to X$.*

Proof. Let K be the image $\rho \circ \alpha(V)$ in Y (K is homeomorphic to V). Let $\pi\colon E \to K$ be the $\mathbf{R}^q$-bundle over K obtained from $T(Y)$ by restricting to the vertical tangent bundle to the q-dimensional fibers over points of K; employing a Riemannian metric on $T(Y)$, $\pi\colon E \to K$ is a Euclidean vector bundle. In what follows one identifies, via the exponential map, (small) neighbourhoods of K in Y with the corresponding neighbourhoods of K in E. Let $\mu = \alpha \circ p\colon K \to X$. Note that $\rho \circ \mu = \mathrm{id}_K$ i.e. μ is a lift of K into X. Applying Theorem 5.1 to the inclusion $i\colon \mathfrak{Op}_E(K) \to Y$ and to the lift $\mu\colon K \to X$, there is a continuous lift $\nu\colon U \to X$, U a disk bundle neighbourhood of K in E, such that $\rho \circ \nu = \mathrm{id}_U$, $\nu = \mu$ on K. In particular, $\nu \circ \rho \circ \alpha = \alpha\colon V \to X$. For U sufficiently small, it follows also that $\nu(U) \subset N$, which completes the proof of the Theorem. $\qquad\square$

Remark 5.4. Theorem 5.3 is applied as follows to lift smooth approximations. Let $f \in \Gamma(Y)$ be a continuous section which lifts to $\alpha \in \Gamma(X)$: $\rho \circ \alpha = f\colon V \to Y$. Employing Theorem 5.3 there is a neighbourhood U of $f(V)$ which lifts into X. Since $p\colon Y \to V$ is a smooth bundle, one can C^0-approximate f in U by a smooth section $g\colon V \to Y$, and there is a homotopy, $H\colon [0,1] \to \Gamma(Y)$, such that $H_0 = f$, $H_1 = g$, and $\mathrm{ev}\, H([0,1] \times V) \subset U$. The lifted homotopy $G = \nu \circ H\colon [0,1] \times V \to N$ satisfies $G_0 = \nu \circ f = \alpha$, $G_1 = \nu \circ H_1$. Consequently, up to a small C^0-homotopy of sections $G\colon [0,1] \to \Gamma(X)$, one may replace the pair of sections (α, f), where $f \in \Gamma(Y)$ is a continuous section which lifts to α, by the pair (β, g), $\beta = G_1 \in \Gamma(X)$, such that $\rho \circ \beta = g \in \Gamma(Y)$ is a smooth section. Furthermore if f is smooth on $\mathfrak{Op}\, K$ one may assume in addition that for all $t \in [0,1]$, $G_t = \alpha$ (constant homotopy) on $\mathfrak{Op}\, K$.

Corollary 5.5. *Let $K \subset V$ be closed, $f \in \Gamma(Y)$, $\alpha \in \Gamma(X)$ such that f lifts to α along $K\colon \rho \circ \alpha = f\colon K \to Y$. Let N be a neighbourhood of the image $\alpha(V)$ in X. Up to a small homotopy $\mathrm{rel}\,K$ of α in N, one may assume that $\rho \circ \alpha = f$ on $\mathfrak{Op}\, K$. Explicitly, there is a homotopy $H\colon [0,1] \to \Gamma(X)$ such that:*

(i) $H_0 = \alpha$; ev $H([0,1] \times V) \subset N$.

(ii) $\rho \circ H_1 = f \colon \mathfrak{Op}\, K \to Y$.

In addition if f is smooth on $V \setminus K$ then the homotopy H can be chosen so that also $\rho \circ H_1$ is smooth on $V \setminus K$.

Proof. Employing Theorem 5.3, there is a neighbourhood U of the image $\rho \circ \alpha(V)$ in Y and a continuous lift $\nu \colon U \to X$ such that $\nu(U) \subset N$, $\nu \circ \rho \circ \alpha = \alpha$. Employing the $\mathbf{R}^q$-bundle $\pi \colon E \to L$ as in the proof of Theorem 5.3, where L is the image $\rho \circ \alpha(V)$ in Y (small q-disk neighbourhoods of L in E are identified with small neighbourhoods of L in Y), the sections $\rho \circ \alpha, f \in \Gamma_{\mathfrak{Op}\, K}(Y)$ are homotopic in U, rel K. Employing standard arguments with bump functions one shows that there is a homotopy rel K, $G \colon [0,1] \to \Gamma(Y)$, G is constant $= \rho \circ \alpha$ on $V \setminus \mathfrak{Op}\, K$, and such that:

(iii) $G_0 = \rho \circ \alpha$; ev $G([0,1] \times V) \subset U$.

(iv) $G_1 = f \colon \mathfrak{Op}_1 K \to Y$ where $\overline{\mathfrak{Op}_1 K} \subset \mathfrak{Op}\, K$.

In case f is smooth on $V \setminus K$, employing smooth approximation theory, I Theorem 1.1, up to a small homotopy on G_1 in U rel $\mathfrak{Op}\, K$, one may assume in addition that G_1 is smooth on $V \setminus K$. Setting $H = \nu \circ G \colon [0,1] \to \Gamma(X)$, the corollary is proved. $\qquad \square$

Corollary 5.6 (Parameters). *Let $\rho \colon X \to Y$ be a microfibration. Let $\pi \colon E \to K$ be a Euclidean vector bundle, as in Theorem 5.1. Let $g \colon A \times \mathfrak{Op}_E(K) \to Y$, $\mu \colon A \times K \to X$, A a compact polyhedron (parameter space), be continuous maps such that $\rho \circ \mu = g \colon A \times K \to Y$. There is a continuous lift $\nu \colon A \times \mathfrak{Op}_E(K) \to X$, such that $\nu = \mu \colon A \times K \to X$, and $\rho \circ \nu = g \colon A \times \mathfrak{Op}_E(K) \to Y$.*

Proof. We consider the $\mathbf{R}^q$-bundle $\mathrm{id} \times \pi \colon A \times E \to A \times K$. Since A is compact it follows that neighbourhoods of $A \times K$ in $A \times E$ of the form $A \times \mathfrak{Op}_E(K)$ are cofinal with neighbourhoods $\mathfrak{Op}_{A \times E}(A \times K)$. Since also $A \times K$ is a locally finite CW-complex, Theorem 5.1 applies to the Euclidean bundle $\mathrm{id} \times \pi \colon A \times E \to A \times K$ and the corollary is proved.

A microfibration $\rho \colon X \to Y$, Y a manifold, is an open map. Indeed, if $\rho(a) = y \in Y$ then small ball neighbourhoods of y lift back to (path connected) sets X that contain a. One therefore obtains the following result (cf. II §1 for the following convex hull notation).

Lemma 5.7. *Let $\rho \colon X \to \mathbf{R}^q$ be a microfibration. Then $\rho(A) \subset \mathbf{R}^q$ is open for each path component A of X. Consequently for each $x \in X$, $\mathrm{Conv}(X, x) = \mathrm{IntConv}(X, x) \subset \mathbf{R}^q$.*

§2. *C*-Structures for Relations over Affine Bundles

5.2.1. Let $p\colon E \to B$ be an affine $\mathbf{R}^q$-bundle over a base manifold B, $\dim B = n \geq 1$, and let $\rho\colon \mathcal{R} \to E$ be a microfibration. $\Gamma(E)$, respectively $\Gamma(\mathcal{R})$, is the space of continuous sections, in the compact-open topology, of the map $p\colon E \to B$, respectively $p \circ \rho\colon \mathcal{R} \to B$.

$$\mathcal{R} \xrightarrow{\ \ \rho\ \ } E$$
$$\downarrow p$$
$$B$$

An important special case, studied in Chapter II, is the inclusion $i\colon \mathcal{R} \to E$, where $\mathcal{R} \subset E$ is open. $\Gamma_K(E)$, $\Gamma_K(\mathcal{R})$, denote the corresponding spaces of sections over the subspace $K \subset B$. For each $b \in B$, $\mathcal{R}_b$ denotes the fiber over b of the map $p \circ \rho\colon \mathcal{R} \to B$. In particular, $\mathcal{R}_b = \rho^{-1}(E_b)$ where $E_b = p^{-1}(b)$ is the fiber over b of the affine bundle $p\colon E \to B$ ($E_b = \mathbf{R}^q$).

If $\phi\colon A \to \Gamma_K(\mathcal{R})$ is a map (A a space of parameters) then $\operatorname{ev}\phi\colon A \times K \to \mathcal{R}$ is the evaluation map, $\operatorname{ev}\phi(x,y) = \phi(x)(y) \in \mathcal{R}$. For each $b \in K$, $\phi_b\colon A \to \mathcal{R}_b$ denotes the induced map, $\phi_b(x) = \phi(x)(b) \in \mathcal{R}_b$. As a particular example, if $\phi\colon [0,1] \to \Gamma_K(\mathcal{R})$ is a path then for each $b \in K$, $\phi_b\colon [0,1] \to \mathcal{R}_b$ is the induced path in the fiber $\mathcal{R}_b$. If $L \subset K$, then $\phi\colon A \to \Gamma_L(\mathcal{R})$ (same notation) denotes the restriction of the map ϕ to sections over the subspace L. By definition, $\rho \circ \phi\colon A \to \Gamma(E)$ is smooth if $\operatorname{ev}\rho \circ \phi \in C^\infty(A \times B, E)$.

5.2.2. For each $(b,a) \in B \times \mathcal{R}_b$ let $\operatorname{Conv}(\mathcal{R}_b, a) \subset E_b$ denote the convex hull (in the affine fiber E_b) of the ρ-image of the path component in $\mathcal{R}_b$ to which a belongs. Applying Corollary 5.7 to the microfibration $\rho\colon \mathcal{R}_b \to E_b$, for each $b \in B$, $\operatorname{Conv}(\mathcal{R}_b, a)$ is open in E_b ($= \mathbf{R}^q$). $\mathcal{R}$ is *ample* over E if for all $(b,a) \in B \times \mathcal{R}_b$, $\operatorname{Conv}(\mathcal{R}_b, a) = E_b$.

Let $\beta \in \Gamma(\mathcal{R})$, $f \in \Gamma(E)$ be continuous sections such that for all $b \in B$,

$$f(b) \in \operatorname{Conv}(\mathcal{R}_b, \beta(b)).$$

Recall, Chapter II, §2, that a *C*-structure over a subset $K \subset B$ with respect to f, β consists of a pair (g, G) where $g\colon [0,1] \to \Gamma_K(\mathcal{R})$ is continuous, $g(0) = g(1) = \beta_K$ (the restriction of β to K), such that for all $b \in K$ the composed path $\rho \circ g_b\colon [0,1] \to E_b$ strictly surrounds $f(b)$, and $G\colon [0,1]^2 \to \Gamma_K(\mathcal{R})$ is a (fiberwise) base point preserving homotopy which contracts the strictly surrounding path g to the base point β_K: for all $(t,s) \in [0,1]^2$,

$$G(t,1) = g(t)\,; \quad G(t,0) = \beta_K\,; \quad G(0,s) = G(1,s) = \beta_K.$$

Since g is a loop at $\beta \in \Gamma_K(\mathcal{R})$ and G is a base point preserving contraction of g to β, we employ indiscriminately throughout the equivalent formulation that $g\colon (S^1, 1) \to (\Gamma_K(\mathcal{R}), \beta)$ and that $G\colon (D^2, 1) \to (\Gamma_K(\mathcal{R}), \beta)$, such that $G = g\colon S^1 \to \Gamma_K(\mathcal{R})$ ($1 \in S^1$ is the base point).

Applying II Lemma 2.2, the space of *C*-structures over K with respect to f, β is contractible. The next result, based on II Proposition 2.3, establishes the existence of *C*-structures which are globally defined over the base manifold B.

Proposition 5.8. *Let* $\beta \in \Gamma(\mathcal{R})$, $f \in \Gamma(E)$ *such that for all* $b \in B$, $f(b) \in$ Conv$(\mathcal{R}_b, \beta(b))$. *There is a C-structure* (ψ, H) *globally defined over* B *with respect to* f, β. *Explicitly, there is a continuous map* $\psi\colon S^1 \to \Gamma(\mathcal{R})$, $\psi(1) = \beta$, *such that for all* $b \in B$, *the* ρ-*image of the loop* $\psi_b\colon S^1 \to \mathcal{R}_b$ *strictly surrounds* $f(b) \in E_b$. *Furthermore, there is a homotopy* $H\colon D^2 \to \Gamma(\mathcal{R})$ *which contracts* ψ *to* $\beta = H(1)$.

Proof. In the special case that $\mathcal{R} \subset E$ is open and $\rho\colon \mathcal{R} \to E$ is the inclusion map, the proposition is equivalent to II Proposition 2.3 (over a base manifold B). For general microfibrations $\rho\colon \mathcal{R} \to E$, the inductive process of Proposition 2.3 is employed to construct suitable sections of $\Gamma(E)$, which then lift back to *C*-structures in $\Gamma(\mathcal{R})$ via the microfibration $\rho\colon \mathcal{R} \to E$, in accordance with Theorem 5.1. To begin this inductive process, for each $b \in B$, we construct *C*-structures over $\mathfrak{Op}\, b \subset B$.

Lemma 5.9. *For all* $b \in B$, *each C-structure over* $\{b\}$ *with respect to* $f \in \Gamma(E)$, $\beta \in \Gamma(\mathcal{R})$, *extends to a C-structure over* $\mathfrak{Op}\, b \subset B$ *with respect to* f, β.

Proof. For the purposes of the lemma we may assume $\mathfrak{Op}\, b$ is homeomorphic to $\mathbf{R}^n$, $n = \dim B$ (It is here that we employ the hypothesis that B is a manifold). Let (g, G) be a *C*-structure over $\{b\}$ with respect to f, β. Note that the pair $(\rho \circ g, \rho \circ G)$ is a *C*-structure (in E) over $\{b\}$ with respect to $f, \rho \circ \beta$. Applying II Lemma 2.4, the pair $(\rho \circ g, \rho \circ G)$ extends to a *C*-structure (h, H) over $\mathfrak{Op}\, b$ with respect to $f, \rho \circ \beta$:

$$h\colon (S^1, 1) \to (\Gamma_{\mathfrak{Op}\, b}(E), \rho \circ \beta); \quad H\colon (D^2, 1) \to (\Gamma_{\mathfrak{Op}\, b}(E), \rho \circ \beta),$$

such that at $b \in \mathfrak{Op}\, b$, the induced maps $h_b = \rho \circ g\colon S^1 \to E_b$, $H_b = \rho \circ G\colon D^2 \to E_b$, and at the base point $1 \in S^1$, $h(1) = \rho \circ \beta \in \Gamma_{\mathfrak{Op}\, b}(E)$. Consequently, with respect to the product bundle, $\pi\colon F = D^2 \times \mathbf{R}^n \to D^2$, the subcomplex $L = \{1\} \subset D^2$, and an identification $\mathfrak{Op}\, b = \mathbf{R}^n$ via a homeomorphism of pairs $(\mathfrak{Op}\, b, \{b\}) \to (\mathbf{R}^n, \{0\})$, there is a commutative diagram (ev $H\colon D^2 \times \mathfrak{Op}\, b \to E$),

$$
\begin{array}{ccc}
D^2 \times \{0\} \cup \mathfrak{Op}_F(L) & \xrightarrow{\text{ev } G \cup \beta} & \mathcal{R} \\
\downarrow{\scriptstyle i} & & \downarrow{\scriptstyle \rho} \\
\mathfrak{Op}_F(D^2) & \xrightarrow[\text{ev } H]{} & E
\end{array}
$$

Applying Theorem 5.1 there is a continuous map $\nu\colon D^2 \to \Gamma_{\mathfrak{Op}\,b}(\mathcal{R})$ whose corresponding evaluation map $\mathrm{ev}\,\nu\colon \mathfrak{Op}_F(D^2) \to \mathcal{R}$ is a continuous lift in the above commutative diagram: (i) $\rho \circ \mathrm{ev}\,\nu = \mathrm{ev}\,H\colon \mathfrak{Op}_F(D^2) \to E$, (ii) $\mathrm{ev}\,\nu = \mathrm{ev}\,G \cup \mathrm{ev}\,\beta\colon D^2 \times \{0\} \cup \mathfrak{Op}_F(L) \to \mathcal{R}$. Consequently the map $\nu\colon D^2 \to \Gamma_{\mathfrak{Op}\,b}(\mathcal{R})$ defines a C-structure $(\nu|_{S^1}, \nu)$ over $\mathfrak{Op}\,b$ with respect to f, β that extends the given C-structure (g, G) over $\{b\}$. $\square$

Employing Lemma 5.9 one patches together the local C-structures constructed above over $\mathfrak{Op}\,b$ for all $b \in B$. The inductive step for this process is as follows.

Lemma 5.10. *Let $K, L \subset B$ be closed subsets and (g_0, G_0), (g_1, G_1) be C-structures respectively over $\mathfrak{Op}\,K$, $\mathfrak{Op}\,L$ with respect to $f \in \Gamma(E)$, $\beta \in \Gamma(\mathcal{R})$. There is a C-structure (h, H) over $\mathfrak{Op}\,(K \cup L)$ with respect to f, β and a neighbourhood N of K whose closure lies in $\mathfrak{Op}\,K$ such that $(h, H) = (g_0, G_0)$ over N and $(h, H) = (g_1, G_1)$ over $\mathfrak{Op}\,L \setminus \mathfrak{Op}\,K$.*

The proof of the lemma is formally the same as the proof of the corresponding II Lemma 2.5 in the case that $\mathcal{R} \subset E$ is open, and is therefore omitted. Returning to the proof of Proposition 5.8, there is a C-structure over $\mathfrak{Op}\,b$ with respect to f, β for each $b \in B$. Hence there are countable, locally finite closed covers $\{W_i\}$, $\{U_i\}$ of the base manifold B such that for all i, $W_i \subset \mathrm{int}\,U_i$, and there is a C-structure over $\mathfrak{Op}\,U_i$ for each index i. Inductively on n, let $K = \bigcup_{i=1}^{i=n} W_i$, a closed set in B, and suppose (ψ_n, G_n) is a C-structure over $\mathfrak{Op}\,K$. Applying Lemma 5.10 to the closed sets K, $L = U_{n+1}$, there is a C-structure (ψ_{n+1}, G_{n+1}) over $\mathfrak{Op}\,(K \cup U_{n+1})$ such that $(\psi_{n+1}, G_{n+1}) = (\psi_n, G_n)$ over $\mathfrak{Op}\,K$. Since the cover $\{W_i\}$ is locally finite it follows that the limit maps,

$$\psi = \lim_{n \to \infty} \psi_n \colon S^1 \to \Gamma(\mathcal{R}); \quad G = \lim_{n \to \infty} G_n \colon D^2 \to \Gamma(\mathcal{R}),$$

are well-defined and continuous. Thus (ψ, G) is a C-structure over B with respect to $f \in \Gamma(E)$, $\beta \in \Gamma(\mathcal{R})$ (cf. the analogous proof of II, Proposition 2.3). $\square$

The existence of C-structures, Proposition 5.8, admits several refinements, analogous to the refinements of the corresponding existence of C-structures, II Proposition 2.3, in the case that $\mathcal{R} \subset E$ is open. We indicate only how the relevant C-structures in E are lifted back to $\mathcal{R}$ in the general case that $\rho\colon \mathcal{R} \to E$ is a microfibration. In case $p\colon E \to B$ is smooth, employing Theorem 5.3, for each neighbourhood N of $\beta(B)$ in $\mathcal{R}$ there is a continuous lift $\nu\colon U \to N$, U a neighbourhood of $\rho \circ \beta(B)$ in E, such that: (i) $\rho \circ \nu = \mathrm{id}_U$, (ii) $\nu \circ \rho \circ \beta = \beta \in \Gamma(\mathcal{R})$.

Complement 5.11 (Relative Theorem). *Suppose $p\colon E \to B$ is a smooth affine $\mathbf{R}^q$-bundle and let $K \subset B$ be a closed subspace such that $f = \rho \circ \beta \in \Gamma_K(E)$. Assume also the following data:*

(i) *a neighbourhood N of the image $\beta(B)$ in $\mathcal{R}$ and a continuous lift $\nu: U \to N$ as above.*

(ii) *A C-structure (m, M) in $U \subset E$ over $\mathfrak{Op}\, K$ with respect to $f, \rho \circ \beta$.*

There is a C-structure (h, H) over B with respect to f, β such that over $\mathfrak{Op}\, K \subset B$, $(h, H) = (\nu \circ m, \nu \circ M)$ (a continuous lift of C-structures over $\mathfrak{Op}\, K$). In particular, over $\mathfrak{Op}\, K$ the image of (the evaluation of the map) H lies in N:

$$\mathrm{ev}\, H(D^2 \times \mathfrak{Op}\, K) \subset N.$$

Proof. Employing II Complement 2.6, C-structures (m, M) in (ii) exist. The Complement follows from Lemma 5.10 applied to the lifted C-structure $(\nu \circ m, \nu \circ M)$ over $\mathfrak{Op}\, K$ and to any C-structure over B with respect to f, β (which exists by Proposition 5.8). Thus one obtains a C-structure (h, H) over B which equals $(\nu \circ m, \nu \circ M)$ over (a smaller) $\mathfrak{Op}\, K$ in B.   □

Complement 5.12 (C^∞-structures). *Suppose $p: E \to B$ is a smooth affine $\mathbf{R}^q$-bundle. Let (h, H) be a C-structure over B with respect to f, β. Up to a small homotopy one may assume $\rho \circ H$ (and hence $\rho \circ h$) is a smooth map i.e. the evaluation map $\rho \circ \mathrm{ev}\, H: D^2 \times B \to E$ is a smooth map. Furthermore this homotopy can be chosen rel $\{1\}$ in case $\rho \circ H(1) = \rho \circ \beta$ is smooth.*

Proof. The complement follows from Remark 5.4 applied to the microfibration $\mathrm{id} \times \rho: D^2 \times \mathcal{R} \to D^2 \times E$, the smooth bundle $\mathrm{id} \times p: D^2 \times E \to D^2 \times B$, and the section of $\Gamma(D^2 \times \mathcal{R})$ induced by the C-structure H. In case $\rho \circ \beta$ is smooth one may assume $H = \beta$ on $\mathfrak{Op}\, 1 \subset D^2$, and Remark 5.4 applies rel $\mathfrak{Op}\, 1$. The C-structure condition is preserved throughout a sufficiently small perturbation of (h, H) since "strictly surrounding" is an open condition.   □

The study of the homotopy groups $\pi_i(\Gamma(\mathcal{R}))$ in connection with the weak homotopy equivalence theorem, VIII Theorem 8.15, employs the following parametrized version of the existence of C-structures.

Complement 5.13 (Parameters). *Let P be an auxiliary manifold (a parameter space). Let $f: P \to \Gamma(E)$, $\beta: P \to \Gamma(\mathcal{R})$ be continuous maps such that for all $(p, b) \in P \times B$,*

$$f(p, b) \in \mathrm{Conv}(\mathcal{R}_b, \beta(p, b)).$$

There is a P-parameter family of C-structures (h, H) over B with respect to f, β. That is, there are continuous maps,

$$h: P \times S^1 \to \Gamma(\mathcal{R}); \quad H: P \times D^2 \to \Gamma(\mathcal{R}),$$

such that for each $p \in P$, (h_p, H_p) is a C-structure over B with respect to the sections $f(p), \beta(p)$ (here $h_p(t) = h(p, t)$; $H_p(t, s) = H(p, t, s)$).

Proof. Associated to the parameter space P are the affine bundle $\mathrm{id} \times p\colon P \times E \to P \times B$, fiber $\mathbf{R}^q$, and the microfibration over $P \times E$, $\mathrm{id} \times \rho\colon P \times \mathcal{R} \to P \times E$. The parametrized families of maps f, β induce sections $\pi_1 \times \mathrm{ev}\, f \in \Gamma(P \times E)$, respectively $\pi_1 \times \mathrm{ev}\, \beta \in \Gamma(P \times \mathcal{R})$, where $\pi_1\colon P \times B \to P$ is the projection onto P. Applying Theorem 5.8 to the above induced sections, the complement is proved (the required convex hull conditions on the induced sections are equivalent to the stated conditions on f, β). $\qquad\square$

Recall that the relation $\rho\colon \mathcal{R} \to E$ is ample if for all $(b, a) \in B \times \mathcal{R}_b$, $\mathrm{Conv}(\mathcal{R}_b, a) = E_b(= \mathbf{R}^q)$. In the case of ample relations C-structures exist for all sections $f \in \Gamma(E)$. We state this as a separate result.

Corollary 5.14 (Ample Relations). *Let $\rho\colon \mathcal{R} \to E$ is ample and suppose that $\mathcal{R}$ admits a section $\beta \in \Gamma(\mathcal{R})$. For each section $f \in \Gamma(E)$ there is a C-structure (h, H) over B with respect to f, β.*

5.2.3. Integral Representation Theorem. In Chapter II the theory of C-structures was employed, II Theorem 2.12, to represent a section $f \in \Gamma(E)$ as a (Riemann) integral, integrand in $\mathcal{R}$, in the case that $\mathcal{R} \subset E$ is open. Essentially this integral representation is obtained by suitably reparametrizing a strictly surrounding path. Lifting this reparametrization to $\mathcal{R}$, the integral representation theorem is proved in the general case of a microfibration $\rho\colon \mathcal{R} \to E$.

Theorem 5.15 (Integral Representation). *Let $p\colon E \to B$ be an affine $\mathbf{R}^q$-bundle over a manifold B and let $\rho\colon \mathcal{R} \to E$ be a microfibration. Suppose $\beta \in \Gamma(\mathcal{R})$, $f \in \Gamma(E)$ are sections such that for all $b \in B$, $f(b) \in \mathrm{Conv}(\mathcal{R}_b, \beta(b))$. Each C-structure (g, G) over B with respect to f, β can be reparametrized to a C-structure (h, H) such that for all $b \in B$ (recall $h\colon [0,1] \to \Gamma(\mathcal{R})$) :*

$$f(b) = \int_0^1 \rho \circ h(t, b)\, dt.$$

Proof. Let (g, G) be a C-structure over B with respect to f, β. Projecting into E, $(\rho \circ g, \rho \circ G)$ is a C-structure on E with respect to $f, \rho \circ \beta$. The integral representation of f is obtained from the strictly surrounding path $\rho \circ g\colon [0,1] \to \Gamma(E)$ by a suitable *internal* reparametrization of the domain of g: $f(b) = \int_0^1 \rho \circ g(\mu(s, b), b)\, ds$ for a suitable function $\mu\colon [0,1] \times B \to [0,1]$ (cf. II Theorem 2.12 for details). Let (h, H) be the C-structure on $\mathcal{R}$ with respect to f, β obtained by the internal reparametrization, $h(t, b) = g(\mu(t, b), b)$, $H(t, s, b) = G(\mu(t, b), s, b)$, for all $(t, s, b) \in [0,1]^2 \times B$. Evidently, for all $b \in B$, $f(b) = \int_0^1 \rho \circ h(t, b)\, dt$, and the theorem is proved. $\qquad\square$

Remark 5.16. The various complements to Proposition 5.8 on the existence of C-structures: relative theorem; C^∞-structures; parameters; ample relations (Complements 5.11–5.14) in turn yield respective complements to the Integral Representation Theorem 5.15, by making further requirements on the surrounding path that appears in the integrand of the integral representation. To illustrate, we state only a complement on C^∞-structures, leaving the statement of the other complements to the reader (cf. II Complements 2.13–2.15, in case $\mathcal{R} \subset E$ is open).

Complement 5.17 (C^∞-structures). *Suppose that $p\colon E \to B$ is a smooth affine $\mathbf{R}^q$-bundle. If $f, \rho \circ \beta \in \Gamma(E)$ are smooth sections then there is a smooth C-structure (h, H) with respect to f, β such that for all $b \in B$, $f(b) = \int_0^1 \rho \circ h(t, b)\, dt$.*

§3. The $C^\perp$-Approximation Theorem

5.3.1. To conclude this chapter, we generalize the main analytic results of Chapter III to the case of microfibrations $\rho\colon \mathcal{R} \to E$. The key point here is that the main approximation results, The One Dimensional Theorem, (III Theorem 3.4), and the $C^\perp$-Approximation Theorem, (III Theorem 3.8), are obtained from the integral representation theorem by further internal reparametrizations of the surrounding path in the integrand, based on the rapidly oscillating function $\theta_\epsilon\colon [0, 1] \to [0, 1]$ introduced in III Proposition 3.1. Lifting back to $\mathcal{R}$, the corresponding approximation theorems are proved for general microfibrations $\rho\colon \mathcal{R} \to E$.

For example, let $p\colon E = B \times \mathbf{R}^q \to B$ be a product $\mathbf{R}^q$-bundle over a split space $B = C \times [0, 1]$. Recall that the section $g \in \Gamma(E)$ is C^1 in $t \in [0, 1]$ if the corresponding evaluation map $\mathrm{ev}\, g \in C^0(B, \mathbf{R}^q)$ is continuously differentiable in $t \in [0, 1]$; $\partial_t g \in \Gamma(E)$ denotes the corresponding section. The above discussion of the generalization of the analytic results in Chapter III to the microfibration case applies to prove Theorem 5.18 and Complement 5.19 below. Details, left to the reader, are adapted easily from the analogous proofs provided below for the $C^\perp$-Approximation Theorem 5.20 and Complement 5.21 in the microfibration case.

One-Dimensional Theorem 5.18. *Let $p\colon E \to B$ be a Euclidean $\mathbf{R}^q$-bundle over a split compact Hausdorff space $B = C \times [0, 1]$. Let $\rho\colon \mathcal{R} \to E$ be a microfibration and suppose $\beta \in \Gamma(\mathcal{R})$, $f_0 \in \Gamma(E)$, f_0 is C^1 in $t \in [0, 1]$, are sections such that for all $b \in B$,*

$$\partial_t f_0(b) \in \mathrm{Conv}(\mathcal{R}_b, \beta(b)).$$

For each $\epsilon > 0$ there is a section $f_\epsilon \in \Gamma(E)$, which is C^1 in $t \in [0, 1]$, and a homotopy $F = F_\epsilon\colon [0, 1] \to \Gamma(\mathcal{R})$ such that the following properties obtain:

(i) $\lim_{\epsilon \to 0} \|f_\epsilon - f_0\| = 0$ *($\|\ \|$ is the sup-norm on $C^0(B, \mathbf{R}^q)$).*

(ii) $F_0 = \beta$; $\rho \circ F_1 = \partial_t f_\epsilon \in \Gamma(E)$ *i.e. the section* $\partial_t f_\epsilon$ *lifts to* $F_1 \in \Gamma(\mathcal{R})$.

(iii) *Relative Theorem: Suppose in addition* $K \subset B$ *is closed such that* $\partial_t f_0 = \rho \circ \beta$ *on* $\mathfrak{Op}\, K$. *Let* N *be a neighbourhood of the image* $\beta(B)$ *in* $\mathcal{R}$. *One may choose* F *so that over* $\mathfrak{Op}\, K$ *the image of* F *lies in* N: $\mathrm{ev}\, F: [0,1] \times \mathfrak{Op}\, K \to N$.

Complement 5.19 (Strong Relative Theorem). *Suppose in addition* $K \subset B$ *is closed such that* $\partial_t f_0 = \rho \circ \beta$ *on* $\mathfrak{Op}\, K$. *For* $\epsilon > 0$ *sufficiently small then, up to a small perturbation on* $\mathfrak{Op}\, K$, *the section* $f_\epsilon \in \Gamma(E)$ *and the homotopy* $F = F_\epsilon: [0,1] \to \Gamma(\mathcal{R})$ *can be chosen to satisfy the additional properties (i)* $f_\epsilon = f_0$ *on* $\mathfrak{Op}\, K$; *(ii)* F *is constant on* $\mathfrak{Op}\, K$: *for all* $t \in [0,1]$, $F_t = \beta$ *on* $\mathfrak{Op}\, K$.

Let $B = [0,1]^n$, and recall the spaces $C^{s,r}(B, \mathbf{R}^q)$, $s \geq r \geq 0$, and the norms $\| \ \|^{s,r}$ introduced in Chapter III. In the usual manner, the space of C^r-sections of the product bundle $\pi: B \times \mathbf{R}^q \to B$ is identified with $C^r(B, \mathbf{R}^q)$.

$C^\perp$-Approximation Theorem 5.20. *Let* $p: E \to B$ *be the product* $\mathbf{R}^q$-*bundle over the n-cube* B, $n \geq 1$, *and let* $\rho: \mathcal{R} \to E$ *be a microfibration. Let also* P *be a compact smooth manifold (a parameter space) and let* $s \geq r \geq 1$ *be integers. Suppose* $g_0: P \to \Gamma(E)$ *is smooth (i.e. the evaluation map is smooth) and* $\beta: P \to \Gamma(\mathcal{R})$ *is continuous such that* $\rho \circ \beta: P \to \Gamma(E)$ *is smooth and such that for all* $(p,b) \in P \times B$:

$$\partial_t^r g_0(p,b) \in \mathrm{Conv}(\mathcal{R}_b, \beta(p,b)).$$

For each $\epsilon > 0$ *there is a* C^∞-*map* $g_\epsilon: P \to \Gamma(E)$ *and a homotopy* $F = F_\epsilon: [0,1] \times P \to \Gamma(\mathcal{R})$ *such that for all* $p \in P$ *the following properties obtain:*

(i) $\lim_{\epsilon \to 0} \|g_\epsilon(p) - g_0(p)\|^{s,r} = 0$, *uniformly in* $p \in P$.

(ii) $F_0(p) = \beta(p)$; $\rho \circ F_1(p) = \partial_t^r g_\epsilon(p)$; $\rho \circ F: [0,1] \times P \to \Gamma(E)$ *is smooth.*

(iii) *(Relative Theorem)* : *Suppose in addition* $K \subset P \times B$ *is closed such that, as maps in* $C^0(P \times B, \mathbf{R}^q)$, $\partial_t^r g_0 = \rho \circ \beta$ *on* $\mathfrak{Op}\, K$. *Let* N *be a neighbourhood in* $P \times \mathcal{R}$ *of the image* $(\pi_1 \times \mathrm{ev}\, \beta)(P \times B)$ $(\pi_1: P \times B \to P$ *is the projection map). One may choose* F *so that over* $\mathfrak{Op}\, K$ *the image of* F *lies in* N: *for all* $t \in [0,1]$, $(\pi_1 \times \mathrm{ev}\, F_t)(\mathfrak{Op}\, K) \subset N$.

Proof. The approximation (i) excludes the pure rth order derivatives ∂_t^r. However the approximation (i) is C^0-close in the derivatives ∂_t^j, $0 \leq j \leq r-1$, and C^s-close in the derivatives in the coordinates $u \in I^{n-1}$.

Employing the microfibration $\mathrm{id} \times \rho: P \times \mathcal{R} \to P \times E$ and the $\mathbf{R}^q$-bundle $\mathrm{id} \times p: P \times E \to P \times B$, there is no loss of generality in supposing that P is a point and that K is closed in B. Thus $\beta \in \Gamma(\mathcal{R})$ and N is a neighbourhood of $\beta(B)$ in $\mathcal{R}$. Applying the Integral Representation Theorem 5.15 there is a C-structure

(h, H) with respect to $\partial_t^r g_0, \beta$ such that $\rho \circ H\colon [0,1]^2 \to \Gamma(E)$ is smooth and such that for all $b \in B$,

$$\partial_t^r g_0(b) = \int_0^1 \rho \circ h(s, b)\, ds; \quad \operatorname{ev} H([0,1]^2 \times \mathfrak{Op}\, K) \subset N.$$

Thus the smooth parametrized family of paths $\rho \circ h\colon [0,1] \to \Gamma(E)$ strictly surrounds the derivative $\partial_t^r g_0$ and provides the above integral representation. The theorem now formally follows from the $C^\perp$-Approximation Theorem III 3.8 by employing $\rho \circ h$ instead of h. Thus for each $\epsilon > 0$ there is a C^∞-function $\theta_\epsilon\colon [0,1] \to [0,1]$ and a smooth section $g_\epsilon \in \Gamma(E)$ such that conclusion (i) is satisfied and $(b = (c, t) \in I^{n-1} \times [0,1] = B)$,

$$\partial_t^r g_\epsilon(c, t) = \rho \circ h(\theta_\epsilon(t), (c, t)). \tag{5.1}$$

Furthermore, let $F = F_\epsilon\colon [0,1] \to \Gamma(\mathcal{R})$ be the homotopy obtained by the internal reparametrization of H,

$$F(s, c, t) = H(\theta_\epsilon(t), s, c, t), \quad 0 \le s \le 1. \tag{5.2}$$

Since $H(t, 0, b) = \beta(b)$, $H(t, 1, b) = h(t, b)$, it follows that $F_0 = \beta \in \Gamma(\mathcal{R})$ and $\rho \circ F_1 = \partial_t^r g_\epsilon$, which proves conclusion (ii). Conclusion (iii) follows from the above properties of H, relative to $\mathfrak{Op}\, K$ (cf. Remark 5.16). $\qquad\square$

Complement 5.21 (Strong Relative Theorem). *Suppose in addition $K \subset P \times B$ is closed such that, as maps in $C^0(P \times B, \mathbf{R}^q)$, $\partial_t^r g_0 = \rho \circ \beta$ on $\mathfrak{Op}\, K$. For $\epsilon > 0$ sufficiently small then, up to a small perturbation on $\mathfrak{Op}\, K$, the map $g_\epsilon\colon P \to \Gamma(E)$ and the homotopy $F = F_\epsilon\colon [0,1] \times P \to \Gamma(\mathcal{R})$ can be chosen to satisfy the additional properties (i) $g_\epsilon = g_0$ on $\mathfrak{Op}\, K$; (ii) F is constant along $\mathfrak{Op}\, K$: for all $t \in [0,1]$, $F_t = \beta$ on $\mathfrak{Op}\, K$.*

Proof. Again we may assume that the parameter space P is a point. Let $g_\epsilon \in \Gamma(E)$, $F = F_\epsilon\colon [0,1] \to \Gamma(\mathcal{R})$, as in conclusion (iii) (relative theorem). Let U be a neighbourhood of the image $\rho \circ \beta(B)$ in E and let $\nu\colon U \to \mathcal{R}$ be a continuous lift such that $\rho \circ \nu = \operatorname{id}_U$, $\nu(U) \subset N$, $\nu \circ \rho \circ \beta = \beta$. Employing Complements 5.11, 5.12 one may assume that over $\mathfrak{Op}\, K$ the C-structure (h, H) above is a continuous lift of the form $(h, H) = (\nu \circ m, \nu \circ M)$ where (m, M) is a smooth C-structure over $\mathfrak{Op}\, K$ in E with respect to $\partial_t^r g_0, \rho \circ \beta$ such that,

$$\operatorname{ev} M([0,1]^2 \times \mathfrak{Op}\, K) \subset U.$$

In particular for all $(b, t) \in \mathfrak{Op}\, K \times [0,1]$ $(b = (c, t) \in I^{n-1} \times [0,1])$,

$$\rho \circ F(u, (c, t)) = M(\theta_\epsilon(t), u, (c, t)), \quad 0 \le u \le 1. \tag{5.3}$$

Applying the $C^\perp$-Approximation Theorem III 3.8 (iii), for $\epsilon > 0$ sufficiently small, there is a smooth section $g'_\epsilon \in \Gamma(E)$ and a smooth homotopy $G = G_\epsilon \colon [0,1] \to \Gamma(E)$ such that the following properties obtain:

(iv) $g'_\epsilon = g_0$ on (a smaller) $\mathfrak{Op}_1 K$; $g'_\epsilon = g_\epsilon$ on $B \setminus \mathfrak{Op}_2 K$, where $\overline{\mathfrak{Op}_1 K} \subset \mathfrak{Op}_2 K$; $\overline{\mathfrak{Op}_2 K} \subset \mathfrak{Op}\, K$.

(v) $G_0 = \rho \circ \beta \in \Gamma(E)$; $G_1 = \partial_t^r g'_\epsilon$; for all $t \in [0,1]$ $G_t = \rho \circ \beta$ on $\mathfrak{Op}_1 K$ (constant homotopy); the image $\mathrm{ev}\, G([0,1] \times \mathfrak{Op}\, K) \subset U$.

(vi) For all $t \in [0,1]$ the homotopy $G_t = \rho \circ F_t$ on $B \setminus \mathfrak{Op}_2 K$.

Since $H = \nu \circ M \colon [0,1]^2 \to \Gamma_{\mathfrak{Op}\, K}(\mathcal{R})$ it follows from (vi), (5.2) and the above expression (5.3) for $\rho \circ F$ over $\mathfrak{Op}\, K$ that $\nu \circ G = F \colon [0,1] \to \Gamma_{\mathfrak{Op}\, K \setminus \mathfrak{Op}_2 K}(\mathcal{R})$ i.e. the lift of the perturbed homotopy G matches up with F on $B \setminus \mathfrak{Op}_2 K$. The Complement follows by replacing the section g_ϵ with the section $g'_\epsilon \in \Gamma(E)$ and, over $[0,1] \times \mathfrak{Op}\, K$, by replacing the homotopy F with the lifted homotopy $\nu \circ G$. $\qquad\square$

Remark 5.22. Employing (5.1), (5.2) above, except for a small perturbation over $\mathfrak{Op}\, K$, the images of the smooth section $g_\epsilon \in \Gamma(E)$ and of the homotopy $F \colon [0,1] \to \Gamma(\mathcal{R})$ (the parameter space P is here identified to a point) are contained in the image of the C-structure (h, H). This small perturbation in the Strong Relative Theorem is an essential feature of the inductive proof of the h-principle over a system of charts: employing Theorem 5.2.1, holonomic sections and homotopies are constructed relative to constructions caried out inductively in previous charts. Thus the geometry must be modified over $\mathfrak{Op}\, K$ by suitable cut-off functions in order to ensure that $g'_\epsilon = g_0$ and the homotopy F is constant $(= \beta)$ on $\mathfrak{Op}\, K$. These modifications by cut-off functions are controlled to lie in a preassigned neighbourhood N of $\beta(B)$ in $\mathcal{R}$.

Corollary 5.23 (Ample Relations). *Suppose in addition* $\rho \colon \mathcal{R} \to E$ *is ample. Let* $\beta \colon P \to \Gamma(\mathcal{R})$ *be continuous, where* P *is a compact smooth manifold, and let* $s \geq r \geq 1$ *be integers. For each smooth map* $g_0 \colon P \to \Gamma(E)$ *the following properties obtain:*

For each $\epsilon > 0$ *there is a smooth map* $g_\epsilon \colon P \to \Gamma(E)$ *and a homotopy* $F = F_\epsilon \colon [0,1] \times P \to \Gamma(\mathcal{R})$ *such that for all* $p \in P$, *conclusions (i), (ii), (iii) of the* $C^\perp$*-Approximation Theorem 5.20, and also Complement 5.21 are satisfied.*

Proof. Since $\rho \colon \mathcal{R} \to E$ is ample then for all $(p,b) \in P \times B$,

$$\partial_t^r g_0(p,b) \in \mathrm{Conv}(\mathcal{R}_b, \beta(p,b)).$$

Thus the main hypothesis of the $C^\perp$-Approximation Theorem 5.20 are satisfied, from which the corollary follows. $\qquad\square$

CHAPTER 6

THE GEOMETRY OF JET SPACES

§1. The Manifold $X^\perp$

6.1.1. Let $p \colon X \to V$ be a smooth fiber bundle, fiber dimension q, over an n-dimensional manifold V. Let $\tau \subset T(V)$ be a codimension 1 hyperplane field on V ($\dim \tau = n - 1$). Recall the smooth affine bundle of jet spaces $p^r_{r-1} \colon X^{(r)} \to X^{(r-1)}$, $r \geq 1$. Associated to the hyperplane field τ is a manifold $X^\perp$ and a natural affine $\mathbf{R}^q$-bundle $p^r_\perp \colon X^{(r)} \to X^\perp$, defined below, whose local structure provides the natural geometrical setting for applications of the main analytic approximation results of Chapter III, in particular the $C^\perp$-Approximation Theorem 3.8. This bundle "factors" the affine bundle $p^r_{r-1} \colon X^{(r)} \to X^{(r-1)}$ in the following sense. There is a natural affine bundle $p^\perp_{r-1} \colon X^\perp \to X^{(r-1)}$ such that,

$$
\begin{array}{ccc}
X^{(r)} & \xrightarrow{\;p^r_\perp\;} & X^\perp \\[4pt]
{\scriptstyle p^r_{r-1}}\big\downarrow & & \big\downarrow{\scriptstyle p^\perp_{r-1}} \\[4pt]
X^{(r-1)} & =\!=\!=\!= & X^{(r-1)}
\end{array}
\tag{6.1}
$$

The manifold $X^\perp$ is constructed in terms of the following "$\perp$-jet" relation on the space of sections $\Gamma^r(X)$ of the smooth bundle $p \colon X \to V$. Let $f \in \Gamma^r(X)$. Note that the $(r-1)$-jet extension $j^{r-1} f \in \Gamma(X^{(r-1)})$ is a C^1 map $j^{r-1} f \colon V \to X^{(r-1)}$. We define the following equivalence relation on sections $f, g \in \Gamma^r(X)$ at each $x \in V$ with respect to the hyperplane $\tau_x \subset T_x(V)$. For each $x \in V$ sections $f, g \in \Gamma^r(X_U)$, U a neighbourhood of x in V, have the same $\perp$-jet at $x \in V$ if the following conditions are satisfied in some (hence all) local coordinates:

(i) $j^{r-1} f(x) = j^{r-1} g(x) (= y) \in X^{(r-1)}$ i.e. f, g have the same $(r-1)$-jet at x.

(ii) $D(j^{r-1} f) = D(j^{r-1} g) \colon \tau_x \to T_y(X^{(r-1)})$ i.e. the derivatives at x of the $(r-1)$-jet extensions of f, g are equal when restricted to the hyperplane $\tau_x \subset T_x(V)$.

The $\perp$-jet relation is an equivalence relation, for which the set of all equivalence classes $[f]^\perp(x)$, $x \in V$, is denoted by $X^\perp$. Evidently the r-jet relation on $\Gamma^r(X_U)$ is compatible with the $\perp$-jet relation at each $x \in V$. Hence there is a natural

D. Spring, *Convex Integration Theory: Solutions to the h-principle in geometry and topology*, Modern Birkhäuser classics, DOI 10.1007/978-3-0348-0060-0_6, © Springer Basel AG 2010

projection map $p_\perp^r \colon X^{(r)} \to X^\perp$, $[f]_r(x) \mapsto [f]^\perp(x)$. Similarly there is a natural projection map $p_{r-1}^\perp \colon X^\perp \to X^{(r-1)}$, $[f]^\perp(x) \mapsto [f]_{r-1}(x)$.

Let $v = (v_1, \ldots, v_n)$ be a basis of $\mathbf{R}^n$. A polynomial $p = \sum_\alpha h^\alpha c_\alpha \in \mathcal{H}_r(n,q)$, $c_\alpha \in \mathbf{R}^q$, $|\alpha| = r$, $h = (h_1, \ldots, h_n) \in \mathbf{R}^n$ (cf. I §1.2.1 for this notation), is written $p = p^\perp + p_n$ where $p_n(h) = p(0, \ldots, 0, h_n) \in \mathcal{H}_r(n,q)$; $p^\perp = p - p_n$. Thus $p_n(h) = h_n^r c$, $c \in \mathbf{R}^q$; analytically, $r! p_n(h) = h_n^r (\partial/\partial h_n)^r p(h)$. Thus, with respect to the basis v, there is an induced splitting,

$$\mathcal{H}_r(n,q) = \mathcal{H}_r^\perp(n,q) \times L^q \quad p \mapsto (p^\perp, p_n).$$

We employ weighted coordinates on $\mathcal{H}_r(n,q)$ (cf. I §1.2.1) in what follows. In these coordinates, we identify $L^q \equiv \mathbf{R}^q$: $p_n(h) = h_n^r c \in L^q$ has coordinates $r!c \in \mathbf{R}^q$. Let $f \in C^r(U,W)$ where $U \subset \mathbf{R}^n$, $W \subset \mathbf{R}^q$ are open. With respect to the above splitting, for each $x \in U$,

$$\frac{1}{r!} D^r f(x)(h) = (D^\perp f(x)(h), \frac{h_n^r}{r!} \partial_n^r f(x)) \in \mathcal{H}_r^\perp(n,q) \times L^q.$$

Explicitly, the polynomial $D^\perp f(x)(h)$ has weighted coordinates $(\partial_v^\alpha f(x)) \in \mathbf{R}^{q d_r - q}$, where ∂_v^α runs over all multi-index rth order partial derivatives such that $\partial_v^\alpha \neq \partial_n^r$ (the pure derivative ∂_n^r is excluded).

Let $J^\perp(U,V) = J^{r-1}(U,V) \times \mathcal{H}_r^\perp(n,q)$. Thus $J^r(U,V) = J^\perp(U,V) \times L^q$. There is a continuous map, $\pi_r \colon C^r(U,W) \to C^0(U, J^\perp(U,V))$, $f \mapsto j^\perp f$, the perp-jet extension of f, such that for all $x \in U$,

$$j^\perp f(x) = (j^{r-1} f(x), D^\perp f(x)) \in J^\perp(U,V).$$

Thus in weighted coordinates, $j^r f(x) = (j^\perp f(x), \partial_n^r f(x))$. Let $\sim$ be the relation on $U \times C^r(U,V)$: $(x,f) \sim (y,g)$ if and only if $x = y$ and $j^\perp f(x) = j^\perp g(x)$ i.e. f, g have the same perp-jet extensions at x. Let $X = \mathbf{R}^n \times \mathbf{R}^q \to \mathbf{R}^q$ be the product $\mathbf{R}^q$ bundle; let $\tau = \ker dh_n \subset T(\mathbf{R}^n)$ be the integrable codimension 1 hyperplane field, with respect to coordinates $(h_1, \ldots, h_n)$ in the basis v of $\mathbf{R}^n$. Then $\sim$ is an equivalence relation on germs of functions at x. Furthermore, the map $[f]^\perp(x) \mapsto j^\perp f(x)$ is well-defined and induces a bijection $X^\perp \to J^\perp(U,V)$. In this way we identify $X^\perp \equiv J^\perp(U,V)$.

Returning to the bundle $p \colon X \to V$ and to the hyperplane field τ above, let $x \in V$ and let $v = (v_1, v_2, \ldots, v_n)$ be a basis of $T_x(V)$ that is adapted to τ: $v_i \in \tau_x$, $1 \leq i \leq n-1$ i.e. $(v_1, v_2, \ldots, v_{n-1})$ is a basis for τ_x. Let $f \in \Gamma^r(X)$. With respect to local coordinates in $V \times X$ at $(x, f(x))$ we assume $f \in C^r(U,W)$ where U, W are open sets respectively in $\mathbf{R}^n$, $\mathbf{R}^q$. We write ∂_{v_i} for the derivative in the direction $v_i \in \mathbf{R}^n$, $1 \leq i \leq n$ ($\partial_{v_i} = \partial/\partial u_i$ in local coordinates $(u_1, \ldots, u_n)$ in the basis v of $\mathbf{R}^n$). We employ directional derivatives since in general τ is not integrable on a neighbourhood $\mathfrak{Op}\, x$ i.e. global smooth coordinates on $\mathfrak{Op}\, x$ for an adapted basis at each point of $\mathfrak{Op}\, x$ do not exist in general. Thus with respect to the adapted basis v of $T_x(V)$,

$$j^r f(x) = (j^{r-1} f(x), \frac{1}{r!} D^r f(x))$$
$$= (j^{r-1} f(x), \partial_v^\alpha f(x)) \in J^{r-1}(U, W) \times \mathcal{H}_r(n, q) = J^r(U, W),$$

where $\partial_v^\alpha = \partial_{v_1}^{p_1} \circ \partial_{v_2}^{p_2} \circ \cdots \circ \partial_{v_n}^{p_n}$ runs over all rth order derivatives; $|\alpha| = p_1 + p_2 + \cdots + p_n = r$. For each index α, $\partial_v^\alpha f(x) \in \mathbf{R}^q$. In these coordinates define,

$$j^\perp f(x) = (j^{r-1} f(x), D^\perp f(x))$$
$$= (j^{r-1} f(x), \partial_v^\alpha f(x)) \in J^{r-1}(U, W) \times \mathcal{H}_r^\perp(n, q) = J^\perp(U, W),$$

where ∂_v^α runs over all rth order derivatives such that $\partial_v^\alpha \neq \partial_{v_n}^r$ i.e. only the pure rth order derivative $\partial_{v_n}^r$ is excluded. In particular, $j^r f(x) = (j^\perp f(x), \partial_{v_n}^r f(x))$. Note that $[f]^\perp(x) = [g]^\perp(x)$ implies that $j^{r-1} f(x) = j^{r-1} g(x)$ and from (ii), $\partial_v^\alpha f(x) = \partial_v^\alpha g(x)$ for all rth order derivatives $\partial_v^\alpha \neq \partial_{v_n}^r$ (the pure derivative $\partial_{v_n}^r$ is the only rth order derivative that does not occur in condition (ii) for the $\perp$-jet relation). Thus if sections $f, g \in \Gamma^r(X_U)$ have the same $\perp$-jet at x then $j^\perp f(x) = j^\perp g(x)$ in the above coordinate representation with respect to the adapted n-frame v i.e. the coordinate notation $j^\perp f(x)$ above is well-defined on $\perp$-jet equivalence classes.

The set $X^\perp$ is naturally topologized as a manifold, the manifold of $\perp$-jets of germs of C^r-sections of the bundle $p \colon X \to V$, whose charts are equivalent to $J^\perp(U, W)$ above. $f \in \Gamma^r(X)$ induces a section $j^\perp f \in \Gamma(X^\perp)$, $x \mapsto [f]^\perp(x)$, where $[f]^\perp(x) \in X^\perp$ is represented in these charts by the above coordinates $j^\perp(f_U)(x) \in J^\perp(U, W)$ with respect to an adapted basis v of $T_x(V)$. In particular $\dim X^\perp + q = \dim X^{(r)}$.

Let $v = (v_1, v_2, \cdots, v_n)$, $w = (w_1, w_2, \cdots, w_n)$ be continuous tangent n-frame fields adapted to τ on $\mathfrak{Op}\, x_0$. Thus each derivative ∂_{v_i}, $1 \leq i \leq n - 1$, is a linear combination, with coefficients that are continuous functions in the base space coordinates in $\mathfrak{Op}\, x_0$, of the derivatives ∂_{w_j}, $1 \leq j \leq n - 1$. Consequently, a change of adapted n-frames from v to w induces an invertible transformation, represented by a square matrix whose entries are continuous functions of the base space coordinates in $\mathfrak{Op}\, x$, from the coordinates of $j^\perp f(x)$ with respect to the n-frame v to the coordinates of $j^\perp f(x)$ with respect to the n-frame w. If τ is a smooth hyperplane field then the changes of coordinates with respect to smooth adapted n-frames is smooth.

The bundle $p \colon X \to V$ includes smooth transition maps $g_{AB} \colon A \cap B \to \mathrm{Diff}(F)$ that are associated to overlapping charts A, B in the base manifold

V. Let $f \in \Gamma^r(X)$. Recall the change of coordinates formula, I (1.5), where $g = \mathrm{ev} \, g_{AB}$: for all $x \in A \cap B$,

$$j^r f|_B(x) = j^r g \circ (\mathrm{id}, f|_A)(x).$$

Restricting to $\perp$-jet extensions, $j^\perp f|_B(x) = j^\perp g \circ (\mathrm{id}, f|_A)(x)$, from which it follows that the induced change of coordinates map for overlapping charts $J^\perp(A, W_1)$, $J^\perp(B, W_2)$, on $X^\perp$ is smooth if τ is a smooth hyperplane field. Consequently, if τ is smooth then $X^\perp$ is a smooth manifold; the projection maps $p^r_\perp : X^\perp \to X^{(r-1)}$, $[f]^\perp \mapsto [f]_{r-1}$, $p^r_\perp : X^{(r)} \to X^\perp$, $[f]_r \mapsto [f]^\perp$ are smooth and are equivalent respectively to the corresponding natural projection maps of charts, $p^r_\perp : J^r(U, W) \to J^\perp(U, W)$, $p^\perp_{r-1} : J^\perp(U, W) \to J^{r-1}(U, W)$.

The Affine Bundle $\mathbf{p}^r_\perp : \mathbf{X}^{(r)} \to \mathbf{X}^\perp$. The projection map $p^r_\perp : X^{(r)} \to X^\perp$ is an $\mathbf{R}^q$-bundle for which, in local coordinates in a chart $J^r(U, \mathbf{R}^q)$ on $X^{(r)}$, the fiber over $j^\perp h(x_0) \in X^\perp$, $h \in \Gamma^r(X)$, $x_0 \in U$, consists of all r-jet extensions $j^r(h + p_n(x - x_0))(x_0)$ where $p_n(h) = h^r_n c \in L^q \equiv \mathbf{R}^q$ ($c \in \mathbf{R}^q$) in an adapted basis v of $T_{x_0}(V)$. Furthermore, applying the change of coordinates formula to $f = g_{AB} \circ (h + p_n(x - x_0))$ it follows from the homogeneity of polynomials in $\mathcal{H}_r(n, q)$ that,

$$j^\perp f(x_0) = j^\perp(g_{AB} \circ h)(x_0);$$
$$\partial^r_{v_n} f(x_0) = \partial^r_{v_n} g_{AB} \circ (h + p_n(x - x_0))(x_0)$$

Thus a change of coordinates on charts of $X^{(r)}$ induces an affine transformation of the $\mathbf{R}^q$-fiber over the base point $j^\perp h(x_0)$ of the bundle $p^r_\perp : X^{(r)} \to X^\perp$. The "translation" term of the affine transformation is induced from the Chain Rule by the derivatives $D^j h(x_0)$, $0 \le j \le r - 1$, in the formula above for $\partial^r_{v_n} f(x_0)$. In addition if $w = (w_1, \ldots, w_n)$ is a change of adapted basis of $T_{x_0}(V)$ then ∂_{v_n} is a linear combination of the derivatives ∂_{w_j}, $1 \le j \le n$. Hence the pure derivative $\partial^r_{v_n}$ is a linear combination of all the derivatives ∂^α_w, $|\alpha| = r$. Since $j^\perp h(x_0)$ includes all derivatives $\partial^\alpha_w \ne \partial^r_{w_n}$ (in the adapted basis w) it follows that a change of adapted basis in $T_{x_0}(V)$ also induces an affine transformation on the $\mathbf{R}^q$-fiber over the base point $j^\perp h(x_0)$ of the bundle $p^r_\perp : X^{(r)} \to X^\perp$. One concludes therefore that the $\mathbf{R}^q$-bundle $p^r_\perp : X^{(r)} \to X^\perp$ is affine. Similar considerations show that the bundle $p^\perp_{r-1} : X^\perp \to X^{(r-1)}$ is an affine bundle and that the affine structures on the bundles $p^r_\perp : X^{(r)} \to X^\perp$, $p^\perp_{r-1} : X^\perp \to X^{(r-1)}$ are both compatible with the affine structure on the bundle $p^r_{r-1} : X^{(r)} \to X^{(r-1)}$.

$\mathbf{X}^\perp$ for Split Manifolds. Suppose that $V = V' \times [0, 1]$ i.e. V is a split smooth manifold, and $\tau \subset T(V)$ is the codimension 1 hyperplane field that is tangent to the slices $V \times \{t\}$, $0 \le t \le 1$. In particular τ is an integrable hyperplane field.

Conversely, if τ is an integrable codimension 1 hyperplane field then locally V splits as above such that τ is the hyperplane field that is tangent to the slices. Since V splits there are preferred locally adapted tangent n-frames v such that $\partial_{u_n} = \partial_t$. Hence, in local coordinates with respect to a preferred adapted n-frame,

$$j^r f(x) = (j^\perp f(x), \partial_t^r f(x)) \in X^\perp \times \mathbf{R}^q.$$

If in addition $p \colon X \to V$ is the product $\mathbf{R}^q$-bundle over a split manifold V, $X = V \times \mathbf{R}^q$, then the above coordinate representation with respect to a preferred adapted n-frame induces a splitting on r-jet spaces,

$$X^{(r)} = X^\perp \times \mathbf{R}^q.$$

As an example, let U be a chart on the base manifold V such that $U \cap \partial V = \emptyset$. Then U is diffeomorphic to the n-cube $[0,1]^n$. Hence U is a split manifold with respect to each of the coordinates in $[0,1]^n$. Over U the bundle $p \colon X \to V$ is locally a product $\mathbf{R}^q$-bundle, for which the above splitting of r-jet spaces obtains.

§2. Principal Decompositions in Jet Spaces

6.2.1. Principal Subspaces in $X^{(r)}$. The affine bundles $p_{r-1}^r \colon X^{(r)} \to X^{(r-1)}$, $p_\perp^r \colon X^{(r)} \to X^\perp$ have the same total space $X^{(r)}$. Employing the commutative diagram (6.1), it follows that each affine fiber $X_w^{(r)}$, $w \in X^{(r-1)}$, of the affine bundle $p_{r-1}^r \colon X^{(r)} \to X^{(r-1)}$ is foliated by a family of parallel q-planes, the fibers of the affine $\mathbf{R}^q$-bundle $p_\perp^r \colon X^{(r)} \to X^\perp$ that lie over $w \in X^{(r-1)}$. This family of parallel affine q-planes in $X_w^{(r)}$ is called the *principal direction* induced by the hyperplane field τ in the fiber $X_w^{(r)}$. Each of these q-planes is a *principal subspace* of the family. Consequently, each fiber $X_w^{(r)}$ is foliated by a principal direction, which in turn is parametrized by the codimension 1 hyperplane fields $\tau \subset T(V)$. Principal subspaces play an important role in the general theory. We prove below an algebraic lemma which has the geometrical consequence that any two r-jets $z_1, z_2 \in X_w^{(r)}$ can be joined by a continuous path in $X_w^{(r)}$, continuous also with respect to the end points z_1, z_2, that is composed of (piecewise linear) segments, each of which lies in a principal subspace with respect to some codimension 1 hyperplane field defined locally on $\mathfrak{Op}\, x \subset V$, $x = s(w) \in V$ ($s \colon X^{(r-1)} \to V$ is the source map). This geometrical property underlies the theory of iterated Convex Hull extensions developed in Chapter VIII.

6.2.2. Let $f \in \Gamma^r(X)$. Suppose $f(U) \subset W$, where U, W are charts respectively on the base V and the fiber F. In local coordinates,

$$j^r f(x) = \left(j^{r-1} f(x), \frac{1}{r!} D^r f(x) \right) \in J^{r-1}(U, W) \times \mathcal{H}_r(n, q). \qquad (6.2)$$

where $\mathcal{H}_r(n, q)$ is the fiber of the affine bundle $p_{r-1}^r \colon X^{(r)} \to X^{(r-1)}$.

Let $z, z_1 \in X_w^{(r)}$ be r-jets in the same affine fiber, $w \in X^{(r-1)}$. In terms of the local coordinates (6.2), throughout this chapter we employ the succinct additive notation: $z = z_1 + p$ where $p \in \mathcal{H}_r(n, q)$ (addition in the second factor). Explicitly, if $z = j^r f(x)$, $z_1 = j^r g(x)$ then $p = D^r(f - g)(x) \in \mathcal{H}_r(n, q)$.

Let $p_\perp^r : X^{(r)} \to X^\perp$ be the affine $\mathbf{R}^q$-bundle associated to a codimension 1 hyperplane field $\tau \subset T(V)$. In what follows we give an algebraic characterization of the principal subspaces of the affine space $X_w^{(r)}$ associated to the hyperplane field τ. In local coordinates on V let $\ell = \sum a_i \, dx_i$ be a 1-form at $x \in V$ such that $\tau_x = \ker \ell$. For each $r \geq 1$, the linear function $\ell \in \mathcal{H}_1(n, 1)$ induces a homogeneous polynomial function (a monomial) $\ell^r \in \mathcal{H}_r(n, 1)$ (symmetric r-fold tensor product of ℓ), such that for all $h \in \mathbf{R}^n$,

$$\ell^r(h) = (\ell(h))^r \in \mathbf{R}.$$

Explicitly in contracted notation $\ell^r = \sum_{|\alpha|=r} r!/\alpha! \, a_\alpha \, dx^\alpha$. For each vector $c = (c_1, c_2, \cdots, c_q) \in \mathbf{R}^q$ let $\ell^r \cdot c \in \mathcal{H}_r(n, q)$ be the polynomial function, with values along the line through $c \in \mathbf{R}^q$ such that for all $h \in \mathbf{R}^n$,

$$\ell^r \cdot c(h) = (\ell(h))^r c \in \mathbf{R}^q. \qquad (6.3)$$

Lemma 6.1. *Suppose $f, g \in \Gamma^r(X)$ have the same $(r-1)$-jet extensions at $x \in V$: $j^{r-1}f(x) = j^{r-1}g(x) \in X^{(r-1)}$. In the local coordinates (6.2), f, g have the same $\perp$-jet at $x \in V$ i.e. $j^\perp f(x) = j^\perp g(x) \in X^\perp$, if and only if there is a vector $c \in \mathbf{R}^q$ such that,*

$$D^r f(x) = D^r g(x) + \ell^r \cdot c.$$

Proof. Let $v = (v_1, v_2, \cdots, v_n)$ be a basis of $T_x(V)$ adapted to τ_x: $v_i \in \tau_x$, $1 \leq i \leq n - 1$. Employing I (1.2) with respect to the adapted basis v of $\mathbf{R}^n$, for all $h \in \mathbf{R}^n$,

$$D^r(f - g)(x)(h) = \sum_{|\alpha|=r} \frac{r!}{\alpha!} h^\alpha \, \partial_v^\alpha (f - g)(x).$$

Thus $j^\perp f(x) = j^\perp g(x)$ if and only if $\partial_v^\alpha (f - g)(x) = 0$ for all derivatives in the above sum such that $\partial_v^\alpha \neq \partial_{v_n}^r$. Consequently the equality of $\perp$-jets is equivalent to,

$$D^r(f - g)(x)(h) = h_n^r \partial_{v_n}^r (f - g)(x) \in \mathbf{R}^q.$$

where $h = (h_1, h_2, \cdots, h_n)$ in the adapted basis v. Since $\ell(h) = h_n \ell(v_n)$, $\ell(v_n) \neq 0$, one obtains the equivalent condition ($\kappa = 1/\ell(v_n)$),

$$D^r f(x) = D^r g(x) + \ell^r \cdot c,$$

where $c = \kappa^r \partial_{v_n}^r (f - g)(x) \in \mathbf{R}^q$, which proves the lemma. $\qquad\square$

Remark 6.2. Let $z = j^r f(x)$, $z_1 = j^r g(x)$ be r-jets in $X^{(r)}$ that lie in the same principal subspace. Thus $j^\perp f(x) = j^\perp g(x)$ and, by the lemma, there is a vector $c \in \mathbf{R}^q$ such that in local coordinates $D^r f(x) = D^r g(x) + \ell^r \cdot c$. Employing the local coordinates (6.2) in $X^{(r)}$,

$$z = (j^{r-1} g(x), D^r g(x) + \ell^r \cdot c) \in J^{r-1}(\mathbf{R}^n, \mathbf{R}^q) \times \mathcal{H}_r(n, q).$$

Employing the additive notation above, we write $z = z_1 + \ell^r \cdot c$ in the local co-ordinates (6.2). In this notation, one has the following explicit algebraic characterization of principal subspaces. An r-jet $z \in X^{(r)}$ lies in the principal subspace through z_1 if and only if, in the above local coordinates, there is a vector $c \in \mathbf{R}^q$ such that,

$$z = z_1 + \ell^r \cdot c.$$

Corollary 6.3. *Let $X_w^{(r)}$ be an affine fiber of the affine bundle $X^{(r)} \to X^{(r-1)}$, $w \in X^{(r-1)}$. Recall the principal direction on $X_w^{(r)}$ with respect to the affine $\mathbf{R}^q$-bundle $X^{(r)} \to X^\perp$. For all $z_1, z_2 \in X_w^{(r)}$, the translation $T \colon X_w^{(r)} \to X_w^{(r)}$, $T(z) = z + (z_2 - z_1)$, induces an affine isomorphism from the principal subspace through z_1 to the principal subspace through z_2.*

Proof. Let $y_1 \in X_w^{(r)}$ be in the principal subspace through z_1. In the above local coordinates, $y_1 = z_1 + \ell^r \cdot c$, where $c \in \mathbf{R}^q$. Since $T(y_1) = z_2 + \ell^r \cdot c$, it follows that T induces an affine isomorphism of principal subspaces. □

6.2.3. An important geometrical problem in $X^{(r)}$ for the general theory of Convex Hull extensions, discussed below, is the problem of sufficiently many principal directions: for all r-jets z_1, z_2 in $X_w^{(r)}$ the problem is to construct r-jets $y_j \in X_w^{(r)}$, $1 \le j \le m$, such that $y_1 = z_1$, $y_m = z_2$ and such that the r-jets y_{j-1}, y_j lie in a principal subspace R_j associated to a codimension 1 hyperplane field $\tau_j \subset T_x(V)$ $(x = s(w) \in V)$, $1 \le j \le m$. If the successive r-jets y_{j-1}, y_j in this sequence are joined by a continuous path in the principal subspace $R_j = \mathbf{R}^q$, one obtains a *principal path* in $X_w^{(r)}$ joining z_1, z_2, in terms of which we reformulate as follows: for all r-jets z_1, z_2 in $X_w^{(r)}$ the problem is to construct a principal path in $X_w^{(r)}$ joining z_1, z_2. Theorem 6.7 below solves this problem in a continuous manner i.e. that depends continuously on the r-jets z_1, z_2.

Proposition 6.4. *There is a basis of $\mathcal{H}_r(n, 1)$ consisting of monomials of the form ℓ_μ^r, $1 \le \mu \le \dim \mathcal{H}_r(n, 1)(= d_r(n))$, where $\ell_\mu \in \mathcal{H}_1(n, 1)$ is linear and for all $h \in \mathbf{R}^n$,*

$$\ell_\mu^r(h) = (\ell_\mu(h))^r \in \mathbf{R}, \quad 1 \le \mu \le d_r(n).$$

Proof. The proposition is a well-known result in algebra that real homogeneous polynomials of degree r in n variables are represented as sums of rth powers of linear forms in n variables. A modern reference is Reznick [35], p. 30. This proposition extends easily to the following corollary on homogeneous polynomials with values in $\mathbf{R}^q$.

Corollary 6.5. *Let ℓ_μ^r, $1 \leq \mu \leq d_r(n)$, be a basis of $\mathcal{H}_r(n,1)$ consisting of monomials. For each polynomial map $p \in \mathcal{H}_r(n,q)$ there are vectors $c_\mu \in \mathbf{R}^q$, $1 \leq \mu \leq d_r(n)$, such that,*

$$p = \sum_{\mu=1}^{d_r(n)} \ell_\mu^r \cdot c_\mu.$$

Proof. In the basis (ℓ_μ^r), $p = (p_1, p_2 \ldots, p_q)$, where $p_i = \sum_\mu c_\mu^i \ell_\mu^r$, $1 \leq i \leq q$. Employing (6.3) it follows that,

$$p = \Big(\sum_\mu c_\mu^1 \ell_\mu^r, \ldots, \sum_\mu c_\mu^q \ell_\mu^r \Big)$$

$$= \sum_\mu (c_\mu^1 \ell_\mu^r, c_\mu^2 \ell_\mu^r, \ldots, c_\mu^q \ell_\mu^r)$$

$$= \sum_{\mu=1}^{d_r(n)} \ell_\mu^r \cdot c_\mu,$$

where $c_\mu = (c_\mu^1, c_\mu^2, \ldots, c_\mu^q) \in \mathbf{R}^q$, $1 \leq \mu \leq d_r(n)$, which proves the Corollary. $\square$

Corollary 6.6. *Let $f \in C^0(Z, \mathcal{H}_r(n,q))$ be a continuous map, Z a topological space. There are continuous map $F_\mu \in C^0(Z, \mathbf{R}^q)$, $1 \leq \mu \leq d_r(n)$, such that,*

$$f = \sum_{\mu=1}^{d_r(n)} \ell_\mu^r \cdot F_\mu.$$

In addition F_μ is smooth, $1 \leq \mu \leq d_r(n)$, if Z is a smooth manifold and f is a smooth map.

Theorem 6.7. *Let $f, g \in \Gamma_{\mathfrak{Op}\, x}(X^{(r)})$ be continuous sections of r-jets over $\mathfrak{Op}\, x \subset V$ such that $p_{r-1}^r \circ f = p_{r-1}^r \circ g \in \Gamma_{\mathfrak{Op}\, x}(X_U^{(r-1)})$ i.e. f, g induce the same $(r-1)$-jet sections over $\mathfrak{Op}\, x$. There are smooth, integrable, codimension 1 tangent hyperplane fields τ_j on $\mathfrak{Op}\, x \subset V$, $1 \leq j \leq d_r(n)$, and continuous sections $y_j \in \Gamma_{\mathfrak{Op}\, x}(X^{(r)})$, $0 \leq j \leq d_r(n)$, such that $y_0 = f$, $y_{d_r(n)} = g$ and such that:*

(i) $p_{r-1}^r \circ y_j = p_{r-1}^r \circ f(= w) \in \Gamma_{\mathfrak{Op}\,x}(X^{(r-1)})$, $1 \le j \le d_r(n)$ *i.e. the sections* f, y_j *induce the same* $(r-1)$*-jet sections.*

(ii) *For all* $u \in \mathfrak{Op}\,x$, *successive* r*-jets* $y_{j-1}(u), y_j(u)$ *lie in a principal subspace* $R_j(u) \subset X_{w(u)}^{(r)}$ *associated to the hyperplane* $\tau_j(u) \subset T_u(V)$, $1 \le j \le d_r(n)$.

(iii) *The sections* $y_j \in \Gamma_{\mathfrak{Op}\,x}(X^{(r)})$, $1 \le j \le d_r(n)$, *are smooth if* f, g *are smooth sections.*

Proof. Let N be a neighbourhood of $j^{r-1}f(x)$ in $X^{(r-1)}$ such that the restriction bundle $p_{r-1}^r \colon X_N^{(r)} \to N$ is trivial. One may assume that $N \equiv J^{r-1}(\mathbf{R}^n, \mathbf{R}^q)$ and that,

$$X_N^{(r)} = J^{r-1}(\mathbf{R}^n, \mathbf{R}^q) \times \mathcal{H}_r(n, q).$$

In these coordinates, since f, g induce the same $(r-1)$-jet sections over $\mathfrak{Op}\,x$ it follows that over $\mathfrak{Op}\,x$, in additive notation (addition in the factor $\mathcal{H}_r(n,q)$), $f = g + p$ where $p \in C^0(\mathbf{R}^n, \mathcal{H}_r(n,q))$. Employing Corollary 6.6, it follows that there are continuous functions $G_\mu \in C^0(\mathbf{R}^n, \mathbf{R}^q)$, $1 \le \mu \le d_r(n)$, such that

$$f = g + \sum_{\mu=1}^{d_r(n)} \ell_\mu^r \cdot G_\mu. \tag{6.4}$$

Let $y_0 = g$ and in the above local coordinates let $y_\mu \in \Gamma_{\mathfrak{Op}\,x}(X^{(r)})$ be the section,

$$
\begin{aligned}
y_\mu &= g + \sum_{i=1}^\mu \ell_i^r \cdot G_i, \quad 1 \le \mu \le d_r(n) \\
&= y_{\mu-1} + \ell_\mu^r \cdot G_\mu, \quad 1 \le \mu \le d_r(n).
\end{aligned}
\tag{6.5}
$$

Employing Corollary 6.6, the functions G_μ, hence the sections y_μ defined above are smooth if in addition f, g are smooth, $1 \le \mu \le d_r(n)$. Since the 1-forms ℓ_μ on $\mathbf{R}^n$ are constant, hence closed, they pull back to closed smooth 1-forms (same notation) on $\mathfrak{Op}\,x$ for which the corresponding codimension 1 hyperplane fields $\tau_\mu = \ker \ell_\mu$ on $\mathfrak{Op}\,x$ are smooth and integrable, $1 \le \mu \le d_r(n)$. Employing (6.5) and the algebraic characterization of principal subspaces provided in Remark 6.2, it follows that for all $u \in \mathfrak{Op}\,x$ successive r-jets $y_{\mu-1}(u), y_\mu(u) \in X_w^{(r)}$, $w = p_{r-1}^r \circ f(u) \in X^{(r-1)}$, lie in a principal subspace $R_\mu(u)$ associated to the hyperplane $\tau_\mu(u) = \ker \ell_\mu(u) \subset T_u(V)$. Setting $\mu = d_r(n)$ in (6.5) it follows from (6.4) that $y_{d_r(n)} = f$, which completes the proof of the theorem. $\quad\square$

Remark 6.8. The proof of Theorem 6.7 implies the following pointwise result which we record separately: Let $z_0, z_1 \in X_w^{(r)}$, $w \in X^{(r-1)}$. Let $z_1 = z_0 + p$, where $p = \sum_j \ell_j^r \cdot c_j \in \mathcal{H}_r(n, q)$. The sequence of r-jets $y_j \in X_w^{(r)}$, $1 \le j \le d_r(n)$, defined inductively by $y_0 = z_0$, $y_j = y_{j-1} + \ell_j^r \cdot c_j$, $1 \le j \le d_r(n)$, satisfies the property that $y_{d_r(n)} = z_1$ and that successive r-jets y_{j-1}, y_j lie in a principal subspace R_j associated to the hyperplane $\tau_j(x) = \ker \ell_j(x)$ $(x = s(w) \in V)$.

6.2.4. Principal Paths. Let $h \colon [a, b] \to \mathbf{R}^q$ be continuous. Associated to each partition, $a = t_0 \le t_1 \le \cdots \le t_m = b$, of the interval $[a, b]$ is the piecewise linear path h' which affinely interpolates between $h(t_{i-1}), h(t_i)$ on the interval $[t_{i-1}, t_i]$, $1 \le i \le m$: for all $t \in [t_{i-1}, t_i]$,

$$h'(t) = \frac{t_i - t}{t_i - t_{i-1}} h(t_{i-1}) + \frac{t - t_{i-1}}{t_i - t_{i-1}} h(t_i), \quad 1 \le i \le m.$$

In case $t_i = t_{i-1}$ then $h'(t_i) = h(t_i)$. This construction is uniformly continuous in the sense that for each $\epsilon > 0$ there is a $\delta > 0$ such that for all partitions of mesh $< \delta$ the corresponding piecewise linear path h' satisfies $\|h - h'\| < \epsilon$ in the sup-norm topology. Let r-jets $z_0, z_1 \in X_w^{(r)}$, $w \in X^{(r-1)}$. A *piecewise principal path* in $X_w^{(r)}$ joining z_0 to z_1 is a piecewise linear path $p \colon [a, b] \to X_w^{(r)}$, $p(a) = z_0$, $p(b) = z_1$, such that, with respect to some subdivision $a = t_0 \le t_1 \le t_2 \le \cdots \le t_m = b$ of $[a, b]$, successive pairs of points $p(t_{j-1}), p(t_j)$ lie in a principal subspace R_j, $1 \le j \le m$. For example, employing Remark 6.8, there is a piecewise principal path in $X_w^{(r)}$ joining z_0 to z_1 which interpolates linearly, as above, between the points y_{i-1}, y_i on the interval $[t_{i-1}, t_i]$, $1 \le i \le d_r(n)$, with respect to a partition $a = t_0 \le t_1 \le \cdots \le t_{d_r(n)} = b$, of the interval $[a, b]$. Employing Theorem 6.7, the piecewise principal path $p \equiv p(z_0, z_1)$ constructed above is continuous in the end points z_0, z_1.

Lemma 6.9. *Let d be a metric on $X^{(r)}$ and let $\epsilon > 0$. For each path $h \colon [0, 1] \to X_w^{(r)}$ there is a piecewise principal path $p \colon [0, 1] \to X_w^{(r)}$ which is an ϵ-approximation to h: for all $t \in [0, 1]$, $d(h(t), p(t)) < \epsilon$.*

Proof. With respect to a subdivision $0 = t_0 \le t_1 \le t_2 \le \cdots \le t_m = 1$ of $[0, 1]$, there is path $p(t)$ in $X_w^{(r)}$ joining $h(0)$ to $h(1)$ such that on each subinterval $[t_{i-1}, t_i]$, $p(t)$ is a piecewise principal path joining $h(t_{i-1}), h(t_i)$, $1 \le i \le m$. Let $h(t_i) = h(t_{i-1}) + g_i$, where $g_i \in \mathcal{H}_r(n, q)$, $1 \le i \le m$. Evidently, g_i approaches $0 \in \mathcal{H}_r(n, q)$, $1 \le i \le m$, as the mesh of the partition of $[0, 1]$ approaches 0. The metric d restricted to the fiber $X_w^{(r)} \equiv \mathcal{H}_r(n, q)$ induces a metric on the linear space $\mathcal{H}_r(n, q)$. Since metrics on finite dimensional linear spaces are equivalent, it follows that, for a sufficiently small mesh, $d(h(t), p(t)) < \epsilon$ for all $t \in [0, 1]$. $\square$

Let $f, g \in \Gamma(X^{(r)})$ be continuous sections such that $p_{r-1}^r \circ f = p_{r-1}^r \circ g \in \Gamma(X^{(r-1)})$ i.e. the sections f, g have the same $(r-1)$-jet components. A *piecewise principal homotopy* of sections that joins f to g over a subspace A of the base manifold V is a homotopy $H\colon [0,1] \to \Gamma_A(X^{(r)})$ such that $H_0 = f|_A$, $H_1 = g|_A$, and such that for all $x \in A$, $H_t(x)$ is a piecewise principal path in the fiber $X^{(r)}_{w(x)}$; $w(x) = p^r_{r-1} \circ f(x) \in X^{(r-1)}$. Employing Theorem 6.7 it follows that locally for each $x \in V$ there is a piecewise principal homotopy of sections $H\colon [0,1] \to \Gamma_{\mathfrak{Op}\,x}(X^{(r)})$ that joins f to g over $\mathfrak{Op}\,x$.

Although Theorem 6.7 and the local construction of piecewise principal paths above are sufficient for our purposes it is interesting to extend the above construction of a local piecewise principal homotopy H over $\mathfrak{Op}\,x$ to a piecewise principal homotopy of sections that joins f to g globally over the base manifold V. Indeed, this global picture is the one advocated in Gromov [18]. The point here is that locally in each fiber $X^{(r)}_w$ the space of piecewise principal paths is contractible. Proceeding inductively over a covering by charts of V, as in the proof of the existence of global C-structures, II Proposition 2.3, this contractibility result is employed to construct a piecewise principal homotopy of sections globally over V which joins f to g. Details are sketched as follows.

Lemma 6.10. *Let $\mathcal{P} \subset C^0([0,1], X^{(r)}_w)$ denote the subspace of piecewise principal paths in a fiber $X^{(r)}_w$ which join $f(x)$ to $g(x)$ $(x = s(w) \in V)$. The space $\mathcal{P}$ is contractible.*

Proof. Let $F\colon [0,1] \to X^{(r)}_w$ be a piecewise principal path, $F_0 = f(x)$, $F_1 = g(x)$, which is constructed as above in §6.2.4 with respect to the fixed family of principal subspaces specified in Remark 6.8. We show that $\mathcal{P}$ strongly deformation retracts to F. To this end let $G \in \mathcal{P}$. Let $H\colon [0,1]^2 \to X^{(r)}_w$ be the homotopy defined as follows: For each $s \in [0,1]$,

   (i) $H(s,u) = G(u)$, $s \le u \le 1$.

   (ii) $H(s,u)$, $0 \le u \le s$, is a piecewise principal path in $X^{(r)}_w$, constructed as above in §6.2.4, that joins $f(x)$ to $G(s)$ on $[0,s]$ (to ensure continuity of H, one rescales a fixed partition of $[0,1]$ to $[0,s]$).

Thus H is continuous; for all $s \in [0,1]$ the path $H_s(\cdot) \in \mathcal{P}$; for each s, the path $H(s,u)$, $u \in [0,s]$, is piecewise principal with respect to the fixed family of principal subspaces specified in Remark 6.8. For all $u \in [0,1]$ $H(1,u) = F(u)$; $H(0,u) = G(u)$ i.e. the homotopy continuously deforms the path G to the path F through piecewise principal paths in $\mathcal{P}$. Furthermore the deformation H is continuous in $G \in \mathcal{P}$.   □

Employing Theorem 6.7 and the existence of fiberwise piecewise principal paths as constructed above in §6.4, Lemma 6.10 extends as follows to charts of the form $\mathfrak{Op}\,x$ on V. Details are left to the reader.

Corollary 6.11. *Let $f, g \in \Gamma_{\mathfrak{Op}\,x}(X^{(r)})$ be continuous sections such that $p^r_{r-1} \circ f = p^r_{r-1} \circ g \in \Gamma_{\mathfrak{Op}\,x}(X^{(r-1)})$. Let $\mathcal{P} \subset C^0([0,1], \Gamma_{\mathfrak{Op}\,x}(X^{(r)})$ be the subspace* *of piecewise principal homotopies which join f to g over $\mathfrak{Op}\, x$. The space $\mathcal{P}$ is contractible.*

Proposition 6.12. *Let $f, g \in \Gamma(X^{(r)})$ satisfy $p_{r-1}^r \circ f = p_{r-1}^r \circ g \in \Gamma(X^{(r-1)})$ i.e. the sections f, g have the same $(r-1)$-components. There is a global piecewise principal homotopy $H \colon [0,1] \to \Gamma(X^{(r)})$ which joins f to g over V.*

Proof. We follow the structure of the proof for the analogous construction of a global C-structure, II Proposition 2.3, which is based on the contractibility of the space of C-structures. Employing Theorem 6.7 and the construction of piecewise principal paths in §6.2.4, for each $x \in V$ a piecewise principal homotopy of sections exists which joins f to g over $\mathfrak{Op}\, x$. In order to patch together these piecewise principal homotopies on overlapping charts the contractibility of the space of piecewise principal paths, Corollary 6.11, is employed inductively over a covering of the base V by charts, the inductive step of which is as follows. The proof of the following lemma is analogous to II Lemma 2.5 and therefore is omitted.

Lemma 6.13. *Let $K, L \subset V$ be closed and let G_0 respectively G_1 be piecewise principal homotopies of sections which join f to g over $\mathfrak{Op}\, K$, respectively $\mathfrak{Op}\, L$. There is a piecewise principal homotopy of sections H which joins f to g over $\mathfrak{Op}\,(K \cup L)$ such that $H = G_0$ over (a smaller) $\mathfrak{Op}_1 K$ and $H = G_1$ over $\mathfrak{Op}\,(L \setminus \mathfrak{Op}\,(K \cap L))$.*

Returning to the proof of the proposition, there are countable locally finite open covers $\{W_i\}$, $\{U_i\}$ of the base manifold V such that for all i, $\overline{W}_i \subset U_i$, and there is a piecewise principal homotopy of sections which joins f to g over each U_i ($U_i \equiv \mathfrak{Op}\, x_i$). Inductively on n, let $K = \bigcup_{i=1}^n \overline{W}_i$, a closed subset of V, and suppose H_n is a piecewise principal homotopy of sections which joins f to g over $\mathfrak{Op}\, K$. Applying Lemma 6.13 to the closed set K and to $\overline{W}_{n+1}$, there is a piecewise principal homotopy H_{n+1} of sections which joins f to g over $\mathfrak{Op}\,(K \cup \overline{W}_{n+1})$ such that $H_{n+1} = H_n$ over $\mathfrak{Op}\, K$. Since the cover $\{W_i\}$ is locally finite it follows that the homotopy $H = \lim_{n \to \infty} H_n$ is well-defined, continuous, and also $H \colon [0,1] \to \Gamma(X^{(r)})$ is a piecewise principal homotopy of sections that joins f to g over the base manifold V. $\square$

Theorem 6.14. *Let $f, g \in \Gamma(X^{(r)})$ satisfy $p_{r-1}^r \circ f = p_{r-1}^r \circ g \in \Gamma(X^{(r-1)})$. There are sections $f_i \in \Gamma(X^{(r)})$, $1 \leq i \leq m$, $f_1 = f$, $f_m = g$, such that for all i, $p_{r-1}^r \circ f_i = p_{r-1}^r \circ f$, and such that for all $x \in V$, successive sections $f_i(x), f_{i+1}(x)$ lie in a principal subspace, $1 \leq i \leq m - 1$.*

Proof. The proof is by induction over a finite cover of V by charts, each of which is at most a countable union of disjoint charts of the form $\mathfrak{Op}\, x$, based on the following construction. Let K, L be closed in V, L a chart of the form $\overline{\mathfrak{Op}\, x}$.

Suppose the theorem holds over $\mathfrak{Op}\, K$: there are sections $f_i \in \Gamma_{\mathfrak{Op}\, K}(X^{(r)})$, $1 \le i \le p$, $f_1 = f$, $f_p = g$, such that for all $x \in \mathfrak{Op}\, K$ successive sections $f_i(x), f_{i+1}(x)$ lie in a principal subspace. We show that the theorem extends over $\mathfrak{Op}\, K \cup L$. In case $K \cap L = \emptyset$, employing Theorem 6.7 on L, the extension to $\mathfrak{Op}\, K \cup L$ is trivial. In case $K \cap L$ is not empty, employing Tietse's Theorem, one extends f_i (same notation) across the chart L to $\mathfrak{Op}\, K \cup L$, $2 \le i \le m - 1$. Applying Theorem 6.7 to the sections f_i, f_{i+1} over L, one employs bump functions in the chart L to show that for each i the section f_i "bifurcates" over L to sections $g_1^i, \ldots, g_{d_r(n)}^i$, $g_1^i = f_i$, $g_{d_r(n)}^i = f_{i+1}$, such that $g_j^i = f_i$ over $\mathfrak{Op}\, K$, $1 \le j \le d_r(n)$, and such that for all $x \in L$ successive sections $g_j^i(x), g_{j+1}^i(x)$ lie in a principal subspace. Details are left to the reader. $\qquad\qquad\qquad\qquad\qquad\qquad\square$

Remark 6.15. A unified and simpler proof of Theorems 6.12, 6.14 is possible if Proposition 6.4 extends to establish the connectivity of the space of monomial bases of $\mathcal{H}_r(n, q)$: up to a change of sign of a linear form, any two monomial bases are connected by a homotopy of monomial bases. The algebraic literature does not seem to address this question.

CHAPTER 7

CONVEX HULL EXTENSIONS

§1. The Microfibration Property

7.1.1. The h-principle. Let $p\colon X \to V$ be a smooth bundle, q-dimensional fibers, over a smooth n-dimensional manifold V, $n \geq 1$; $s\colon X^{(r)} \to V$ is the source map. Let $\rho\colon \mathcal{R} \to X^{(r)}$ be a microfibration. We recall the notation introduced in I §3. A section $\alpha \in \Gamma(\mathcal{R})$ ($s \circ \rho \circ \alpha = \mathrm{id}_V$) is *holonomic* if there is a C^r-section $f \in \Gamma^r(X)$ such that $j^r f = \rho \circ \alpha \in \Gamma(X^{(r)})$. The relation $\mathcal{R}$ satisfies the *h-principle* if for each $\alpha \in \Gamma(\mathcal{R})$ there is a homotopy of sections $H\colon [0, 1] \to \Gamma(\mathcal{R})$, $H_0 = \alpha$, such that the section H_1 is holonomic. The h-principle is required to be a relative condition in the following sense. Let $K \subset V$ be closed and suppose α is holonomic on K: there is a C^r-section $g \in \Gamma^r(X)$ such that $\rho \circ \alpha = j^r g \in \Gamma_K(X^{(r)})$. Then in addition we require that for all $t \in [0, 1]$, $H_t = \alpha \in \Gamma_K(\mathcal{R})$ (constant homotopy over K).

Let $\tau \subset T(V)$ be a codimension 1 hyperplane field on V. Associated to τ is the affine $\mathbf{R}^q$-bundle $p_\perp^r\colon X^{(r)} \to X^\perp$ (cf. VI §6.1.1). Thus we have data as in Chapter V, §5.2.1:

$$
\begin{array}{ccc}
\mathcal{R} & \xrightarrow{\ \rho\ } & X^{(r)} \\
& & \big\downarrow{\scriptstyle p_\perp^r} \\
& & X^\perp
\end{array}
$$

Let $\alpha \in \Gamma(\mathcal{R})$ and let $f \in \Gamma^r(X)$ be a C^r-section such that for all $x \in V$,

$$j^r f(x) \in \mathrm{Conv}(\mathcal{R}_{z(x)}, \alpha(x)). \tag{7.1}$$

where $z(x) = j^\perp f(x) \in X^\perp$. Thus in this notation for all $x \in V$, the r-jets $\rho \circ \alpha(x), j^r f(x)$ both lie in the principal subspace given by the $\mathbf{R}^q$-fiber over $j^\perp f(x)$ in the bundle $p_\perp^r\colon X^{(r)} \to X^\perp$. The condition (7.1) states (cf. V §5.2.2) that for each $x \in V$, the r-jet $j^r f(x)$ lies in the convex hull (in this principal subspace) of the ρ-image of the path component in $\mathcal{R}_{z(x)}$ to which $\alpha(x)$ belongs. The above hypothesis (7.1) on α, f is very strong and implies that, in local coordinates, the r-jets $\rho \circ \alpha, j^r f(x)$ agree on all the "$\perp$"-derivatives. Employing the topological and analytic machinery of Chapter V, the h-Stability Theorem

D. Spring, *Convex Integration Theory: Solutions to the h-principle in geometry and topology*, Modern Birkhäuser classics, DOI 10.1007/978-3-0348-0060-0_7, © Springer Basel AG 2010

7.2 proves that, under the hypothesis (7.1), the h-principle applies to α if in addition $\tau \subset T(V)$ is a smooth, integrable, codimension 1 hyperplane tangent field. In fact a more refined result is proved, including a relative theorem and a $C^{\perp}$-density result. The entire Chapter VII is devoted to the proof of the h-stability theorem. The homotopy in the h-stability theorem can be chosen to be holonomic at each stage, thus generalizing the corresponding results of Gromov [18].

The h-stability theorem is the key structural result that is employed inductively in the general theory of convex hull extensions. The h-stability theorem is used in Chapter VIII to prove the C^{r-1}-dense h-principle for short maps (Theorem 8.4), and also in Chapter IX to solve non-linear systems of PDEs. Thus the h-stability theorem lays the groundwork for the applications of convex integration theory to the global solution of non-linear systems of PDEs studied in Chapter IX. These applications of the h-stability theorem do not require that the relation $\mathcal{R}$ be ample, thus generalizing considerably the corresponding results proved by Gromov for ample relations $\mathcal{R}$. The results proved in Chapter IV for open ample relations in spaces of 1-jets $X^{(1)}$ are consequences of the theory of ample relations in Chapter VIII.

7.1.2. Convex Hull Extensions. Let $p \colon E \to B$ be an affine $\mathbf{R}^q$-bundle over a manifold B and let $\rho \colon \mathcal{R} \to E$ be continuous (not necessarily a microfibration). Let $\pi \colon E^* \to \mathcal{R}$ be the pullback bundle, fiber $\mathbf{R}^q$, along the map $p \circ \rho \colon \mathcal{R} \to B$. Thus $E^* = \{(a, x) \in \mathcal{R} \times E \mid p \circ \rho(a) = p(x) \in B\}$; $\pi(a, x) = a \in \mathcal{R}$. The induced bundle map $\rho^* \colon E^* \to E$ is the projection $(a, x) \mapsto x \in E$. Let $\mathcal{R}^* \subset E$ be the "convex hull relation" in E: $\mathcal{R}^* = \{(a, x) \in E^* \mid x \in \mathrm{Conv}(\mathcal{R}_b, a)$ where $b = p(x) \in B\}$. There is an inclusion map $i \colon \mathcal{R} \to \mathcal{R}^*$, $a \mapsto (a, \rho(a))$, and an induced projection map $\pi \colon \mathcal{R}^* \to \mathcal{R}$, $(a, x) \mapsto a$. The bundle projection map $\rho^* \colon E^* \to E$ restricts to a projection map (same notation) $\rho^* \colon \mathcal{R}^* \to E$.

$$
\begin{array}{ccc}
E^* & \xrightarrow{\;\rho^*\;} & E \\
{\scriptstyle \pi}\big\downarrow & & \big\downarrow{\scriptstyle p} \\
\mathcal{R} & \xrightarrow[p \circ \rho]{} & B
\end{array}
$$

We identify $\mathcal{R} \subset \mathcal{R}^*$ via the inclusion $i \colon \mathcal{R} \to \mathcal{R}^*$. Note that a section $\beta \in \Gamma(\mathcal{R}^*)$ consists of a pair of sections $\beta = (\alpha, f)$ where $\alpha = \pi \circ \beta \in \Gamma(\mathcal{R})$, $f = \rho^* \circ \beta \in \Gamma(E)$. Thus associated to the relation $\rho \colon \mathcal{R} \to E$ is the convex hull relation $\mathcal{R}^* \subset E^*$, about which we have the following important microfibration property.

Lemma 7.1. *Suppose* $\rho \colon \mathcal{R} \to E$ *is a microfibration. The projection map* $\rho^* \colon \mathcal{R}^* \to E$ *is a microfibration (known as the convex hull extension of $\mathcal{R}$).*

Proof. Let $F\colon P \times [0,1] \to E$, $g\colon P \to \mathcal{R}^*$ be continuous maps, P a compact polyhedron, such that $\rho^* \circ g = F_0 \colon P \to E$. In particular, for all $p \in P$ ($b = p \circ F \colon P \times [0,1] \to B$):

$$F_0(p) \in \mathrm{Conv}(\mathcal{R}_{b(p,0)}, \pi \circ g(p)). \tag{7.2}$$

Since $p\colon E \to B$ is a bundle it follows that the composition $p \circ \rho\colon \mathcal{R} \to B$ is a microfibration. We therefore have the following commutative diagram:

$$
\begin{array}{ccc}
P \times \{0\} & \xrightarrow{\pi \circ g} & \mathcal{R} \\
\downarrow{\scriptstyle i} & & \downarrow{\scriptstyle p\circ\rho} \\
P \times [0,1] & \xrightarrow[p\circ F]{} & B
\end{array}
$$

By the microfibration property, there is an $\epsilon \in (0,1]$ and a continuous lift $G\colon P \times [0,\epsilon] \to \mathcal{R}$ such that: (*i*) $p \circ \rho \circ G = p \circ F\colon P \times [0,\epsilon] \to B$; (*ii*) $G_0 = \pi \circ g$. Let $H\colon P \times [0,\epsilon] \to E^*$ be the continuous map, $H(p,t) = (G(p,t), F(p,t))$. Evidently, $\rho^* \circ H = F\colon P \times [0,\epsilon] \to E$; $H_0 = (G_0, F_0) = g$.

To complete the proof of the lemma, we show that there is a $\delta \in (0,\epsilon]$ such that in addition $H(P \times [0,\delta]) \subset \mathcal{R}^*$: for all $(p,t) \in P \times [0,\delta]$,

$$F(p,t) \in \mathrm{Conv}(\mathcal{R}_{b(p,t)}, G(p,t)). \tag{7.3}$$

This is a somewhat delicate point whose proof requires the local existence theory of C-structures established in Chapter V. Since $G_0 = \pi \circ g$ it follows from (7.2) that the convex hull condition (7.3) holds along $P \times \{0\}$. Applying V Lemma 5.9 (the local existence of C-structures) to the microfibration $\mathrm{id} \times \rho\colon P \times \mathcal{R} \to P \times E$, for each $q \in P$ there is a neighbourhood $\mathfrak{Op}\,(q,0)$ in $P \times [0,1]$ such that (7.3) is satisfied for all $(p,t) \in \mathfrak{Op}\,(q,0)$. since P is compact there is a $\delta \in (0,1]$ such that (7.3) is satisfied on $P \times [0,\delta]$, and the lemma is proved. $\qquad\square$

Thus the lemma provides a microfibration $\rho^*\colon \mathcal{R}^* \to E$, the *convex hull extension* of the relation $\mathcal{R}$ with respect to the affine bundle $p\colon E \to B$. We consider now the case of a microfibration $\rho\colon \mathcal{R} \to X^{(r)}$ and the affine $\mathbf{R}^q$-bundle $p_\perp^r\colon X^{(r)} \to X^\perp$ associated to a codimension 1 tangent field $\tau \subset T(V)$. Let $\mathrm{Conv}_\tau \mathcal{R}$ denote the corresponding convex hull extension of the relation $\mathcal{R}$ over $X^{(r)}$. There is a natural inclusion $i\colon \mathcal{R} \to \mathrm{Conv}_\tau \mathcal{R}$ and a projection map $\rho^*\colon \mathrm{Conv}_\tau \mathcal{R} \to X^{(r)}$, $(a,z) \mapsto z$. Applying the lemma, $\rho^*\colon \mathrm{Conv}_\tau \mathcal{R} \to X^{(r)}$ is a microfibration that depends on the tangent field $\tau \subset T(V)$.

$$
\begin{array}{ccc}
\mathrm{Conv}_\tau \mathcal{R} & \xrightarrow{\ \rho^*\ } & X^{(r)} \\
& & \downarrow{\scriptstyle p_\perp^r} \\
& & X^\perp
\end{array}
$$

Geometrically $\mathrm{Conv}_\tau \mathcal{R}$ is described as follows: $(a, z) \in \mathrm{Conv}_\tau \mathcal{R}$ if and only if the r-jets $\rho(a), z \in X_w^{(r)}$, $w = p_{r-1}^r(z) \in X^{(r-1)}$, both lie in a principal subspace (an $\mathbf{R}^q$-fiber) of the bundle $p_\perp^r : X^{(r)} \to X^\perp$, such that the following convex hull condition obtains:

$$z \in \mathrm{Conv}(\mathcal{R}_b, a); \quad b = p_\perp^r(z) = p_\perp^r \circ \rho(a) \in X^\perp.$$

Thus z lies in the convex hull of the ρ-image of the path component in $\mathcal{R}_b$ to which a belongs (cf. V §5.2.2 for this notation). This complicated microfibration encodes the convex hull information required for the general theory. We identify $\mathcal{R} \subset \mathrm{Conv}_\tau \mathcal{R}$ via the inclusion map $i \colon \mathcal{R} \to \mathrm{Conv}_\tau \mathcal{R}$, $a \mapsto (a, \rho(a))$.

§2. The h-Stability Theorem

7.2.1. A section $\varphi \in \Gamma(\mathrm{Conv}_\tau \mathcal{R})$ consists of a pair of sections (α, g) where $\alpha = \pi \circ \varphi \in \Gamma(\mathcal{R})$, $g = \rho^* \circ \varphi \in \Gamma(X^{(r)})$. In particular, with respect to the bundle $p_\perp^r \colon X^{(r)} \to X^\perp$, $p_\perp^r \circ \rho \circ \alpha = p_\perp^r \circ g \in \Gamma(X^\perp)$. The section φ is holonomic if $g = j^r f$ for some C^r-section $f \in \Gamma^r(X)$. Explicitly, if $\varphi = (\alpha, j^r f) \in \Gamma(\mathrm{Conv}_\tau \mathcal{R})$ is holonomic then for all $x \in V$,

$$j^r f(x) \in \mathrm{Conv}(\mathcal{R}_{b(x)}, \alpha(x)); \quad b(x) = j^\perp f(x) = p_\perp^r \circ \rho \circ \alpha(x) \in X^\perp. \qquad (7.4)$$

Employing the inclusion $\mathcal{R} \subset \Gamma(\mathrm{Conv}_\tau \mathcal{R})$, $\varphi = (\alpha, j^r f) \in \Gamma(\mathcal{R})$ is holonomic if and only if $\alpha \in \Gamma(\mathcal{R})$ is holonomic ($\rho \circ \alpha = j^r f$)).

With these preliminaries, the work of this chapter is the proof of the following result which is the main inductive step for the proof of the h-principles in Chapter VIII.

Theorem 7.2 (The h-Stability Theorem). *Let $\rho \colon \mathcal{R} \to X^{(r)}$ be a microfibration and let $\tau \subset T(V)$ be a codimension 1 tangent field on V that is smooth and integrable. Let $\varphi = (\alpha, j^r f_0) \in \Gamma(\mathrm{Conv}_\tau \mathcal{R})$ be a holonomic section as in (7.4) for a C^r-section $f_0 \in \Gamma^r(X)$. Let $\mathcal{N}$ be a neighbourhood of $j^\perp f_0(V)$ in $X^\perp$. There is a homotopy of holonomic sections $H \colon [0,1] \to \Gamma(\mathrm{Conv}_\tau \mathcal{R})$, $H_t = (\alpha_t, j^r g_t)$, $t \in [0,1]$, such that the following properties obtain:*

(i) *$H_0 = \varphi$; $H_1 \in \Gamma(\mathcal{R})$, hence α_1 is holonomic: $\rho \circ \alpha_1 = j^r g_1 \in \Gamma(X^{(r)})$. Thus, each holonomic section of $\mathrm{Conv}_\tau \mathcal{R}$ can be continuously deformed through holonomic sections in $\Gamma(\mathrm{Conv}_\tau \mathcal{R})$ to a holonomic section of $\mathcal{R}$. In particular, employing the homotopy α_t, the h-principal holds for the section $\alpha \in \Gamma(\mathcal{R})$.*

(ii) *($C^\perp$-Dense h-Principle): For all $t \in [0,1]$, the image $j^\perp g_t(V) \subset \mathcal{N}$. In particular $j^\perp \circ \rho \circ \alpha_1(V) = j^\perp g_1(V) \subset \mathcal{N}$.*

(iii) *(Relative Theorem)*: *Furthermore, let* $K = \{x \in V \mid \rho \circ \alpha(x) = j^r f_0(x)\}$ *i.e. α is holonomic over K. Then one can choose the homotopy H so that for all $t \in [0, 1]$, $H_t = \varphi$ on K (constant homotopy on K).*

Proof. In order to apply the analytical and topological machinery of Chapter V, the C^r-smoothness of the data $f_0 \in \Gamma^r(X)$ needs to be refined, rel K, to C^∞-smoothness on $V \setminus K$, and then the set K in the relative theorem needs to be replaced by $\mathfrak{Op}\, K$. The smoothness problem is treated first.

Lemma 7.3. *There is a (small) homotopy rel K of holonomic sections $\varphi_t = (\alpha_t, j^r g_t) \in \Gamma(\mathrm{Conv}_\tau \mathcal{R})$, $t \in [0, 1]$, $\varphi_0 = \varphi$, such that g_1 is smooth on $V \setminus K$.*

Proof. Let $B \subset X^{(r)}$ be the embedded submanifold $B = j^r f_0(V)$. Since φ is holonomic then $\rho^* \circ \varphi = j^r f_0 \in \Gamma(X^{(r)})$. Applying V Theorem 5.3 to the microfibration $\rho^* \colon \mathrm{Conv}_\tau \mathcal{R} \to X^{(r)}$ and to the smooth bundle $s \colon X^{(r)} \to V$, there is a neighbourhood U of B in $X^{(r)}$ and a continuous lift $\nu \colon U \to \mathrm{Conv}_\tau \mathcal{R}$ such that: (i) $\rho^* \circ \nu = \mathrm{id}_U$; (ii) $\nu \circ j^r f_0 = \varphi \in \Gamma(\mathrm{Conv}_\tau \mathcal{R})$.

Applying smooth approximation theory I Theorem 1.1, there is a homotopy rel K, $g_t \in \Gamma^r(X)$, $t \in [0, 1]$, $g_0 = f_0$, such that g_1 is smooth on $V \setminus K$ and such that for all $t \in [0, 1]$, $j^r g_t(V) \subset U$. Setting $\varphi_t = \nu \circ j^r g_t \in \Gamma(\mathrm{Conv}_\tau \mathcal{R})$, $t \in [0, 1]$, it follows from property (i) for the lift ν that $\rho^* \circ \varphi_t = j^r g_t$ i.e. φ_t is a homotopy of holonomic sections. Furthermore $j^r(g_t)(w) = j^r f_0(w)$ for all $w \in K$. Hence for all $t \in [0, 1]$, $\varphi_t = \varphi \in \Gamma_K(\mathrm{Conv}_\tau \mathcal{R})$ (constant homotopy over K). Since $\rho^* \circ \varphi_1 = j^r g_1$ is smooth on $V \setminus K$, the lemma is proved. $\qquad\square$

Employing Lemma 7.3, the data $\varphi = (\alpha, j^r f_0) \in \Gamma(\mathrm{Conv}_\tau \mathcal{R})$ is assumed from now on to satisfy the additional hypothesis that f_0 is smooth on $V \setminus K$. In general the closed set K in the relative theorem is not subject to *a priori* control. In order to apply the analytical machinery of Chapter V, the closed set K in the relative theorem is replaced with $\mathfrak{Op}\, K$. This is stated as a separate lemma.

Lemma 7.4. *There is a homotopy rel K of holonomic sections, $\varphi_t = (\alpha_t, j^r f_0) \in \Gamma(\mathrm{Conv}_\tau \mathcal{R})$, $t \in [0, 1]$, $\varphi_0 = \varphi$, such that $\rho \circ \alpha_1 = j^r f_0$ on $\mathfrak{Op}\, K$, and $\rho \circ \alpha_1$ is smooth on $V \setminus K$.*

Proof. As explained in the introduction, employing V Corollary 5.5, up to a small homotopy rel K, $\alpha_t \in \Gamma(\mathcal{R})$, $t \in [0, 1]$, $\alpha_0 = \alpha$, one may assume $\rho \circ \alpha_1 = j^r f_0$ on $\mathfrak{Op}\, K$, and that $\rho \circ \alpha$ is smooth on $V \setminus K$. However, the condition $\varphi_t = (\alpha_t, j^r f_0) \in \Gamma(\mathrm{Conv}_\tau \mathcal{R})$, $t \in [0, 1]$, requires also the convex hull property (7.4) with respect to the bundle $p_\perp^r \colon X^{(r)} \to X^\perp$: for all $(t, x) \in [0, 1] \times V$,

$$j^r f_0(x) \in \mathrm{Conv}(\mathcal{R}_{b(x)}, \alpha_t(x)). \tag{7.5}$$

To ensure (7.5), the homotopy α_t is constructed more carefully. Since τ is smooth, it follows that $p_\perp^r \colon X^{(r)} \to X^\perp$ is a smooth bundle. With respect to the microfibration $\rho\colon \mathcal{R} \to X^{(r)}$ and the smooth bundle $p^r_\perp\colon X^{(r)} \to X^\perp$, let $\beta = \alpha \circ s \in \Gamma(\mathcal{R})$, $g = j^r f_0 \circ s \in \Gamma(X^{(r)})$, where $s\colon X^\perp \to V$ is the source map. Let $L = j^\perp f_0(K) \subset X^\perp$; $M = g(X^\perp) \subset X^{(r)}$. Since $\rho \circ \alpha = j^r f_0$ on K it follows that $\rho \circ \beta = g$ on L.

$$\mathcal{R} \xrightarrow{\ \rho\ } X^{(r)} \xleftarrow{\ i\ } M$$

$$\downarrow{\scriptstyle p^r_\perp}$$

$$X^\perp \xleftarrow[\ i\ ]{} L$$

Applying Theorem 5.3 to the above data, there is a q-disk bundle neighbourhood U of M in $X^{(r)}$ and a continuous lift $\nu\colon U \to \mathcal{R}$ such that $\rho \circ \nu = \mathrm{id}_U$; $\nu \circ \rho \circ \beta = \beta$. Employing V Corollary 5.5 and the lift $\nu\colon U \to \mathcal{R}$, there is a small homotopy rel L, $\beta_t \in \Gamma(\mathcal{R})$, $t \in [0,1]$, such that:

(i) $\beta_0 = \beta$; $\rho \circ \beta_1 = g$ on $\mathfrak{Op}\, L$.

(ii) $\rho \circ \beta_1$ is smooth on $X^\perp \setminus L$.

(iii) for all $(t,y) \in [0,1] \times X^\perp$, $\rho \circ \beta_t(y) \in U$.

Let $\alpha_t = \beta_t \circ j^\perp f_0 \in \Gamma(\mathcal{R})$, $t \in [0,1]$. Employing (i), (ii), (iii), the homotopy α_t satisfies the properties: $\alpha_0 = \alpha$; $\rho \circ \alpha_1 = j^r f_0$ on $\mathfrak{Op}\, K$; $\rho \circ \alpha_1$ is smooth on $V \setminus K$.

For all $y \in X^\perp$, $\beta_t(y) \in \mathcal{R}_y$, $t \in [0,1]$, is a path from $\beta(y)$. Hence, for all $x \in V$, $\alpha_t(x) \in \mathcal{R}_{b(x)}$, $t \in [0,1]$, $b(x) = j^\perp f_0(x)$, is a path from $\alpha(x)$. Employing (7.4) at $\varphi = (\alpha, j^r f_0) \in \Gamma(\mathrm{Conv}_\tau \mathcal{R})$, it follows that for all $(t,x) \in [0,1] \times V$,

$$j^r f_0(x) \in \mathrm{Conv}(\mathcal{R}_{b(x)}, \alpha_t(x)), \quad b(x) = j^\perp f_0(x) \in X^\perp.$$

Setting $\varphi_t = (\alpha_t, j^r f_0) \in \Gamma(\mathrm{Conv}_\tau \mathcal{R})$, $t \in [0,1]$, the lemma is proved. $\qquad\square$

Returning to the proof of the h-stability theorem, employing lemmas 7.3, 7.4, the holonomic data $\varphi = (\alpha, j^r f_0) \in \Gamma(\mathrm{Conv}_\tau \mathcal{R})$ is assumed to satisfy the conditions: $f_0 \in \Gamma^r(X)$ is smooth on $V \setminus K$; $\rho \circ \alpha \in \Gamma(X^{(r)})$ is smooth on $V \setminus K$; $\rho \circ \alpha = j^r f_0$ on $\mathfrak{Op}\, K$.

Since the codimension 1 tangent field τ is smooth and integrable, then locally at each point $x \in V$ there is a chart $\mathfrak{Op}\, x = U$ which is a split manifold diffeomorphic to $U_0 \times [0,1] \equiv I^{n-1} \times [0,1]$, and the restriction of τ to U is tangent to the submanifolds $U_0 \times \{t\}$, $0 \le t \le 1$. Employing VI §6.1, in local coordinates adapted to τ over the split manifold U $((u,t) \in U_0 \times [0,1])$,

$$X^{(r)}_U = X^\perp_U \times \mathbf{R}^q,$$

where the $\mathbf{R}^q$-factor in this product decomposition corresponds to the pure derivative ∂^r_t. Thus in the above local coordinates, $f_0 \in C^\infty(I^n, \mathbf{R}^q)$ in case $U \subset V \setminus K$.

Following the general proof procedure of Chapter IV, let $(W_i)_{i\geq 1}$, $(U_i)_{i\geq 1}$ be locally finite coverings by closed charts of $\overline{V \setminus K_1}$ where K_1 is a closed neighbourhood of K in $\mathfrak{Op}\, K$ such that for all $i \geq 1$: (i) $W_i \subset \mathrm{int}\, U_i$; $U_i \equiv I^{n-1} \times [0,1]$ is adapted as above to the integrable codimension 1 tangent field τ; $U_i \subset V \setminus K$. In particular the sections, $f_0 \in \Gamma(X)$, $\rho \circ \alpha \in \Gamma(X^{(r)})$, are smooth on each chart U_i of the cover.

The proof procedure is an inductive process over successive charts U_i, $i \geq 1$, the main step of which is the following local extension lemma. Let $U \subset V$ be a closed chart which is a split manifold as above with respect to the integrable tangent field τ. Let $\mathcal{R}_U$, $X_U^{(r)}$, $X_U^{\perp}$ denote the subspaces over U of these spaces. Thus $\rho \colon \mathcal{R}_U \to X_U^{(r)}$ is a microfibration, $p_{\perp}^r \colon X_U^{(r)} \to X_U^{\perp}$ is smooth affine $\mathbf{R}^q$-bundle, and in local coordinates adapted to τ, $X_U^{(r)} = X_U^{\perp} \times \mathbf{R}^q$.

Local Extension Lemma 7.5. *Let $\varphi = (\alpha, j^r g) \in \Gamma(\mathrm{Conv}_\tau \mathcal{R})$ be holonomic such that g, $\rho \circ \alpha$ are smooth sections over U. Let $A, W \subset V$ be closed sets, $W \subset \mathrm{int}\, U$, where U is a closed chart on V that is a split manifold as above for the smooth integrable codimension 1 field τ. Suppose that the following properties obtain:*

(i) *$\rho \circ \alpha = j^r g$ on $\mathfrak{Op}\, A$ i.e. $\alpha \in \Gamma(\mathcal{R})$ is holonomic on $\mathfrak{Op}\, A$.*

(ii) *$j^{\perp} g(V) \subset \mathcal{N}$ i.e. g is a $C^{\perp}$-approximation to f_0.*

There is a homotopy of holonomic sections $\varphi_t = (\alpha_t, j^r \ell^t) \in \Gamma(\mathrm{Conv}_\tau \mathcal{R})$, $t \in [0,1]$, $\varphi_0 = \varphi$, $\alpha_t \in \Gamma(\mathcal{R})$, $\ell^t \in \Gamma^r(X)$, such that:

(iii) *For all $t \in [0,1]$, $\ell^t = g$ and $\varphi_t = \varphi$ on $\mathfrak{Op}\, A \cup \mathfrak{Op}\, \overline{V \setminus U}$; the sections $\ell^t \in \Gamma^r(X)$, $\rho \circ \alpha_t \in \Gamma(X^{(r)})$ are smooth over U for all $t \in [0,1]$.*

(iv) *For all $t \in [0,1]$, the image $j^{\perp} \ell^t(V) \subset \mathcal{N}$ i.e. for all $t \in [0,1]$, ℓ^t is a $C^{\perp}$-approximation to f_0.*

(v) *$\rho \circ \alpha_1 = j^r \ell^1$ on $\mathfrak{Op}\, A \cup \mathfrak{Op}\, W$ i.e. $\alpha_1 \in \Gamma(\mathcal{R})$ is holonomic on $\mathfrak{Op}\, A \cup \mathfrak{Op}\, W$. Thus the homotopy $\alpha_t \in \Gamma(\mathcal{R})$ is constant on $\mathfrak{Op}\, A$ and it connects $\alpha_0 = \alpha$, which is holonomic on $\mathfrak{Op}\, A$, to α_1 which is holonomic on $\mathfrak{Op}\, A \cup \mathfrak{Op}\, W$.*

Proof. From (iii), for all $t \in [0,1]$, the functions ℓ^t and the homotopy φ_t are constant on $\overline{V \setminus U}$ i.e. all the changes to g and to φ occur inside the chart U. The purpose of the local extension lemma is to improve the holonomicity of the section α on $\mathfrak{Op}\, A$ (cf. (i)) to holonomicity of the section α_1 on $\mathfrak{Op}\, A \cup \mathfrak{Op}\, W$ (cf. (v)) where W is a closed chart in the interior of U. This is the essential inductive step in the proof procedure.

Let B be the smoothly embedded submanifold $j^{\perp} g(U) \subset X^{\perp}$ and let $p_B \colon Y \to B$ be a tubular m-disk neighbourhood of B in $X^{\perp}$, $m + n = \dim X^{\perp}$, whose m-disk fibers are vertical $(= \ker Ds)$ along B with respect to the source map $s \colon X^{\perp} \to V$. Employing local adapted coordinates, $Y = J^{\perp}(I^n, \mathbf{R}^q)$ (a linear space of "⊥"-derivatives). Employing property (ii) of the data for g one may assume that $Y \subset \mathcal{N}$. For each subspace $Z \subset Y$, one has the microfibration $\rho \colon \mathcal{R}_Z \to X_Z^{(r)}$, where in the above adapted local coordinates, $X_Z^{(r)} = Z \times \mathbf{R}^q$, and $\mathcal{R}_Z = (p_\perp^r \circ \rho)^{-1}(Z)$:

$$\mathcal{R}_Z \xrightarrow{\rho} X_Z^{(r)} = Z \times \mathbf{R}^q \xrightarrow{p_\perp^r} Z \subset X^\perp.$$

Note that the restriction (same notation) $g \in C^\infty(I^n, \mathbf{R}^q)$ in the above local coordinates on the split manifold chart $U \equiv I^n$, where $I^n = I^{n-1} \times [0,1]$ in split coordinates. Let $\pi \colon E \to I^n$ be the product $\mathbf{R}^q$-bundle over $U = I^n$ equivalent to the trivial bundle $p_\perp^r \colon X_B^{(r)} \to B$ that is obtained by pulling back along the embedding $j^\perp g \colon I^n \to B$. Let $\rho \colon \mathcal{S} \to E$ be the microfibration obtained as the pullback of the microfibration $\rho \colon \mathcal{R}_B \to X_B^{(r)}$ along the bundle map $E \to X_B^{(r)}$. Let $\beta \in \Gamma(\mathcal{S})$ be the pullback of the section $\alpha \circ s \in \Gamma(\mathcal{R}_B)$. Let also $K = A \cap I^n$; $L = j^\perp g(K) \subset B$. In what follows, we identify $\Gamma^\infty(E) \equiv C^\infty(I^n, \mathbf{R}^q)$.

$$
\begin{array}{ccc}
E & \longrightarrow & X_B^{(r)} \\
\pi \downarrow & & \downarrow p_\perp^r \\
I^n & \xrightarrow{j^\perp g} & B
\end{array}
$$

Since $\varphi = (\alpha, j^r g) \in \Gamma(\mathrm{Conv}_\tau \mathcal{R})$ it follows that with respect to the pullback bundle $\pi \colon E \to I^n$, for all $x = (u,t) \in I^{n-1} \times [0,1]$,

$$\partial_t^r g(x) \in \mathrm{Conv}(\mathcal{S}_x, \beta(x)). \tag{7.6}$$

Since $\rho \circ \alpha = j^r g$ on $\mathfrak{Op}\, A$, it follows that the pullback $\rho \circ \beta = \partial_t^r g$ on $\mathfrak{Op}\, K$. Applying the $C^\perp$-Approximation Theorem V 5.21 (without parameters) one obtains the following theorem.

Theorem 7.6. *For $\epsilon > 0$ sufficiently small there is a smooth map $g_\epsilon \in C^\infty(I^n, \mathbf{R}^q)$ and a homotopy $F = F_\epsilon \colon [0,1] \to \Gamma(\mathcal{S})$ such that:*

(p_1) $\lim_{\epsilon \to 0} \|g_\epsilon - g\|^{r,r-1} = 0$.

(p_2) $F_0 = \beta$; $\rho \circ F_1 = \partial_t^r g_\epsilon \colon I^n \to \mathbf{R}^q$; *for all $t \in [0,1]$, $\rho \circ F_t \in C^\infty(I^n, \mathbf{R}^q)$.*

(p_3) $g_\epsilon = g$ on $\mathfrak{Op}\, K$. *For all $t \in [0,1]$, $F_t = \beta \in \Gamma_{\mathfrak{Op}\, K}(\mathcal{S})$ (constant homotopy on $\mathfrak{Op}\, K$).*

The entire analytic theory of Chapter V is required to produce the smooth map g_ϵ and the homotopy F_t of Theorem 7.6 with respect to the pullback bundle $\pi \colon E \to I^n$ and the microfibration $\rho \colon \mathcal{S} \to E$. We employ Theorem 7.6 as follows to construct the component $\ell^t \in \Gamma^r(X)$ of the homotopy $\varphi_t = (\alpha_t, j^r \ell^t) \in$

$\Gamma(\mathrm{Conv}_\tau \mathcal{R})$, $t \in [0,1]$, required for the conclusions of the Local Extension Lemma 7.5. Since the closed chart $W \subset \mathrm{int}\, U$ there is a smooth cut-off function $\mu\colon V \to [0,1]$ such that $\mu = 1$ on $\mathfrak{Op}\, W$ and $\mu = 0$ on $\mathfrak{Op}\, \overline{V \setminus U}$ (a neighbourhood of the complement of U). For each $t \in [0,1]$ define a section $\ell^t \equiv \ell^t_\epsilon \in \Gamma^r(X)$ as follows:

$$\ell^t(x) = \begin{cases} g(x) + t\mu(x)(g_\epsilon - g)(x) \in C^\infty(I^n, \mathbf{R}^q) & \text{if } x \in U \\ g(x) & \text{if } x \in \mathfrak{Op}\, \overline{V \setminus U}. \end{cases}$$

Since $\mu = 0$ on $\mathfrak{Op}\, \partial U$ it follows that $\ell^t \in \Gamma^r(X)$, $t \in [0,1]$, is well-defined and smooth on U.

Evidently, $\ell^0 = g$; $\ell^1 = g_\epsilon$ on $\mathfrak{Op}\, W$. Since $g_\epsilon = g$ on $\mathfrak{Op}\, K$ it follows that $\ell^t = g$ on $\mathfrak{Op}\, A$ (constant homotopy). Furthermore over the chart U, employing the norm estimate in Theorem 7.6 for the map g_ϵ, for all $t \in [0,1]$, $\lim_{\epsilon \to 0} \|\ell^t - g\|^{r,r-1} = 0$. Consequently, for $\epsilon > 0$ sufficiently small, it follows that for all $t \in [0,1]$, $j^\perp \ell^t(U) \subset Y$ i.e. $j^\perp \ell^t, j^\perp g$ are C^0-close. More precisely, for each bundle neighbourhood $N(B)$ of B in Y, it follows that for $\epsilon > 0$ sufficiently small, for all $t \in [0,1]$ $j^\perp \ell^t(U) \subset N(B)$, hence $j^\perp \ell^t(U) \subset \mathcal{N}$.

The construction of the component $\alpha_t \in \Gamma(\mathcal{R})$ of the required homotopy $\varphi_t = (\alpha_t, j^r \ell^t)$ is decidedly more complex, and is completed in Lemma 7.13 below. It is for this purpose that the normal bundle $p_B\colon Y \to B$ in $X^\perp$ is introduced. For $\epsilon > 0$ sufficiently small, the sections $j^\perp g_\epsilon, j^\perp g$ are C^0-close in Y. Furthermore, employing the microfibration $\rho\colon \mathcal{R} \to X^{(r)}$, the C-structure in $\mathcal{S}$ employed to construct the homotopy F_ϵ in Theorem 7.6 pushes forward and extends to a C-structure over Y. In particular one obtains a C-structure over the base $j^\perp \ell^t_\epsilon(I^n) \subset Y$ which fiberwise strictly surrounds $j^r \ell^t_\epsilon$, an essential step in the proof of the Local Extension Lemma 7.5. The details of the above steps are delicate and occupy the next several lemmas. The extension of C-structures into Y is proved in Lemma 7.10. The situation is analogous to the push forward problem encountered in the proof of the C^0-dense h-principle, IV Remark 4.5, in the special case of open relations $\mathcal{R} \subset X^{(1)}$: the sections $j^\perp \ell^t_\epsilon, j^\perp g$ are C^0-close; however because of the presence of the rapidly oscillating function θ_ϵ the pure derivatives $\partial^r_t \ell^s_\epsilon, \partial^r_t g$, $s \in [0,1]$, do not satisfy uniform approximations as $\epsilon \to 0$. Special arguments are therefore required to construct a homotopy of sections $\alpha_t \in \Gamma(\mathcal{R})$ such that $(\alpha_t, j^r \ell^t_\epsilon) \in \Gamma(\mathrm{Conv}_\tau \mathcal{R})$, $t \in [0,1]$.

We first recall some details concerning the homotopy F in Theorem 7.6. (cf. V Remark 5.22). Since $\rho \circ \beta = \partial^r_t g$ on $\mathfrak{Op}\, K$, employing V Complement 5.11, there is a smooth C-structure (m, M) with respect to the sections $\partial^r_t y, \beta$, where $m\colon [0,1] \rightarrowtail \Gamma(\mathcal{S})$, $m(0) = \beta$, such that for all $x \in I^n$ the path $\rho \circ m_u(x)$ in E_x, $u \in [0,1]$, strictly surrounds $\partial^r_t g(x)$. Furthermore, there is a q-disk bundle neighbourhood T of the graph of $\rho \circ \beta \in \Gamma(E)$ and a lift $\nu\colon T \to \mathcal{S}$, $\nu \circ \rho \circ \beta = \beta$, $\rho \circ \nu = \mathrm{id}_T$, such that on $\mathfrak{Op}\, K$ (m, M) is a continuous lift: $(m, M) = (\nu \circ m_0, \nu \circ$

M_0), where (m_0, M_0) is a smooth C-structure over $\mathfrak{Op}\, K$ with respect to $\partial_t^r g$, $\rho \circ \beta \in \Gamma(E)$ such that ev $M_0([0,1]^2 \times \mathfrak{Op}\, K) \subset T$. Employing V Theorem 5.21, the homotopy $F \colon [0,1] \to \Gamma(\mathcal{S})$ has the following form ($x = (c,t) \in I^{n-1} \times [0,1]$; $s \in [0,1]$):

$$F_s(c,t) = \begin{cases} M(\theta_\epsilon(t), s, c, t)) \in \mathcal{S}_x & \text{if } x \in I^n \setminus \mathfrak{Op}_2 K \\ \nu \circ F_s'(x) & \text{if } x \in \mathfrak{Op}\, K. \end{cases} \tag{7.7}$$

where $\overline{\mathfrak{Op}_2 K} \subset \mathfrak{Op}\, K$. The homotopy $F' \colon [0,1] \to \Gamma_{\mathfrak{Op}\, K}(T)$ satisfies the properties:

(i) $F_0' = \rho \circ \beta = \partial_t^r g$, $F_1' = \partial_t^r g_\epsilon$.

(ii) $F_s' = \rho \circ \beta$ (constant homotopy) on (a smaller) $\mathfrak{Op}_1 K$.

(iii) $F_s'(x) = M_0(\theta_\epsilon(t), s, x)$ on $\mathfrak{Op}\, K \setminus \mathfrak{Op}_2 K$.

Remark 7.7. Except for a small perturbation in the tube T to ensure that the homotopy $F = \nu \circ F'$ is constant $= \beta$ on $\mathfrak{Op}_1 K$, the image of F is contained in the image of the C-structure M. The construction of F' makes essential use of the contractible q-disk fibers of the bundle T. This split definition for constructing the homotopy F in the tube T so that it is constant on $\mathfrak{Op}\, K$ (cf. (p_3)) is employed again in Lemma 7.12 to construct a suitable extension of the push forward of F over a neighbourhood $N(B)$.

In what follows $B(y; r) \subset \mathbf{R}^q$ is the ball of radius r, center y. The balls obtained in the following technical lemma and corollary are required to establish the convex hull properties of Proposition 7.14.

Lemma 7.8. *Let (m, M) be the C-structure above with respect to $\partial_t^r g, \beta$. There is a $\delta > 0$ such that for all $(t, x) \in [0,1] \times I^n$,*

$$B(\partial_t^r g(x); \delta) \cup B(\rho \circ m(t, x); \delta) \subset \mathrm{Conv}(\mathcal{S}_x, \beta(x)).$$

Proof. The homotopy $m \colon [0,1] \to \Gamma(\mathcal{S})$ satisfies the properties: (i) $m(0) = \beta$; (ii) the projection $\rho \circ m \colon [0,1] \to \Gamma(E)$ is a path that strictly surrounds the section $\partial_t^r g$ in each fiber. By compactness there is a $\delta_1 > 0$ such that for all $x \in I^n$,

$$B(\partial_t^r g(x); \delta_1) \subset \mathrm{Conv}(\mathcal{S}_x, \beta(x)).$$

Again, since $m(0) = \beta$ it follows that for all $(u, x) \in [0,1] \times I^n$,

$$\rho \circ m_u(x) \in \mathrm{Conv}(\mathcal{S}_x, \beta(x)). \tag{7.8}$$

Employing (7.8) it follows from V Proposition 5.8 applied to the bundle $[0,1] \times E \to [0,1] \times I^n$ that there is a parametrized family of C-structures (h, H), with respect to m, β, where $h\colon [0,1]^2 \to \Gamma(\mathcal{S})$, $h(0,0) = \beta$, such that for all $(u, v, x) \in [0,1]^2 \times I^n$, the path $\rho \circ h(u, v, x)$ in E_x, $v \in [0,1]$, strictly surrounds $\rho \circ m(u, x)$. By compactness, there is a $\delta_2 > 0$ such that for all $(t, x) \in [0,1] \times I^n$,

$$B(\rho \circ m(t, x); \delta_2) \subset \mathrm{Conv}(\mathcal{S}_x, \beta(x)),$$

which completes the proof of the lemma.   □

Corollary 7.9. *There is a $\delta > 0$ such that for all $\epsilon > 0$ sufficiently small and all $x \in I^n$,*

$$B(\partial_t^r g_\epsilon(x); \delta) \subset \mathrm{Conv}(\mathcal{S}_x, \beta(x)).$$

Proof. Recall the q-disk bundle $T \subset E$ over the base section $\rho \circ \beta(I^n)$, and the continuous lift $\nu\colon T \to \mathcal{S}$, $\rho \circ \nu = \mathrm{id}_T$, $\nu \circ \rho \circ \beta = \beta$. Joining points in the q-disk fibers of T to the base and then employing the lift ν, it follows that for all $x \in I^n$ (T_x is the q-disk fiber of the projection $\pi\colon T \to I^n$),

$$T_x \subset \mathrm{Conv}(\mathcal{S}_x, \beta(x)).$$

Employing (7.7), for $\epsilon > 0$ sufficiently small, the homotopy F' lies in T and satisfies $F_1' = \partial_t^r g_\epsilon$ over $\mathfrak{Op}\, K$ i.e. for all $x \in \mathfrak{Op}\, K$, $\partial_t^r g_\epsilon(x) \subset T_x$. In fact, from V Theorem 5.20, the homotopy F' lies in the interior of the tubular neighbourhood T (If $T = T(\sigma)$ (radius σ), then the homotopy F' can be chosen to be in $T(\sigma/2)$, provided $\epsilon > 0$ is sufficiently small). Hence the corollary is proved over $\mathfrak{Op}\, K$.

Employing (7.7) it follows that for all $x = (c, t) \in I^n \setminus \mathfrak{Op}\, K$:

$$\rho \circ F_1(x) = \partial_t^r g_\epsilon(x) = \rho \circ M(\theta_\epsilon(t), 1, x).$$

By construction of the C-structure (m, M), the contracting homotopy M runs back and forth along the strictly surrounding homotopy $m\colon [0,1] \to \Gamma(\mathcal{S})$. Employing Lemma 7.8, it follows that there is a $\delta > 0$ such that for all $x \in I^n \setminus \mathfrak{Op}\, K$,

$$B(\partial_t^r g_\epsilon(x); \delta) \subset \mathrm{Conv}(\mathcal{S}_x; \beta(x)),$$

which completes the proof of the Corollary.   □

Returning to the proof of the local extension lemma, since $X_Y^{(r)} = Y \times \mathbf{R}^q$ in local adapted coordinates, the source map $s\colon X^\perp \to V$ induces a bundle map $s \times \mathrm{id}\colon X_Y^{(r)} \to E$, $(y, v) \mapsto (s(y), v) \in I^n \times \mathbf{R}^q = E$, which is an isomorphism on each fiber.

$$
\begin{array}{ccc}
X_Y^{(r)} = Y \times \mathbf{R}^q & \xrightarrow{\ s \times \mathrm{id}\ } & E \\
{\scriptstyle p_\perp^r}\big\downarrow & & \big\downarrow{\scriptstyle \pi} \\
Y & \xrightarrow[\ \ s\ \ ]{} & I^n
\end{array}
$$

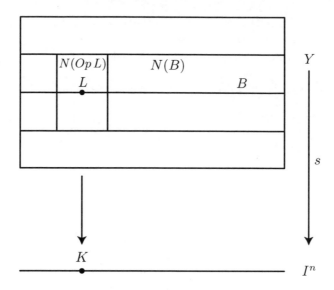

$$\textsc{Figure 7.1}$$

$s^*f \in \Gamma(X_Y^{(r)})$ is the pullback of a section $f \in \Gamma(E)$: for all $y \in Y$, $s^*f(y) = (y, f(s(y))) \in X_Y^{(r)}$. Similarly C-structures in E pull back to C-structures in $X_Y^{(r)}$.

Let $G = s^*\partial_t^r g$, $G_\epsilon = s^*\partial_t^r g_\epsilon \in \Gamma(X_Y^{(r)})$ be the pullbacks respectively of the sections $\partial_t^r g$, $\partial_t^r g_\epsilon \in \Gamma(E)$: for all $y \in Y$,

$$G(y) = \big(y, \partial_t^r g(s(y))\big); \ \ G_\epsilon(y) = \big(y, \partial_t^r g_\epsilon(s(y))\big) \in X_Y^{(r)} = Y \times \mathbf{R}^q. \qquad (7.9)$$

Let (m', M'), $M' : D^2 \to \Gamma(\mathcal{R}_B)$, be the push forward of the C-structure (m, M). Note that the push forward of the section $M(1) = \beta \in \Gamma(\mathcal{S})$ is the section $M'(1) = \alpha \circ s \in \Gamma(\mathcal{R}_B)$. Thus (m', M') is a C-structure with respect to $G, \alpha \circ s$ over $B \subset Y$. Let $N(B) \subset Y$ be an m-disk bundle neighbourhood of B. For notational convenience in what follows, for each $C \subset B$ let $N(C)$ denote the restriction of $N(B)$ over C: $N(C) = s^{-1}(s(C)) \cap N(B)$.

Lemma 7.10. *There is a neighbourhood $N(B)$ of B in Y, a section $\gamma \in \Gamma(\mathcal{R}_{N(B)})$, and a C-structure (j, J) in $\mathcal{R}$ over $N(B)$ with respect to the sections G, γ such that the following properties are satisfied:*

(i) *γ is an extension of the section $\alpha \circ s \in \Gamma(\mathcal{R}_B)$; $\rho \circ \gamma = G$ on $N(\mathfrak{Op}\, L)$.*

(ii) *The C-structure (j, J) extends the C-structure (m', M') over B.*

(iii) *$(\rho \circ j, \rho \circ J) = (s^*(\rho \circ m), s^*(\rho \circ M))$; $\rho \circ J : D^2 \to \Gamma(X_{N(B)}^{(r)})$ is smooth.*

Proof. Applying V Theorem 5.1 to the bundle id $\times p_B \colon D^2 \times Y \to D^2 \times B$, it follows that there is an m-disk bundle neighbourhood $N(B)$ of B in Y and a continuous extension $J \colon D^2 \to \Gamma(\mathcal{R}_{N(B)})$ such that:

(p_4) For all $u \in D^2$, $J_u = M'_u \in \Gamma(\mathcal{R}_B)$ i.e. J restricts to M' over B.

(p_5) The projection $\rho \circ J = s^*(\rho \circ M) \colon D^2 \to \Gamma(X^{(r)}_{N(B)})$. Explicitly, for all $(u, y) \in D^2 \times N(B)$,

$$\rho \circ J_u(y) = \big(y, \pi_2 \circ \rho \circ M_u(s(y))\big),$$

where $\pi_2 \colon E = I^n \times \mathbf{R}^q \to \mathbf{R}^q$ is the projection map.

Since (m, M) is a smooth C-structure, employing (p_5), $\rho \circ J$ is smooth. Let $J(1) = \gamma \in \Gamma(\mathcal{R}_{N(B)})$. Since J extends M', it follows that $\gamma = M'(1) = \alpha \circ s$ along B i.e. γ extends $\alpha \circ s$. Recall that $\rho \circ \beta = \partial_t^r g$ on $\mathfrak{Op}\, K$ and $M(1) = \beta \in \Gamma(\mathcal{S})$. Employing ($p_5$), (7.9), it follows that:

$$\rho \circ \gamma(y) = \big(y, \partial_t^r g(s(y))\big) = G(y) \in X^{(r)}_Y \quad \text{for all } y \in N(\mathfrak{Op}\, L). \tag{7.10}$$

Hence conclusion (i) of the lemma is proved. Let $j \colon S^1 \to \Gamma(\mathcal{R}_{N(B)})$ be the restriction of J to $S^1 = \partial D^2$. From (p_5), $\rho \circ j = s^*(\rho \circ m)$ and hence:

$$(\rho \circ j, \rho \circ J) = (s^*(\rho \circ m), s^*(\rho \circ M)).$$

Since the homotopy $\rho \circ m$ strictly surrounds the section $\partial_t^r g \in \Gamma(E)$ it follows that the homotopy $\rho \circ j \colon S^1 \to \Gamma(X^{(r)}_{N(B)})$ strictly surrounds the section $G = s^* \partial_t^r g \in \Gamma(X^{(r)}_{N(B)})$. Thus the pair (j, J) is a C-structure with respect to G, γ and, employing (7.10), $\rho \circ \gamma = G$ on $N(\mathfrak{Op}\, L)$, which completes the proof of the lemma. □

Recall the lift $\nu \colon T \to \mathcal{S}$, where T is a q-disk bundle neighbourhood of the image $\rho \circ \beta(I^n) \subset E$. Applying V Theorem 5.3 to the section $\gamma \in \Gamma(\mathcal{R}_{N(B)})$ there is a q-disk bundle neighbourhood $\overline{T}$ of the image $\rho \circ \gamma(N(B))$ in $X^{(r)}_{N(B)}$ and a continuous extension (same notation) $\nu \colon \overline{T} \to \mathcal{R}$ such that $\nu \circ \rho \circ \gamma = \gamma$; $\rho \circ \nu = \mathrm{id}_{\overline{T}}$.

Note that $\rho \circ \beta = \partial_t^r g$ on $\mathfrak{Op}\, K$ and from Lemma 7.10 $\rho \circ \gamma = G = s^* \partial_t^r g$ on $N(\mathfrak{Op}\, L) = s^{-1}(\mathfrak{Op}\, K) \cap N(B)$. Let T_g, respectively $\overline{T}_G$, denote the subdisk bundle of T over the base $\partial_t^r g(\mathfrak{Op}\, K)$ in E, respectively the subdisk bundle of $\overline{T}$ over the base $G(N(\mathfrak{Op}\, L))$ in $X^{(r)}_Y$. With respect to the bundle map, $s \times \mathrm{id} \colon N(\mathfrak{Op}\, L) \times \mathbf{R}^q \to \mathfrak{Op}\, K \times \mathbf{R}^q$, employing $G = s^* \partial_t^r g$, if the q-disk bundle T_g has sufficiently small radius,

$$(s \times \mathrm{id})^{-1}(T_g) \subset \overline{T}_G. \tag{7.11}$$

Corollary 7.11. *One may suppose in addition that the C-structure $(j, J) = (\nu \circ j_0, \nu \circ J_0)$ over $N(\mathfrak{Op}\, L)$ where $(j_0, J_0) = (s^* m_0, s^* M_0)$ is a C-structure in $\overline{T}_G$ over $N(\mathfrak{Op}\, L)$ with respect to the sections $G, \rho \circ \gamma$.*

Proof. Recall that over $\mathfrak{Op}\, K$ the C-structure $(m, M) = (\nu \circ m_0, \nu \circ M_0)$, where (m_0, M_0) is a C-structure over $\mathfrak{Op}\, K$ in the q-disk bundle T_g with respect to $\partial_t^r g, \rho \circ \beta$. Setting $(j_0, J_0) = (s^* m_0, s^* M_0)$, employing (7.11), it follows that (j_0, J_0) is a C-structure over $N(\mathfrak{Op}\, L)$ in the q-disk bundle $\overline{T}_G$ with respect to $G, \rho \circ \gamma$.

Employing V Theorem 5.1 one obtains a neighbourhood $N(B)$ of B in Y, a C-structure (same notation) (j, J) over $N(B)$ which satisfies the conclusions of Lemma 7.10 and such that in addition $(j, J) = (\nu \circ j_0, \nu \circ J_0)$ on $N(\mathfrak{Op}\, L)$ where (j_0, J_0) is a C-structure in $\overline{T}_G$ as above. $\qquad\square$

Working in the bundle $\pi \colon E \to I^n$, the smooth map g_ϵ and the homotopy $F = F_\epsilon \colon [0, 1] \to \Gamma(\mathcal{S})$ were constructed in Theorem 7.6 to satisfy the properties (p_i), $1 \le i \le 3$, and for which the homotopy F is given explicitly by the formula (7.7). In the next lemma, which is central for the constructions that follow, employing the C-structure (j, J) given in Lemma 7.10, one extends the lift $s^* F$ over B to a homotopy P in $\mathcal{R}$ over $N(B)$ such that $P_0 = \gamma$ and such that $\rho \circ P_1 = G_\epsilon = s^* \partial_t^r g_\epsilon$. The split definition (cf. (7.7)) of the homotopy F over $\mathfrak{Op}\, K$ and $I^n \setminus \mathfrak{Op}_2 K$ necessitates a corresponding split definition (and verification of properties) of P over $N(\mathfrak{Op}\, L)$ and $N(B) \setminus N(\mathfrak{Op}_2 L)$ where $N(\mathfrak{Op}_2 L) = s^{-1}(\mathfrak{Op}_2 K) \cap N(B)$. Since $F' \colon [0, 1] \to \Gamma_{\mathfrak{Op}\, K}(T_g)$, employing (7.11), note in particular that $s^* F' \colon [0, 1] \to \Gamma_{N(\mathfrak{Op}\, L)}(\overline{T}_G)$. In what follows we employ the equivalent formulation of C-structures on domains $D^2, [0, 1]^2$ (cf. §5.2.2).

Lemma 7.12. *Let $P \colon [0, 1] \to \Gamma(\mathcal{R}_{N(B)})$ be the homotopy (cf. (7.7)): for all $(u, y) \in [0, 1] \times N(B)$,*

$$P_u(y) = \begin{cases} J(\theta_\epsilon(t), u, y) & \text{if } y \in N(B) \setminus N(\mathfrak{Op}_2 L) \\ \nu \circ P'_u(y) & \text{if } y \in N(\mathfrak{Op}\, L). \end{cases} \qquad (7.12)$$

where $P' = s^* F' \colon [0, 1] \to \Gamma_{N(\mathfrak{Op}\, L)}(\overline{T}_G)$. The homotopy P satisfies the following properties: for all $u \in [0, 1]$,

(i) $P_0 = \gamma$; $\rho \circ P_1 = G_\epsilon$; $\rho \circ P \colon [0, 1] \to \Gamma(X_{N(B)}^{(r)})$ *is smooth.*

(ii) $P_u = \gamma \in \Gamma(\mathcal{R}_{N(\mathfrak{Op}_1 L)})$ *(constant homotopy on a smaller $N(\mathfrak{Op}_1 L)$).*

Proof. Employing Corollary 7.11, $J = \nu \circ J_0$ on $N(\mathfrak{Op}\, L)$, and $J_0 = s^* M_0$. By construction (cf. (7.7)), for all $x = (c, t) \in \mathfrak{Op}\, K \setminus \mathfrak{Op}_2 K$,

$$F'_u(x) = M_0(\theta_\epsilon(t), u, x).$$

Since $P' = s^* F'$, it follows that for all $(u, y) \in [0,1] \times N(\mathfrak{Op}\, L) \setminus N(\mathfrak{Op}_2 L)$,

$$P'_u(y) = J_0(\theta_\epsilon(t), u, y),$$

Since $J = \nu \circ J_0$ on $N(\mathfrak{Op}\, L)$ it follows that the homotopy P is well-defined, continuous and that $\rho \circ P$ is smooth.

To complete the proof of conclusion (i), we work in $N(\mathfrak{Op}\, L)$ and then in $N(B) \setminus N(\mathfrak{Op}_2 L)$. Employing (7.7), the homotopy F' satisfies the properties: over $\mathfrak{Op}\, K$,

$$F'_0 = \rho \circ \beta = \partial_t^r g; \quad F'_1 = \partial_t^r g_\epsilon.$$

Since $P' = s^* F'$ over $N(\mathfrak{Op}\, L)$, it follows that over $N(\mathfrak{Op}\, L)$,

$$P'_0 = s^* \partial_t^r g = G; \quad P'_1 = s^* \partial_t^r g_\epsilon = G_\epsilon. \tag{7.13}$$

Employing (7.10), $G = \rho \circ \gamma$ over $N(\mathfrak{Op}\, L)$ i.e. $P'_0 = \rho \circ \gamma$ and hence $P_0 = \nu \circ \rho \circ \gamma = \gamma$ on $N(\mathfrak{Op}\, L)$. Since the C-structure (j, J) satisfies $J(t, 0) = \gamma$ for all $t \in [0,1]$, it follows that $P_0 = \gamma$ on $N(B)$.

From (7.13), $\rho \circ P_1 = P'_1 = G_\epsilon$ on $N(\mathfrak{Op}\, L)$. Employing (7.7) and (p_2) of Theorem 7.6 for the homotopy F, for all $x = (c, t) \in I^n \setminus \mathfrak{Op}_2 K$,

$$\partial_t^r g_\epsilon(x) = \rho \circ F_1(x) = \rho \circ M(\theta_\epsilon(t), 1, x).$$

Since $\rho \circ J = s^*(\rho \circ M)$, $G_\epsilon = s^* \partial_t^r g_\epsilon$, it follows that $\rho \circ P_1 = G_\epsilon$ on $N(B)$, which completes the proof of conclusion (i).

Since $F'_u = \rho \circ \beta = \partial_t^r g$ on (a smaller) $\mathfrak{Op}_1 K$ (constant homotopy), employing (7.10), for all $u \in [0,1]$, $P'_u = s^* F'_u = G = \rho \circ \gamma$ on $N(\mathfrak{Op}_1 L)$. Hence $P_u = \nu \circ P'_u = \gamma$ on $N(\mathfrak{Op}_1 L)$ (constant homotopy) which proves conclusion (ii), and the proof of the lemma is complete. $\qquad\square$

Recall the given data $\varphi = (\alpha, j^r g) \in \Gamma(\mathrm{Conv}_\tau\, \mathcal{R})$. Let $\alpha_t \in \Gamma(\mathcal{R})$, $t \in [0,1]$, be the homotopy:

$$\alpha_t(x) = \begin{cases} \alpha(x) & \text{if } x \in \mathfrak{Op}\, \overline{V \setminus U} \\ P_{\mu(x)t}(j^\perp \ell^t(x)) & \text{if } x \in U. \end{cases} \tag{7.14}$$

Since $\mu = 0$ on $\mathfrak{Op}\, \partial U$, it follows that on U, $\alpha_t = P_0(j^\perp g)$ near the boundary of U. From Lemma 7.12, $P_0 = \gamma$, and from Lemma 7.10, the section $\gamma \in \Gamma(\mathcal{R}_{N(B)})$ is an extension of the section $\alpha \circ s \in \Gamma(\mathcal{R}_B)$. In particular, $\gamma \circ j^\perp g = \alpha \in \Gamma(\mathcal{R}_U)$ which proves that the homotopy α_t is well-defined, continuous, and for all $t \in [0,1]$, $\rho \circ \alpha_t \in \Gamma(X^{(r)})$ is smooth on U.

Lemma 7.13. *The homotopy* $\alpha_t \in \Gamma(\mathcal{R})$, $t \in [0,1]$, *satisfies the following properties: for all* $t \in [0,1]$,

(i) $\alpha_0 = \alpha$; $p_\perp^r \circ \rho \circ \alpha_t = j^\perp \ell^t \in \Gamma(X^\perp)$; $\rho \circ \alpha_t$ *is smooth on* U.

(ii) $\rho \circ \alpha_1 = j^r \ell^1$ *on* $\mathfrak{Op}\, W$.

(iii) $\alpha_t = \alpha$ *on* $\mathfrak{Op}\, A$ *(constant homotopy).*

Proof. As explained above, $\alpha_0 = P_0(j^\perp g) = \alpha$ on U hence $\alpha_0 = \alpha$. Since $P_t \in \Gamma(\mathcal{R}_{N(B)})$ is a section, then for all $y \in N(B)$, $p_\perp^r \circ \rho \circ P_t(y) = y$; setting $y = j^\perp \ell^t(x)$ it follows that $p_\perp^r \circ \rho \circ \alpha_t = j^\perp \ell^t$, which completes the proof of conclusion (i).

Recall that $\ell^1 = g_\epsilon$ on $\mathfrak{Op}\, W$ and, from Lemma 7.12, $\rho \circ P_1 = G_\epsilon$ on $N(B)$. Consequently employing (7.9), for all $x \in \mathfrak{Op}\, W$,

$$\rho \circ \alpha_1(x) = \rho \circ P_1(j^\perp g_\epsilon(x)) = (j^\perp g_\epsilon(x), \partial_t^r g_\epsilon(x)) = j^r \ell^1(x),$$

which proves conclusion (ii). Recall that $\ell^t = g$ on $\mathfrak{Op}\, A$ and, from Lemma 7.12, $P_t = \gamma$ on $N(\mathfrak{Op}\, L)$ (constant homotopy). Also from Lemma 7.10, $\gamma = \alpha \circ s$ along $B = j^\perp g(U)$. Consequently, for all $(t,x) \in [0,1] \times \mathfrak{Op}\, K$ $(K = A \cap U)$,

$$\alpha_t(x) = P_{\mu(x)t}(j^\perp g(x)) = \gamma(j^\perp g(x)) = \alpha(x),$$

which proves conclusion (iii), and the proof of the lemma is complete.   □

Let $\varphi_t = (\alpha_t, j^r \ell^t) \in \Gamma(\mathcal{R}) \times \Gamma^r(X)$, $t \in [0,1]$. Employing Lemma 7.13, one verifies the following properties:

(i) $\varphi_0 = (\alpha, j^r g) = \varphi$.

(ii) $\varphi_t = \varphi$ (constant homotopy) on $\mathfrak{Op}\, A$.

(iii) $\rho \circ \alpha_1 = j^r \ell^1$ on $\mathfrak{Op}\, W$.

(iv) For all $t \in [0,1]$ $j^\perp \ell^t(V) \subset \mathcal{N}$.

Consequently the homotopy φ_t, $t \in [0,1]$, satisfies all the conclusions of the local extension lemma except possibly the convex hull property $\varphi_t \in \Gamma(\mathrm{Conv}_\tau \mathcal{R})$, $t \in [0,1]$, which is proved separately in the following proposition.

Proposition 7.14. *For* $\epsilon > 0$ *sufficiently small,* $\varphi_t = (\alpha_t, j^r \ell_\epsilon^t)$ *is a homotopy in* $\Gamma(\mathrm{Conv}_\tau \mathcal{R})$: *for all* $(t,x) \in [0,1] \times V$,

$$j^r \ell^t(x) \in \mathrm{Conv}(\mathcal{R}_{b_t(x)}, \alpha_t(x)); \quad b_t(x) = j^\perp \ell^t(x) \in X^\perp.$$

Proof. On $\mathfrak{Op}\, \overline{V \setminus U}$, $\ell^t = g$, $\alpha_t = \alpha$ (constant homotopy). By hypothesis $\varphi = (\alpha, j^r g) \in \Gamma(\mathrm{Conv}_\tau \mathcal{R})$ i.e. the lemma is trivial on $\mathfrak{Op}\, \overline{V \setminus U}$.

Lemma 7.15. *Recall that on $U = I^n$, $\ell^s(= \ell^s_\epsilon) = g + s\mu(g_\epsilon - g) \in C^\infty(I^n, \mathbf{R}^q)$. If $\epsilon > 0$ is sufficiently small then for all $(s, x) \in [0, 1] \times I^n$,*

$$\partial_t^r \ell^s_\epsilon(x) \in \mathrm{Conv}(\mathcal{S}_x, \beta(x)).$$

Proof. Calculating successive derivatives in $t \in [0, 1]$, on obtains,

$$\partial_t^r \ell^s = \partial_t^r g + s\mu \partial_t^r(g_\epsilon - g) + \sum_{p=1}^r s\kappa_p \, \partial_t^p(\mu) \, \partial_t^{r-p}(g_\epsilon - g)$$
$$= \partial_t^r g + s\mu \partial_t^r(g_\epsilon - g) + \mathcal{E}(\epsilon),$$

where κ_p is a constant which depends only on p, $1 \leq p \leq r$, and, from the norm estimate (p_1) for g_ϵ in Theorem 7.6, the error term $\lim_{\epsilon \to 0} \mathcal{E}(\epsilon) = 0$. Thus for each $x \in I^n$, up to a small perturbation by the error $\mathcal{E}(\epsilon)$, $\partial_t^r \ell^s(x)$ is a convex combination of the derivatives $\partial_t^r g(x)$, $\partial_t^r g_\epsilon(x)$. Employing Lemma 7.8, Corollary 7.9, there is a $\delta > 0$ such that for all $\epsilon > 0$ sufficiently small, and all $x \in U$,

$$B(\partial_t^r g(x); \delta) \cup B(\partial_t^r g_\epsilon(x); \delta) \subset \mathrm{Conv}(\mathcal{S}_x, \beta(x)).$$

Since any convex linear combination of the derivatives $\partial_t^r g(x)$, $\partial_t^r g_\epsilon(x)$ lies in the convex hull of the above q-balls of radius δ, one concludes that, for $\epsilon > 0$ sufficiently small, $\partial_t^r \ell^s(x) \in \mathrm{Conv}(\mathcal{S}_x, \beta(x))$, and the lemma is proved. $\square$

Returning to the proof of the proposition, let $f^v(=f^v_\epsilon)=s^*\partial_t^r \ell^v_\epsilon \in \Gamma(X^{(r)}_{N(B)})$: for all $(v, y) \in [0, 1] \times N(B)$,

$$f^v(y) = \big(y, \partial_t^r \ell^v(s(y))\big) \in X^{(r)} = X^\perp \times \mathbf{R}^q.$$

In particular, the push forward of the section $\partial_t^r \ell^v \in \Gamma(E)$ to the bundle $p^r_\perp : X^{(r)}_B \to B$ is the restriction of the section f^v along B: $f^v(b) = (j^\perp g(x), \partial_t^r \ell^v_\epsilon(x))$, where $b = j^\perp g(x) \in B$. By construction, the push forward of $\beta \in \Gamma(\mathcal{S})$ is the section $\alpha \circ s \in \Gamma(X^{(r)}_B)$. The C-structure (m, M) employed in the proofs of Lemma 7.8, Corollary 7.9, pushes forward to a C-structure (m', M'), where $M' : [0, 1]^2 \to \Gamma(\mathcal{R}_B)$. With respect to the C-structure (m', M'), it follows from Lemma 7.15 that for all $\epsilon > 0$ sufficiently small and all $(v, b) \in [0, 1] \times B$,

$$f^v(b) \subset \mathrm{Conv}(\mathcal{R}_b, \alpha \circ s(b)).$$

Furthermore, from Lemma 7.10, the section $\gamma \in \Gamma(\mathcal{R}_{N(B)})$ is an extension of the section $\alpha \circ s \in \Gamma(\mathcal{R}_B)$, and also the C-structure (j, J) over $N(B)$ is an extension of the C-structure (m', M'). Since B is compact, it follows that there is a neighbourhood $N(B)$ of B in Y such that for all $\epsilon > 0$ sufficiently small and for all $(v, y) \in [0, 1] \times N(B)$,

$$f^v(y) \in \mathrm{Conv}(\mathcal{R}_y, \gamma(y)).$$

For $\epsilon > 0$ sufficiently small, $j^\perp \ell^v(U) \subset N(B)$. Consequently, setting $y = j^\perp \ell^v(x)$ $(= b_v(x)) \in N(B)$, for $\epsilon > 0$ sufficiently small and for all $(v, x) \in [0, 1] \times U$,

$$f^v(b_v(x)) = (j^\perp \ell^v(x), \partial_t^r \ell^v(x)) = j^r \ell^v(x) \in \mathrm{Conv}(\mathcal{R}_{b_v(x)}, \gamma(b_v(x))). \qquad (7.15)$$

Employing the definition (7.14), for all $x \in U$ $\alpha_t(x) = P_{\mu(x)t}(b_t(x)) \in \mathcal{R}_{b_t(x)}$. Since $P_0 = \gamma$ it follows that the homotopy, $s \mapsto P_{\mu(x)st}(b_t(x))$, $s \in [0, 1]$, is a path in $\mathcal{R}_{b_t(x)}$ that connects $P_0(b_t(x)) = \gamma(b_t(x))$ to $\alpha_t(x)$. Consequently, employing (7.15), for all $(t, x) \in [0, 1] \times U$,

$$j^r \ell^t(x) \in \mathrm{Conv}(\mathcal{R}_{b_t(x)}, \alpha_t(x)),$$

which proves the proposition, and the proof of the local extension lemma is complete. $\qquad \square$

Returning to the proof of the h-stability theorem, employing inductively the Local Extension Lemma 7.5, the proof procedure of Chapter IV applies to the data: (i) $\varphi = (\alpha, j^r f_0) \in \Gamma(\mathrm{Conv}_\tau \mathcal{R})$; (ii) the closed set $K = \{x \in V \mid \rho \circ \alpha(x) = j^r f_0(x)\}$; (iii) the locally finite covers $(W_i)_{i \geq 1}$, $(U_i)_{i \geq 1}$ of $\overline{V \setminus K_1}$, to construct a sequence of holonomic homotopies, $(\varphi_t^i = (\alpha_t^i, j^r \ell_t^i) \in \Gamma(\mathrm{Conv}_\tau \mathcal{R}))_{i \geq 0}$, $t \in [0, 1]$, $\varphi_t^0 = \varphi$ (constant homotopy), ℓ_t^i is smooth on $V \setminus K$ for all i, t, such that the following properties are satisfied for all $i \geq 1$:

(p_1) For all $t \in [0, 1]$, $\varphi_t^i = \varphi_t^{i-1}$ on $\mathfrak{Op}\, \overline{V \setminus U_i}$; $\varphi_t^i = \varphi$ on $\mathfrak{Op}\, K$ (constant homotopy).

(p_2) $\varphi_0^i = \varphi_1^{i-1}$. Thus $\alpha_0^i = \alpha_1^{i-1}$; $\ell_0^i = \ell_1^{i-1}$.

(p_3) For all $t \in [0, 1]$, the image $j^r \ell_t^i(V) \subset \mathcal{N}$.

(p_4) For all $t \in [0, 1]$, $\varphi_t^i = \varphi_1^{i-1}$ on $\bigcup_0^{i-1} \mathfrak{Op}\, W_j$, a constant homotopy $(W_0 = \emptyset)$.

(p_5) $\rho \circ \alpha_1^i = j^r \ell_1^i$ on $\bigcup_0^i \mathfrak{Op}\, W_i$ (thus $\rho \circ \alpha_1^i = j^r \ell_1^i$ on $\mathfrak{Op}\, W_i \subset U_i$).

Concatenating the homotopies φ_t^i, $i \geq 0$, one obtains a homotopy of holonomic sections $\varphi_t = (\alpha_t, j^r f_t) \in \Gamma(\mathrm{Conv}_\tau \mathcal{R})$, $\varphi_0 = \varphi$, such that: (i) $\rho \circ \alpha_1 = j^r f_1$ (α_1 is holonomic); (ii) $\varphi_t = \varphi$ on $\mathfrak{Op}\, K$ (constant homotopy); for all $t \in [0, 1]$ $j^\perp f_t(V) \subset \mathcal{N}$, and hence the proof of the h-stability theorem is complete. $\qquad \square$

7.2.2. Approximations. Let $p: E \to B$ be an affine $\mathbf{R}^q$-bundle and let $\rho: \mathcal{R} \to E$ be a microfibration, where $(\mathcal{R}, d)$ is a metric space. Recall V §5.2.2 the convex hull $\mathrm{Conv}(\mathcal{R}_b, a)$ where $a \in \mathcal{R}_b$. For each $\delta > 0$ and each $(b, a) \in B \times \mathcal{R}_b$ let

$\text{Conv}^\delta(\mathcal{R}_b, a)$ denote the convex hull (in the fiber E_b) of the ρ-image of the path component of a in $\mathcal{R}_b \cap B(a; \delta)$.

Suppose $f \in \Gamma(E)$, $\beta \in \Gamma(\mathcal{R})$ satisfy the convex hull condition: for all $b \in B$, $f(b) \in \text{Conv}^\delta(\mathcal{R}_b, \beta(b))$. Then V Proposition 5.8 generalizes in the obvious way to prove the existence of a C-structure (h, H) with respect to f, β such that in addition, for all $(s, t, b) \in [0, 1]^2 \times B$, $d(\beta(b), H(s, t, b)) < \delta$. Since $\dim V = n$, it follows from dimension theory, Hurewicz and Wallman [24], that there is an open cover $(U_i)_{i \geq 1}$ of V for which each open set in the cover meets non-trivially at most $(n+1)$ open sets of the cover. The following refinement of the h-stability theorem is employed in Chapter IX to solve closed relations in jet spaces; details are left to the reader.

Complement 7.16 Approximations. *Suppose in addition to the hypotheses of the h-stability theorem, $(\mathcal{R}, d)$ is a metric space and there is a $\delta > 0$ such that for all $x \in V$,*

$$j^r f_0(x) \in \text{Conv}^\delta(\mathcal{R}_{b(x)}, \alpha(x)); \quad b(x) = j^\perp f_0(x) \in X^\perp.$$

The homotopy $H_t = (\alpha_t, j^r g_t) \in \Gamma(\text{Conv}_\tau \mathcal{R})$, $t \in [0, 1]$, in the conclusions of the theorem can be chosen so that in addition for all $(t, x) \in [0, 1] \times V$, $d(\alpha_t(x), \alpha(x)) < (n+1)\delta$.

7.2.4. We discuss below a different relative version of the h-Stability Theorem 7.2 that is closer to the approach in Gromov [18]. Let $\rho: \mathcal{R} \to X^{(r)}$ be a microfibration. Let $U \subset V$ be open and let $\tau \subset T(U)$ be a codimension 1 hyperplane tangent field on U. In particular τ need not extend continuously to a tangent field on V. Associated to τ is the affine $\mathbf{R}^q$-bundle $p_\perp^r: X_U^{(r)} \to X_U^\perp$.

We extend the definition of $\text{Conv}_\tau \mathcal{R} \subset \mathcal{R} \times X^{(r)}$ (cf. §7.1.2) to the case that $\tau \subset T(U)$: $(a, z) \in \text{Conv}_\tau \mathcal{R}$ if and only if: (i) $s(z) \in V \setminus U$ and $\rho(a) = z$, or (ii) $s(z) \in U$ and $z \in \text{Conv}(\mathcal{R}_b, a)$, where $b = p_\perp^r(z) \in X_U^\perp$. There is a natural inclusion map $i: \mathcal{R} \to \text{Conv}_\tau \mathcal{R}$, $a \mapsto (a, \rho(a))$, and a natural projection map $\rho^*: \text{Conv}_\tau \mathcal{R} \to X^{(r)}$, $(a, z) \mapsto z$.

A section $(\alpha, g) \in \Gamma(\text{Conv}_\tau \mathcal{R})$ consists of continuous sections $\alpha \in \Gamma(\mathcal{R})$, $g \in \Gamma(X^{(r)})$, such that: (i) for all $x \in V \setminus U$, $\rho \circ \alpha(x) = g(x)$; (ii) for all $x \in U$, $g(x) \in \text{Conv}(\mathcal{R}_{b(x)}, \alpha(x))$, where $b(x) = j^\perp g(x) \in X^\perp$. A section $(\alpha, g) \in \Gamma(\text{Conv}_\tau \mathcal{R})$ is holonomic if $g = j^r f$ for a (unique) section $f \in \Gamma^r(X)$. In particular $\rho \circ \alpha(x) = j^r f(x)$ for all $x \in V \setminus U$.

Theorem 7.17. *Let $U \subset V$ be open and let $\tau \subset T(U)$ be a codimension 1 tangent field on U which is smooth and integrable. Let $\varphi = (\alpha, j^r f_0) \in \Gamma(\text{Conv}_\tau \mathcal{R})$ be a holonomic section. Let $\mathcal{N}$ be a neighbourhood of the image $j^\perp f_0(U)$ in $X_U^\perp$. There is a homotopy of holonomic sections $H: [0, 1] \to \Gamma(\text{Conv}_\tau \mathcal{R})$, $H_t = (\alpha_t, j^r g_t)$, $t \in [0, 1]$, such that:*

(i) $H_0 = \varphi$; $H_1 \in \Gamma(\mathcal{R})$, *hence* α_1 *is holonomic:* $\rho \circ \alpha_1 = j^r g_1 \in \Gamma(X^{(r)})$.

(ii) *For all* $t \in [0, 1]$, *the image* $j^\perp g_t(U) \subset \mathcal{N}$.

(iii) *For all* $(t, x) \in [0, 1] \times (V \setminus U)$, $H_t(x) = \varphi(x)$ *(constant homotopy)*.

Briefly, to prove the theorem, up to a small homotopy rel $V \setminus U$, one may assume that f_0 is smooth on U and that $\rho \circ \alpha$ is smooth on U (cf. Lemmas 7.3, 7.4). One then proceeds as in the h-Stability Theorem 7.2 (non-relative case) with $V = U$. Details are left to the reader.

CHAPTER 8

AMPLE RELATIONS

§1. Short Sections

In this chapter the h-stability theorem VII Theorem 7.2 is applied to prove the C^{r-1}-dense h-principle Theorem 8.12 for ample relations $\rho\colon \mathcal{R} \to X^{(r)}$, a central result of the general theory. Recall that Theorem 7.2 is proved in the strong form i.e. the asserted homotopy is holonomic at each stage. This strong form of h-stability is exploited in §8.1.2 to develop a theory of short sections, which provides a natural context for studying non-ample relations. A principal result is the C^{r-1}-density h-principle Theorem 8.4 for short maps. A strong Weak Homotopy Equivalence Theorem 8.15 is proved, and in §4 we generalize to h-principles relative to an open set $\mathcal{R}_0 \subset X^{(r)}$. The chapter includes several examples covered by the main theorems.

8.1.1. Iterated Convex Hull Extensions. Let $U \subset V$ be open and let $\tau_i \subset T(U)$, $1 \leq i \leq m$, $m \geq 2$, be codimension 1 tangent hyperplane fields on U. Let $p_\perp^r\colon X_U^{(r)} \to X_i^\perp$ be the affine $\mathbf{R}^q$-bundle associated to the hyperplane field τ_i on U, $1 \leq i \leq m$. Let $\rho\colon \mathcal{R} \to X_U^{(r)}$ be continuous (not necessarily a microfibration). For example $\mathcal{R}$ is the restriction over $X_U^{(r)}$ of a relation over $X^{(r)}$. Employing VII §7.1.2, there is an iterated sequence of convex hull relations over $X_U^{(r)}$ defined inductively as follows with respect to the sequence of affine $\mathbf{R}^q$-bundles $p_\perp^r\colon X_U^{(r)} \to X_j^\perp$ associated to $\tau_j \subset T(V)$ ($\mathrm{Conv}_0(\mathcal{R}) = \mathcal{R}$; $\rho_0 = \rho\colon \mathcal{R} \to X_U^{(r)}$):

$$\mathrm{Conv}_j(\mathcal{R}) = \mathrm{Conv}_{\tau_j}(\mathrm{Conv}_{j-1}(\mathcal{R})) \quad 1 \leq j \leq m. \tag{8.1}$$

Thus $\mathrm{Conv}_j(\mathcal{R}) \subset \mathcal{R} \times X_U^{(r)} \times \cdots \times X_U^{(r)}$ ($(j+1)$ factors). There is a natural projection $\rho_j\colon \mathrm{Conv}_j(\mathcal{R}) \to X_U^{(r)}$, $(a, z_1, z_2, \ldots z_j) \mapsto z_j$ (projection onto the last factor), $1 \leq j \leq m$. There is also a natural inclusion $i\colon \mathrm{Conv}_{j-1}(\mathcal{R}) \to \mathrm{Conv}_j(\mathcal{R})$, $a \mapsto (a, \rho_{j-1}(a))$, $1 \leq j \leq m$. The composition of these inclusion maps induces an identification by inclusion $\mathrm{Conv}_j(\mathcal{R}) \subset \mathrm{Conv}_k(\mathcal{R})$, $0 \leq j \leq k \leq m$. In particular, $\mathcal{R} \subset \mathrm{Conv}_j(\mathcal{R})$, $1 \leq j \leq m$, and there is a sequence of commutative

D. Spring, *Convex Integration Theory: Solutions to the h-principle in geometry and topology*, Modern Birkhäuser classics, DOI 10.1007/978-3-0348-0060-0_8, © Springer Basel AG 2010

diagrams, $1 \le j \le m$:

$$
\begin{array}{ccc}
\mathcal{R} & \xrightarrow{\ i\ } & \mathrm{Conv}_j(\mathcal{R}) \\
{\scriptstyle p^r_{r-1}\circ\rho}\Big\downarrow & & \Big\downarrow{\scriptstyle p^r_{r-1}\circ\rho_j} \\
X^{(r-1)} & =\!\!=\!\!= & X^{(r-1)}
\end{array}
\tag{8.2}
$$

The geometric properties of the convex hull extensions $\mathrm{Conv}_j(\mathcal{R})$, $1 \le j \le m$, are somewhat subtle and are deferred to §8.1.3, where illustrative examples are given of iterated convex hull extensions that occur in topology and in PDE theory. In this section we develop the formal properties of convex hull extensions, based on the h-stability theorem, VII Theorem 7.2.

Remark 8.1. As a preliminary remark to these geometrical properties of $\mathrm{Conv}_m(\mathcal{R})$, let $y_{m+1} = (a, z_1, z_2, \ldots, z_m) \in \mathrm{Conv}_m(\mathcal{R})$, where $a \in \mathcal{R}$, $z_i \in X_w^{(r)}$, $w = p^r_{r-1} \circ \rho(a) \in X^{(r-1)}$, $1 \le i \le m$. Thus the r-jets $\rho(a), z_i$, $1 \le i \le m$, all lie in the same affine fiber $X_w^{(r)}$ of the bundle $X^{(r)} \to X^{(r-1)}$. Furthermore, the r-jets $\rho(a), z_1$, respectively z_{i-1}, z_i, both lie in a principal subspace (an $\mathbf{R}^q$-fiber) of the bundle $p^r_\perp \colon X_U^{(r)} \to X_1^\perp$ associated to the tangent hyperplane field τ_1, respectively both lie in a principal subspace of the bundle $p^r_\perp \colon X_U^{(r)} \to X_i^\perp$ associated to the tangent hyperplane field τ_i, $2 \le i \le m$. In particular, employing VI §6.2.4, there is a principal path in $X_w^{(r)}$ connecting z_m to $\rho(a)$. Note that $(a, z_1, z_2, \cdots, z_{m-1}) \in \mathrm{Conv}_{m-1}(\mathcal{R})$, and $\rho_m(y_{m+1}) = z_m \in X_w^{(r)}$. Further geometrical details of (8.1) are provided in §8.1.3.

Since the relation $\rho_j \colon \mathrm{Conv}_j(\mathcal{R}) \to X_U^{(r)}$ is the convex hull extension of the relation $\rho_{j-1} \colon \mathrm{Conv}_{j-1}(\mathcal{R}) \to X_U^{(r)}$ with respect to the tangent hyperplane field τ_j, $1 \le j \le m$, employing VII Lemma 7.1, one obtains by induction the following result.

Lemma 8.2. Let $\rho \colon \mathcal{R} \to X_U^{(r)}$ be a microfibration. Then $\rho_j \colon \mathrm{Conv}_j(\mathcal{R}) \to X_U^{(r)}$ is a microfibration, the iterated jth convex hull extension of $\mathcal{R}$, $1 \le j \le m$.

A section $\varphi \in \Gamma(\mathrm{Conv}_m(\mathcal{R}))$ consists of an m-tuple of sections, $\varphi = (\alpha, g_1, g_2, \ldots, g_m)$, $\alpha \in \Gamma(\mathcal{R})$; $g_j \in \Gamma(X_U^{(r)})$, $1 \le j \le m$, such that for each $x \in U$,

$$
\varphi(x) = (\alpha(x), g_1(x), \ldots g_m(x)) \in \mathrm{Conv}_m(\mathcal{R}).
$$

Employing Remark 8.1, for each $x \in U$ $\rho \circ \alpha(x), g_i(x) \in X_{w(x)}^{(r)}$, $1 \le i \le m$, where $w(x) = p^r_{r-1} \circ \rho \circ \alpha(x) \in X^{(r-1)}$ i.e. all these r-jets lie in the same affine fiber $X_{w(x)}^{(r)}$. In particular if $g = p^r_{r-1} \circ \rho \circ \alpha \in \Gamma(X_U^{(r-1)})$, then $\alpha \in \Gamma(\mathcal{R})$ is a continuous lift of g into $\mathcal{R}$, and $g_j \in \Gamma(X_U^{(r)})$ is a continuous lift of g into $X_U^{(r)}$, $1 \le j \le m$.

Since $\rho_m \circ \varphi = g_m \in \Gamma(X_U^{(r)})$, it follows that $\varphi \in \Gamma(\mathrm{Conv}_m(\mathcal{R}))$ is holonomic if and only if there is a C^r-section $h \in \Gamma^r(X_U)$ such that $g_m = j^r h$: $\rho_m \circ \varphi = j^r h \in \Gamma(X_U^{(r)})$. Furthermore, with respect to the identification above by inclusion $\mathcal{R} \subset \mathrm{Conv}_m(\mathcal{R})$, $\varphi \in \Gamma(\mathcal{R})$ if and only if the m-tuple $\varphi = (\alpha, \rho \circ \alpha, \dots, \rho \circ \alpha)$. In particular, $\varphi \in \Gamma(\mathcal{R})$ is holonomic if and only if α is holonomic on U: $\rho_m \circ \varphi = \rho \circ \alpha = j^r h$.

8.1.2. Short Sections. We assume throughout this section that $\rho \colon \mathcal{R} \to X_U^{(r)}$ is a microfibration and that the tangent hyperplane fields on U are smooth and integrable. The pair $(\alpha, f) \in \Gamma(\mathcal{R}) \times \Gamma^r(X_U)$ is a *formal solution* to the relation $\mathcal{R}$ if $p_{r-1}^r \circ \rho \circ \alpha = j^{r-1} f \in \Gamma(X_U^{(r-1)})$. Hence $\alpha \in \Gamma(\mathcal{R})$ is a continuous lift of $j^{r-1} f$ into $\mathcal{R}$ i.e. α solves the relation $\mathcal{R}$ up to jets of order $(r-1)$. Employing iterated convex hull extensions on U (cf. §8.1.1), a C^r-section $f \in \Gamma^r(X_U)$ is defined to be *short* if for some $m \geq 1$ there is a holonomic section $\varphi \in \Gamma(\mathrm{Conv}_m(\mathcal{R}))$ such that $\rho_m \circ \varphi = j^r f \in \Gamma(X_U^{(r)})$. Thus $\varphi = (\alpha, g_1, \dots, g_{m-1}, j^r f)$, where $\alpha \in \Gamma(\mathcal{R})$; in particular $p_{r-1}^r \circ \rho \circ \alpha = j^{r-1} f \in \Gamma(X^{(r-1)})$ i.e. the pair (α, f) is a *short formal solution* to $\mathcal{R}$. The problem of this chapter is to change a formal solution (α, f) through a homotopy of formal solutions into a solution (β, g) of $\mathcal{R}$ i.e. $\rho \circ \beta = j^r g$. In case f is short, this is accomplished in Theorem 8.4 below.

Lemma 8.3. *The space* $\mathcal{S} = \{f \in \Gamma^r(X_U) \mid f \text{ is short}\}$ *is open in* $\Gamma^r(X_U)$ *in the fine* C^r-*topology.*

Proof. Let $f \in \mathcal{S}$ be short: $j^r f = \rho_m \circ \varphi$, where $\varphi \in \Gamma(\mathrm{Conv}_m(\mathcal{R}))$. Applying Lemma 8.2, $\rho_m \colon \mathrm{Conv}_m(\mathcal{R}) \to X_U^{(r)}$ is a microfibration. Employing V Corollary 5.3, there is a neighbourhood T of the image $j^r f(U) \subset X_U^{(r)}$ and a continuous lift $\nu \colon T \to \mathrm{Conv}_m(\mathcal{R})$, $\rho_m \circ \nu = \mathrm{id}_T$; $\nu \circ \rho_m \circ \varphi = \varphi$. The set $\mathcal{N}_f = \{g \in \Gamma^r(X_U) \mid j^r g(U) \subset T\}$ is a neighbourhood of f in the fine C^r-topology. For each $g \in \mathcal{N}_f$ let $\varphi_g = \nu \circ j^r g \in \mathrm{Conv}_m(\mathcal{R})$. It follows that $\rho_m \circ \varphi_g = j^r g$ i.e. g is short and the lemma is proved. $\qquad\square$

Theorem 8.4. *Let* $f \in \mathcal{S}$ *be short with respect to* $\varphi \in \Gamma(\mathrm{Conv}_m(\mathcal{R}))$:

$$\varphi = (\alpha, g_1, g_2, \cdots, g_{m-1}, j^r f); \quad \alpha \in \Gamma(\mathcal{R}); g_i \in \Gamma(X_U^{(r)}), 1 \leq i \leq m-1.$$

Let $\mathcal{N}$ *be a neighbourhood of the image* $j^{r-1} f(U) \subset X_U^{(r-1)}$. *There is a homotopy of holonomic sections,*

$$\varphi_t = (\alpha_t, g_1^t, g_2^t, \dots, g_{m-1}^t, j^r f_t) \in \Gamma(\mathrm{Conv}_m(\mathcal{R})), \quad t \in [0, 1],$$

such that the following properties obtain:

(i) $\varphi_0 = \varphi$; $\varphi_1 \in \Gamma(\mathcal{R})$, hence α_1 is holonomic: $\rho \circ \alpha_1 = j^r f_1$. Thus the section φ is continuously deformed through holonomic sections of $\mathrm{Conv}_m(\mathcal{R})$ to a holonomic section of $\mathcal{R}$. In particular, employing the homotopy α_t, $t \in [0,1]$, the h-principle holds for $\alpha \in \Gamma(\mathcal{R})$.

(ii) (C^{r-1}-dense principle) : For all $t \in [0,1]$, the image $j^{r-1} f_t(U) \subset \mathcal{N}$.

(iii) (Relative Theorem) : Let $K = \{x \in U \mid \rho \circ \alpha(x) = g_i(x) = j^r f(x), 1 \le i \le m-1\}$. Then one can choose the homotopy φ_t so that for all $t \in [0,1]$, $\varphi_t = \varphi$ on K (constant homotopy on K).

Proof. Applying the h-stability theorem VII Theorem 7.2 to the data consisting of the microfibration $\rho_{m-1}\colon \mathrm{Conv}_{m-1}(\mathcal{R}) \to X_U^{(r)}$ and the section φ, there is a homotopy rel K of holonomic sections $H_t \in \Gamma(\mathrm{Conv}_m(\mathcal{R}))$, $t \in [0,1]$, such that: (i) $H_1 \in \Gamma(\mathrm{Conv}_{m-1}(\mathcal{R}))$; (ii) for all $t \in [0,1]$, $p_{r-1}^r \circ \rho_m \circ H_t(U) \subset \mathcal{N}$ (in fact, the homotopy H_t satisfies a $C^\perp$-density result in $X_m^\perp$); (iii) for all $t \in [0,1]$, $H_t = \varphi$ on K (constant homotopy on K). Since $H_1 \in \Gamma(\mathrm{Conv}_{m-1}(\mathcal{R}))$ is holonomic, the h-stability theorem again applies, and an obvious induction proves the theorem. $\qquad\square$

8.1.3. In order to apply Theorem 8.4 suitable geometric criteria are required in order to recognize short sections. For this purpose a detailed geometric description of $\mathrm{Conv}_m(\mathcal{R})$ is provided in Proposition 8.5 below. The geometric content of iterated convex hull extensions serves also for the analysis of general non-ample relations over jet spaces, in particular for the PDE theory in Chapter IX. It is somewhat remarkable however that in the case of ample relations over jet spaces this rather technical geometric analysis can be avoided. Indeed, following Gromov [18], the theory of iterated convex hull extensions proceeds algebraically, based on Gromov's observation that the ampleness of a relation extends to the ampleness of all of its iterated convex hull extensions. This basic fact is proved in Corollary 8.10. Indeed, Gromov [18] does not discuss the detailed geometry of iterated convex hull extensions. Consequently his main results (for the most part) are stated in terms of ample relations only, which was sufficient for his purposes. Iterated convex hull extensions are studied from three perspectives: analytically in Proposition 8.5; algebraically in Lemma 8.9; affine geometrically in §8.4.2 (Lemma 8.39).

Employing (8.1) an $(m+1)$-tuple $y_{m+1} = (a, z_1, z_2, \ldots, z_m) \in \mathrm{Conv}_m(\mathcal{R})$ if and only if,

$$z_m \in \mathrm{Conv}(R_b, y_m); \quad b = p_\perp^r(z_m) \in X_m^\perp,$$

where $R_b = (p_\perp^r \circ \rho_{m-1})^{-1}(b)$ is the fiber in $\mathrm{Conv}_{m-1}(\mathcal{R})$ over $b \in X_m^\perp$, and $y_m = (a, z_1, z_2, \ldots, z_{m-1}) \in R_b$. In particular, $z_{m-1}(= \rho_{m-1}(y_m))$, z_m lie in the principal subspace over b in the bundle $p_\perp^r \colon X_U^{(r)} \to X_m^\perp$.

Hence there is a path $y_m^t = (a^t, z_1^t, z_2^t, \ldots, z_{m-1}^t)$ in R_b, $y_m^0 = y_m$, $t \in [0,1]$, such that the path of r-jets $z_{m-1}^t \ (= \rho_{m-1}(y_m^t))$, $t \in [0,1]$, surrounds z_m in the $\mathbf{R}^q$-fiber over b with respect to the bundle $p_\perp^r : X_U^{(r)} \to X_m^\perp$. These considerations motivate the following iterative process for characterizing $\mathrm{Conv}_m(\mathcal{R})$.

Proposition 8.5. *Let* $y = (a, z_1, \ldots z_m) \in \mathcal{R} \times \prod_1^m X_U^{(r)}$. *The element* $y \in \mathrm{Conv}_m(\mathcal{R})$ *if and only if there is a sequence of j-parameter families of paths,* $T(j) = (t_1, \ldots, t_j) \in [0,1]^j$ *(continuous in each variable, not necessarily jointly continuous),*

$$y_j^{T(m-j)} = \left(a^{T(m-j)}, z_1^{T(m-j)}, \ldots, z_j^{T(m-j)}\right)$$
$$= \left(c^{T(m-j)}, z_j^{T(m-j)}\right) \in \mathrm{Conv}_j(\mathcal{R}), \quad 0 \le j \le m,$$

where $y_m^{T(0)} = y$; $y_0^{T(m)} = a^{T(m)} \in \mathcal{R}$, *and such that the following properties are satisfied for all* $T(j) \in [0,1]^j$, $1 \le j \le m$:

(p₁) $y_j^{0,\ldots,0} = (a, z_1, \ldots z_j)$, $1 \le j \le m$.

(p₂) $y_{j-1}^{T(m-j),0} = c^{T(m-j)}$, $1 \le j \le m$ ($y_{m-1}^0 = c = (a, z_1, \ldots, z_{m-1})$ *in case* $j = m$).

(p₃) $\rho_j(y_j^{T(m-j)}) = z_j^{T(m-j)} \in X_U^{(r)}$ *(the last component)*, $1 \le j \le m$.

(p₄) *The r-jets* $\rho(a^{T(m)}), z_1^{T(m-1)}$ *both lie in a principal subspace over the base point* $b^{T(m-1)} = p_\perp^r(z_1^{T(m-1)}) \in X_1^\perp$ *of the bundle* $p_\perp^r : X_U^{(r)} \to X_1^\perp$. *The path* $\rho(a^{T(m-1),s})$, $s \in [0,1]$, *surrounds* $z_1^{T(m-1)}$.

(p₅) *Successive r-jets* $z_j^{T(m-j)}, z_{j+1}^{T(m-j-1)}$ *both lie in a principal subspace over the base point* $b^{T(m-j-1)} = p_\perp^r(z_{j+1}^{T(m-j-1)}) \in X_{j+1}^\perp$ *of the bundle* $p_\perp^r : X_U^{(r)} \to X_{j+1}^\perp$, $1 \le j \le m-1$. *The path* $z_j^{T(m-j-1),s}$, $s \in [0,1]$, *surrounds* $z_{j+1}^{T(m-j-1)}$, $1 \le j \le m-1$.

Proof. The element $y = (c, z_m) \in \mathrm{Conv}_m(\mathcal{R})$ if and only if there is path $y_{m-1}^s \in \mathrm{Conv}_{m-1}(\mathcal{R})$, $y_{m-1}^0 = c$, $s \in [0,1]$, of the form,

$$y_{m-1}^s = (a^s, z_1^s, \ldots, z_{m-1}^s)$$
$$= (c^s, z_{m-1}^s) \in \mathrm{Conv}_{m-1}(\mathcal{R}),$$

such that the projected path $\rho_{m-1}(y_{m-1}^s) = z_{m-1}^s$, $s \in [0,1]$, surrounds the r-jet z_m in the principal subspace in the bundle $p_\perp^r : X^{(r)} \to X_m^\perp$ over the base point $b = p_\perp^r(z_m) \in X_m^\perp$ associated to τ_m.

Similarly, the path $y_{m-1}^s = (c^s, z_{m-1}^s) \in \mathrm{Conv}_{m-1}(\mathcal{R})$ if and only if for each $s \in [0,1]$, there is a path $y_{m-2}^{s,t} \in \mathrm{Conv}_{m-2}(\mathcal{R})$, $y_{m-2}^{s,0} = c^s$, $t \in [0,1]$, of the form,

$$y_{m-2}^{s,t} = (a^{s,t}, z_1^{s,t}, \ldots, z_{m-2}^{s,t})$$
$$= (c^{s,t}, z_{m-2}^{s,t}) \in \mathrm{Conv}_{m-2}(\mathcal{R}),$$

such that the projected path $\rho_{m-2}(y_{m-2}^{s,t}) = z_{m-2}^{s,t}$, $t \in [0,1]$, surrounds the r-jet z_{m-1}^s in the principal subspace in the bundle $p_\perp^r : X^{(r)} \to X_{m-1}^\perp$ over the base point $b^s = p_\perp^r(z_{m-1}^s) \in X_{m-1}^\perp$ associated to τ_{m-1}. An obvious induction proves the proposition.   □

Informally the geometrical content of $\mathrm{Conv}_m(\mathcal{R})$ is as follows. $y \in \mathrm{Conv}_m(\mathcal{R})$ comes equipped with a path y_{m-1}^s, $s \in [0,1]$, in $\mathrm{Conv}_{m-1}(\mathcal{R})$ whose projection z_{m-1}^s surrounds the r-jet z_m in a principal subspace of the bundle $p_\perp^r : X_U^{(r)} \to X_m^\perp$; for each point on this path, there is a path $y_{m-2}^{s,t}$, $t \in [0,1]$, in $\mathrm{Conv}_{m-2}(\mathcal{R})$ whose projection $z_{m-2}^{s,t}$ surrounds the r-jet z_{m-1}^s in a principal subspace of the bundle $p_\perp^r : X_U^{(r)} \to X_{m-1}^\perp$, and so on for m steps to obtain: for each point on a $(m-1)$-parameter family of paths $y_1^{t_1,\ldots,t_{m-1}}$ in $\mathrm{Conv}_1(\mathcal{R})$ there is a path $a^{t_1,\ldots,t_m}$ in $\mathcal{R}$, $t_m \in [0,1]$, whose projection $\rho(a^{t_1,\ldots,t_m})$ surrounds the r-jet $z_1^{t_1,\ldots,t_m}$ in a principal subspace of the bundle $p_\perp^r : X_U^{(r)} \to X_1^\perp$.

The construction of C-structures (V Proposition 5.8) restores joint continuity of the variables in the base V and successive parametrized families of surrounding paths i.e. the separate continuity in Proposition 8.5 poses no technical problems.

8.1.3. Examples. We illustrate the geometry of iterated convex hull extensions with an example from non-linear PDE theory (Chapter IX) and an example from local immersion theory.

PDE Example. Let $U \subset \mathbf{R}^2$ be open and $f = (f_1, f_2) \in C^1(U, \mathbf{R}^2)$ such that $((x,y)$ are coordinates in $U)$:

$$(\partial_x f_1)^2 + (\partial_x f_2)^2 = A(j^0 f) \tag{8.3}$$
$$(\partial_y f_1)^2 + (\partial_y f_2)^2 = B(j^0 f, \partial_x f). \tag{8.4}$$

where A, B are continuous positive functions. Let $\tau_1 = \partial_x$, $\tau_2 = \partial_y$. Thus τ_1, τ_2 are codimension 1 tangent fields on U. Let $p : X_U \times \mathbf{R}^2 \to U$ be the product $\mathbf{R}^2$-bundle over U. We identify $\Gamma^r(X) \equiv C^r(U, \mathbf{R}^2)$.

Let $\mathcal{R} \subset X_U^{(1)} = U \times \mathbf{R}^2 \times \mathbf{R}^2 \times \mathbf{R}^2$ be the closed relation defined by the equations (8.3), (8.4):

$$(x, y, u, v, p, q, r, w) \in \mathcal{R} \Leftrightarrow \begin{cases} r^2 + w^2 = A(x,y,u,v) \\ p^2 + q^2 = B(x,y,u,v,r,w) \end{cases}$$

Here $(p, q) \in \mathbf{R}^2$ corresponds to the derivative ∂_y; $(r, w) \in \mathbf{R}^2$ corresponds to the derivative ∂_x. Let $g \in C^1(U, \mathbf{R}^2)$ and let a triple $\varphi = (\alpha, z_1, z_2) \in \mathcal{R} \times X_U^{(1)} \times X_U^{(1)}$ be defined as follows at $(x, y) \in U$: $\alpha = (j^0 g(x, y), p, q, r, w) \in \mathcal{R}$; $z_1 = (j^0 g(x, y), \partial_y g(x, y), r, w) \in X_U^{(1)}$; $z_2 = j^1 g(x, y) = (j^0 g(x, y), \partial_y g(x, y), \partial_x g(x, y)) \in X_U^{(1)}$.

Note that z_1, z_2 lie in a principal subspace of the bundle $X_U^{(1)} \to X_1^{\perp}$ with respect to the tangent field τ_1; α, z_1 lie in a principal subspace of the bundle $X_U^{(1)} \to X_2^{\perp}$ with respect to the tangent field τ_2. Suppose there is a 2-parameter family of paths, $(\alpha^{s,t}, z_1^s, z_2)$, $\alpha^{0,0} = \alpha$, $z_1^0 = z_1$, $(s, t) \in [0, 1]^2$,

$$\alpha^{s,t} = (j^0 g(x, y), p^{s,t}, q^{s,t}, r^s, w^s) \in \mathcal{R}$$

$$z_1^s = (j^0 g(x, y), \partial_y g(x, y), r^s, w^s) \in X_U^{(1)}$$

such that the following properties are satisfied:

(i) The path of 1-jets z_1^s, $s \in [0, 1]$, surrounds z_2 i.e. the path $(r^s, w^s) \in \mathbf{R}^2$, $s \in [0, 1]$, surrounds $\partial_x g(x, y)$ (in a principal subspace of the bundle $X_U^{(1)} \to X_1^{\perp}$).

(ii) For each $s \in [0, 1]$, the path of 1-jets $\alpha^{s,t}$, $t \in [0, 1]$, surrounds z_1^s i.e. the path $(p^{s,t}, q^{s,t}) \in \mathbf{R}^2$, $t \in [0, 1]$, surrounds $\partial_y g(x, y)$ (in a principal subspace of the bundle $X_U^{(1)} \to X_2^{\perp}$).

Let $\mathrm{Conv}_1(\mathcal{R}) = \mathrm{Conv}_{\tau_2}(\mathcal{R})$; $\mathrm{Conv}_2(\mathcal{R}) = \mathrm{Conv}_{\tau_1}(\mathrm{Conv}_{\tau_2}(\mathcal{R}))$. For all $(s, t) \in [0, 1]^2$, let $\varphi^{s,t} = \alpha^{s,t} \in \mathcal{R}$; $\varphi^s = (\alpha^{s,0}, z_1^s) \in \mathcal{R} \times X_U^{(1)}$. Employing (i), (ii), the hypotheses of Proposition 8.5 are easily verified with respect to the 2-parameter family $\varphi^{s,t}, \varphi^s, \varphi$. Consequently, under the hypotheses (i), (ii), for all $s \in [0, 1]$, $\varphi^s \in \mathrm{Conv}_1(\mathcal{R})$; $\varphi \in \mathrm{Conv}_2(\mathcal{R})$.

The restrictions imposed by the conditions (i), (ii) on the first derivatives $\partial_x g(x, y)$, $\partial_y g(x, y)$ determine the conditions for a short section $g \in C^1(U, \mathbf{R}^2)$. Explicitly $g = (g_1, g_2) \in \mathcal{S}$ (the space of short sections) if and only if for all $(x, y, s) \in U \times [0, 1]$ the following conditions obtain:

$$(\partial_x g_1(x, y))^2 + (\partial_x g_2(x, y))^2 \leq A(j^0(g(x, y)) \tag{8.5}$$

$$(\partial_y g_1(x, y))^2 + (\partial_y g_2(x, y))^2 \leq B(j^0 g(x, y), r^s, w^s). \tag{8.6}$$

Thus $g \in C^1(U, \mathbf{R}^2)$ is in $\mathcal{S}$ if and only if: there is a map $(r, w) \in C^0(U, \mathbf{R}^2)$, and for each $(x, y) \in U$ there is a path as above $(r^s(x, y), w^s(x, y)) \in \mathbf{R}^2$, $s \in [0, 1]$, such that $(r^s)^2 + (w^s)^2 = A(j^0 g)$ and conditions (8.5), (8.6) hold.

Since the constant maps from U to $\mathbf{R}^2$ satisfy the short section conditions (8.5), (8.6) it follows that $\mathcal{S}$ is non-empty. Let $g \in \mathcal{S}$ be a short section. There exists a holonomic section $\varphi = (\alpha, z_1, j^1 g) \in \Gamma(\mathrm{Conv}_2(\mathcal{R}))$ as above. The inclusion map $i \colon \mathcal{R} \to X_U^{(1)}$ is not a microfibration since $\mathcal{R}$ is a closed subspace whose interior is empty. However, passing to an open neighbourhood $\mathfrak{Op}\,\mathcal{R}$ in $X_U^{(1)}$, the inclusion is a microfibration. Applying Theorem 8.4, one obtains the following global existence and C^0-density theorem for approximate solutions to the system of PDEs (8.3), (8.4).

Theorem 8.6. *Let $g \in \mathcal{S}$ be a short section for the system* (8.3), (8.4), *and let $\varphi = (\alpha, z_1, j^1 g) \in \Gamma(\mathrm{Conv}_2(\mathcal{R}))$ as above. Let also $\mathfrak{Op}\,\mathcal{R}$ be a neighbourhood of $\mathcal{R}$ in $X_U^{(1)}$ and $\mathcal{N}$ be a neighbourhood of the image $j^0 g(U)$ in $X_U^{(0)}$. Then there is a homotopy of holonomic sections, $\varphi_t = (\alpha_t, z_1^t, j^1 g_t) \in \Gamma(\mathrm{Conv}_2(\mathfrak{Op}\,\mathcal{R}))$, $\alpha_t \in \Gamma(\mathfrak{Op}\,\mathcal{R})$, $t \in [0,1]$, such that the following properties obtain:*

(i) *$\varphi_0 = \varphi$; $\varphi_1 \in \Gamma(\mathfrak{Op}\,\mathcal{R}))$, hence α_1 is holonomic: $\alpha_1 = z_1^1 = j^1 g_1$. Thus, employing the homotopy α_t, $t \in [0,1]$, the h-principle holds for $\alpha \in \Gamma(\mathfrak{Op}\,\mathcal{R})$.*

(ii) *(C^0-dense principle) : For all $t \in [0,1]$, the image $j^0 g_t(U) \subset \mathcal{N}$.*

(iii) *(Relative Theorem) : Let $K = \{(x,y) \in U \mid \alpha(x,y) = j^1 g(x,y)\}$. Then one can choose the homotopy φ_t so that for all $t \in [0,1]$, $\varphi_t = \varphi$ on K (constant homotopy on K).*

The system (8.3), (8.4) is a particular case of triangular systems of PDEs that are studied in Chapter IX. Theorem 8.5 is the first step in an inductive process for constructing C^1-solutions to (8.3), (8.4). In Chapter IX a sequence of open "metaneighbourhoods" of $\mathcal{R}$ in $X_U^{(1)}$, the intersection of whose closures is $\mathcal{R}$, is employed to construct (as in Theorem 8.6) a sequence of approximate C^1-solutions which in the limit converges to a C^1-solution to the system (8.3), (8.4). The C^0-density principle also applies in the limit.

Immersion Example. Let $p \colon X_U = U \times \mathbf{R}^q \to U$ be the product $\mathbf{R}^q$-bundle over an open set $U \subset \mathbf{R}^n$, $1 \leq n \leq q$. We identify $\Gamma^r(X_U) \equiv C^r(U, \mathbf{R}^q)$. The manifold $X_U^{(1)}$ is a product,

$$X_U^{(1)} = U \times \prod_1^{n+1} \mathbf{R}^q.$$

With respect to coordinates $(u_1, u_2, \ldots, u_n) \in U$, a section $f \in \Gamma^1(X)$ induces the 1-jet section $j^1 f \in \Gamma(X_U^{(1)})$ ($\partial_i = \partial/\partial u_i$, $1 \leq i \leq n$):

$$j^1 f(x) = (x, f(x), \partial_1 f(x), \partial_2 f(x), \ldots, \partial_n f(x)) \in X_U^{(1)} \quad \text{for all } x \in U.$$

Let $\tau_i = \ker du_i \subset T(U)$, a codimension 1 integrable tangent hyperplane field on U, $1 \leq i \leq n$, and let $p_\perp^1 \colon X_U^{(1)} \to X_i^\perp$ be the product $\mathbf{R}^q$-bundle associated to the hyperplane field τ_i, $1 \leq i \leq n$. Thus $X_U^{(1)} = X_i^\perp \times \mathbf{R}^q$, where the $\mathbf{R}^q$-factor corresponds to the derivative ∂_i, $1 \leq i \leq n$.

Let $\mathcal{R} \subset X_U^{(1)}$ be the immersion relation that corresponds to the problem of immersing U in $\mathbf{R}^q$: $(x, y, \alpha_1, \alpha_2, \ldots, \alpha_n) \in \mathcal{R}$, $x \in U$ (the source), $y \in \mathbf{R}^q$ (the target), if and only if the vectors $\alpha_i \in \mathbf{R}^q$, $1 \leq i \leq n$, are linearly independent. Thus $\mathcal{R}$ is open in $X_U^{(1)}$ and the inclusion map $i \colon \mathcal{R} \to X_U^{(1)}$ is a microfibration.

In case $q > n$, we show that the immersion relation $\mathcal{R}$ is ample (cf. §8.2.1 for a more general discussion). For each i, $1 \leq i \leq n$, let R_b be the fiber in $\mathcal{R}$ over $b \in X_i^\perp$ with respect to the map $p_\perp^1 \colon \mathcal{R} \to X_i^\perp$ (thus $R_b \subset b \times \mathbf{R}^q \equiv \mathbf{R}^q$). $\mathcal{R}$ is ample in the following sense: for each $b \in X_i^\perp$, R_b is path connected and the convex hull of R_b (in $\mathbf{R}^q$) is the ambient space $\mathbf{R}^q$. Indeed, let $b = (x, y, \alpha_1, \ldots, \alpha_{i-1}, \alpha_{i+1}, \ldots, \alpha_n) \in X_i^\perp$. Thus,

$$(x, y, \alpha_1, \ldots \alpha_{i-1}, \beta, \alpha_{i+1}, \ldots, \alpha_n) \in R_b,$$

if and only if $\beta \in \mathbf{R}^q \setminus L$, where L is the $(n-1)$-dimensional subspace spanned by the remaining vectors α_j, $j \neq i$. In case $q > n$ $L \subset \mathbf{R}^q$ has codimension ≥ 2, from which it follows that $\mathcal{R}$ ample in the above sense.

With respect to the tangent hyperplane fields $\tau_i \subset T(U)$, employing (8.1), let $\mathrm{Conv}_1(\mathcal{R}) = \mathrm{Conv}_{\tau_1}(\mathcal{R})$; $\mathrm{Conv}_j(\mathcal{R}) = \mathrm{Conv}_{\tau_j}(\mathrm{Conv}_{j-1}(\mathcal{R}))$, $1 \leq j \leq n$. Let $g \in C^1(U, \mathbf{R}^q)$. For each $x \in U$ let $\varphi(x) = (\gamma(x), z_1(x), \ldots, z_n(x)) \in \mathcal{R} \times \prod_1^n X_U^{(1)}$ such that:

$$\gamma(x) = (j^0 g(x), \alpha_1, \alpha_2, \ldots, \alpha_n) \in \mathcal{R},$$
$$z_i(x) = (j^0 g(x), \partial_1 g(x), \ldots \partial_i g(x), \alpha_{i+1}, \ldots \alpha_n) \in X_U^{(1)}, \quad 1 \leq i \leq n.$$

In particular, $z_n(x) = j^1 g(x)$. Successive 1-jets $\gamma(x), z_1(x)$, respectively $z_i(x)$, $z_{i+1}(x)$, differ only in the ∂_1-factor, respectively in the ∂_{i+1}-th factor i.e. these successive pairs of 1-jets lie in a principal subspace of the bundle $X_U^{(1)} \to X_i^\perp$, $1 \leq i \leq n$.

In §8.2.1, employing the ampleness of $\mathcal{R}$, it follows formally from Corollary 8.10 that $\varphi \in \Gamma(\mathrm{Conv}_n(\mathcal{R}))$. We prove this fact directly by an explicit construction in order to illustrate both the intricacies of convex hull extensions and also how the geometry of convex hull extensions applies in the classical case of immersion theory.

Lemma 8.7. *Let $q > n$. For all $x \in U$ the element,*

$$\varphi(x) = (\gamma(x), z_1(x), \ldots, z_n(x)) \in \mathrm{Conv}_n(\mathcal{R}) \quad (z_n(x) = j^1 g(x)).$$

Proof. For each $x \in U$, let $\alpha_n^{t_1} \in \mathbf{R}^q$, $\alpha_n^0 = \alpha_n$, $t_1 \in [0,1]$, be a path that surrounds $\partial_n g(x)$ and such that for all $t_1 \in [0,1]$,

$$(j^0 g(x), \alpha_1, \alpha_2, \ldots, \alpha_{n-1}, \alpha_n^{t_1}) \in \mathcal{R}.$$

The existence of such paths $\alpha_n^{t_1}$ follows from the ampleness of $\mathcal{R}$. By downward induction, employing the ampleness of $\mathcal{R}$, for each $x \in U$ there is a parametrized family of paths $\alpha_i^{t_1, t_2, \ldots, t_{n-i+1}} \in \mathbf{R}^q$, $1 \leq i \leq n$, such that the following properties are satisfied:

(i) $\alpha_i^{t_1, \ldots, t_{n-i}, 0} = \alpha_i$; the path $\alpha_i^{t_1, \ldots, t_{n-i}, s}$, $s \in [0,1]$, surrounds $\partial_i g(x)$.

(ii) For all $(t_1, \ldots, t_{n-i+1}) \in [0,1]^{n-i+1}$,

$$(j^0 g(x), \alpha_1, \ldots, \alpha_{i-1}, \alpha_i^{t_1, \ldots, t_{n-i+1}}, \alpha_{i+1}^{t_1, \ldots, t_{n-i}}, \ldots, \alpha_n^{t_1}) \in \mathcal{R}.$$

Employing the above parametrized family of paths, associated to $\varphi = (\gamma, z_1, \ldots, j^1 g)$ is the auxiliary n-parameter family, $(t_1, t_2, \ldots t_n) \in [0,1]^n$:

$$(\gamma^{t_1, \ldots, t_n}(x), z_1^{t_1, \ldots, t_{n-1}}(x), \ldots, z_{n-1}^{t_1}(x), j^1 g(x)) \in \mathcal{R} \times \prod_i^n X_U^{(1)},$$

such that for all $x \in U$,

$$\gamma^{t_1, \ldots, t_n}(x) = (j^0 g(x), \alpha_1^{t_1, \ldots, t_n}, \alpha_2^{t_1, \ldots, t_{n-1}}, \ldots, \alpha_n^{t_1}) \in \mathcal{R},$$
$$z_i^{t_1, \ldots, t_{n-i}}(x) = (j^0 g(x), \partial_1 g(x), \ldots \partial_i g(x), \alpha_{i+1}^{t_1, \ldots, t_{n-i}}, \ldots \alpha_n^{t_1}) \in X_U^{(1)},$$
$$1 \leq i \leq n \quad (z_n(x) = j^1 g(x)).$$

This auxiliary parametrized family satisfies the following properties:

(iii) $\gamma^{0, \ldots, 0} = \gamma$; $z_i^{0, \ldots, 0} = z_i$, $1 \leq i \leq n$.

(iv) $\gamma^{t_1, \ldots, t_n}, z_1^{t_1, \ldots, t_{n-1}}$ differ only by the elements $\alpha_1^{t_1, \ldots, t_n}$, $\partial_1 g(x)$ in the ∂_1-component. Hence $\gamma^{t_1, \ldots, t_n}, z_1^{t_1, \ldots, t_{n-1}}$ lie in a principal subspace R_{b_1} of the bundle $X_U^{(1)} \to X_1^\perp$. Furthermore, employing (i) above, the path $\gamma^{t_1, \ldots, s}$, $s \in [0,1]$, is a path in R_{b_1} which surrounds $z_1^{t_1, \ldots, t_n}$ i.e. $\alpha_1^{t_1, \ldots, t_{n-1}, s}$, $s \in [0,1]$, surrounds $\partial_1 g(x)$.

(v) Successive elements $z_i^{t_1, \ldots, t_{n-i}}, z_{i+1}^{t_1, \ldots, t_{n-i-1}}$ differ only by the elements $\alpha_{i+1}^{t_1, \ldots, t_{n-i}}$, $\partial_{i+1} g(x)$ in the ∂_{i+1}-component, $1 \leq i \leq n-1$. Hence these successive elements lie in a principal subspace $R_{b_{i+1}}$ in the bundle $X_U^{(1)} \to X_{i+1}^\perp$. Furthermore, employing (i) above, the path $z_i^{t_1, \ldots, t_{n-i}}$, $t_{n-i} \in [0,1]$, surrounds $z_{i+1}^{t_1,\ldots,t_{n-i-1}}$ in $R_{b_{i+1}}$ i.e. the path $\alpha_{i+1}^{t_1,\ldots,t_{n-i}}$, $t_{n-i} \in [0,1]$, surrounds $\partial_{i+1} g(x)$, $1 \le i \le n-1$.

Returning to the proof of the lemma, for each $(t_1, \ldots t_n) \in [0,1]^n$ let $y_0^{t_1,\ldots,t_n} = \gamma^{t_1,\ldots,t_n} \in \mathcal{R}$; $y_n = \varphi$, and let,

$$y_i^{t_1,\ldots,t_{n-i}} = (\gamma^{t_1,\ldots,t_{n-i},0,\ldots,0}, z_1^{t_1,\ldots,t_{n-i},0,\ldots,0}, \ldots, z_i^{t_1,\ldots,t_{n-i}})$$

$$\in \mathcal{R} \times \prod_1^i X_U^{(1)}, \quad 1 \le i \le n-1.$$

The sequence $y_i^{t_1,\ldots,t_{n-i}}$, $1 \le i \le n$, satisfies the following properties.

(vi) Employing (iii), $y_j^{0,\ldots,0} = (\gamma, z_1, z_2, \ldots, z_j)$, $1 \le j \le n-1$.

(vii) $\rho_i(y_i^{t_1,\ldots,t_{n-i}}) = z_i^{t_1,\ldots,t_{n-i}} \in X_U^{(1)}$, $1 \le i \le n$.

(viii) $y_i^{t_1,\ldots,t_{n-i}} = (y_{i-1}^{t_1,\ldots,t_{n-i},0}, z_i^{t_1,\ldots,t_{n-i}})$, $1 \le i \le n$.

Employing properties (iv) to (viii), one verifies that for all $(t_1, \ldots, t_n) \in [0,1]^n$, the sequence $y_i^{t_1,\ldots,t_{n-i}}$, $1 \le i \le n$, satisfies the hypotheses (p_1) to (p_5) of Proposition 8.5. One concludes that $y_i^{t_1,\ldots,t_{n-i}} \in \mathrm{Conv}_i(\mathcal{R})$, $0 \le i \le n$. In particular, $y_n = \varphi = (\gamma, z_1, z_2, \ldots, z_n) \in \mathrm{Conv}_n(\mathcal{R})$, which proves the lemma. $\qquad\square$

Applying Theorem 8.4 one obtains the following local immersion theorem.

Theorem 8.8. *Let* $\varphi = (\alpha, z_1, z_2, \ldots, z_{n-1}, j^1 g) \in \Gamma(\mathrm{Conv}_n(\mathcal{R}))$ *as in Lemma 8.7; $g \in C^1(U, \mathbf{R}^q)$, $\alpha \in \Gamma(\mathcal{R})$. Let $\mathcal{N}$ be a neighbourhood of the image $j^0 g(U) \subset X_U^{(0)}$. There is a homotopy of holonomic sections $\varphi_t = (\alpha_t, z_1^t, \ldots, z_{n-1}^t, j^1 g_t) \in \Gamma(\mathrm{Conv}_n(\mathcal{R}))$, $t \in [0,1]$, such that the following properties obtain:*

(i) *$\varphi_0 = \varphi$; $\varphi_1 \in \Gamma(\mathcal{R})$, hence α_1 is holonomic: $\alpha_1 = j^1 g_1$ i.e. the h-principle holds for $\alpha \in \Gamma(\mathcal{R})$.*

(ii) *For all $t \in [0,1]$, $j^0 g_t(U) \subset \mathcal{N}$.*

(iii) *Let $K = \{ x \in U \mid \alpha(x) = j^1 g(x) \}$. Then one can choose the homotopy φ_t so that for all $t \in [0,1]$, $\varphi_t = \varphi$ on K (constant homotopy on K).*

Let $f \in C^1(V, W)$, and let $p \colon X = V \times W \to V$ be the product bundle over the base manifold V, where V, W are smooth manifolds, $\dim V = n$, $\dim W = q$, $1 \le n < q$. One identifies $\Gamma^r(X) \equiv C^r(V, W)$. Let $\mathcal{R} \subset X^{(1)}$ be the immersion relation. Suppose there is a section $\alpha \in \Gamma(\mathcal{R})$ such that $p_\perp^1 \circ \alpha = j^0 f = f \in \Gamma(X)$. Thus the pair (α, f) is a formal solution to the immersion problem of immersing V into W, and α agrees with f on the 0-jet level i.e. α is a continuous lift of f into $\mathcal{R}$. Employing a locally finite cover of V by charts (cf. the proof procedure, Chapter IV), the local immersion theorem 8.8 applies locally in each chart to prove the h-principle for α: there is a homotopy $\alpha_t \in \Gamma(\mathcal{R})$, $t \in [0,1]$, $\alpha_0 = \alpha$, such that α_1 is holonomic. Furthermore, there is a C^0-density result with respect to $f \in \Gamma(X)$. Thus, in case $q > n$, convex hull extension theory, applied to the ample immersion relation $\mathcal{R} \subset X^{(1)}$, reproves the main immersion result, IV Theorem 4.9. This application is a special case of the general theory of ample relations $\rho \colon \mathcal{R} \to X^{(r)}$, $r \geq 1$, that is developed in the next section.

§2. h-Principle for Ample Relations

8.2.1. Let $\rho \colon \mathcal{R} \to X^{(r)}$ be continuous. The relation $\mathcal{R}$ is *ample* if for all $x \in V$ and all codimension 1 tangent hyperplane fields $\tau \subset T(\mathfrak{Op}\, x)$, the following convex hull condition obtains with respect to the affine $\mathbf{R}^q$-bundle $p_\perp^r \colon X_U^{(r)} \to X_U^\perp$ associated to τ ($U = \mathfrak{Op}\, x$): For each $b \in X_U^\perp$ let $\mathcal{R}_b = (p_\perp^r \circ \rho)^{-1}(b) \subset \mathcal{R}$, the fiber over b for the map $p_\perp^r \circ \rho \colon \mathcal{R}_U \to X_U^\perp$. Then for all $(b,a) \in X_U^\perp \times \mathcal{R}_b$ (cf. V §5.2.2 for this notation):

$$X_b^{(r)} = \operatorname{Conv}(\mathcal{R}_b, a). \tag{8.7}$$

Thus for each r-jet z in the principal subspace over the base point b in the bundle $p_\perp^r \colon X_U^{(r)} \to X_U^\perp$, there is a path $a_t \in \mathcal{R}_b$, $a_0 = a$, whose image $\rho(a_t)$, $t \in [0,1]$, surrounds z. In case the map ρ is not onto $X^{(r)}$, by convention (8.7) is vacuously true if $\mathcal{R}_b = \emptyset$.

The condition (8.7) is affine invariant, hence independent of the local adapted coordinates in $\mathfrak{Op}\, x$ employed to define $X_U^\perp$. Hence (8.7) is verified pointwise at $x \in V$ and depends only on the hyperplane $\tau(x) \subset T_x(V)$.

An equivalent formulation of ampleness, closer to Gromov [18], is as follows. The (first) *principal extension* of $\mathcal{R}$ with respect to the hyperplane field τ is the subset $\operatorname{Pr}_\tau(\mathcal{R}) \subset \mathcal{R} \times X^{(r)}$ such that $(a,z) \in \operatorname{Pr}_\tau(\mathcal{R})$ if and only if $\rho(a), z$ lie in a principal subspace of the bundle $p_\perp^r \colon X_U^{(r)} \to X_U^\perp$ associated to τ. Let τ_i, $1 \leq i \leq m$, $m \geq 2$, be a sequence of codimension 1 tangent hyperplane fields on an open set U in V. The iterated extension relations $\operatorname{Pr}_j(\mathcal{R})$, $\operatorname{Conv}_j(\mathcal{R}) \subset \mathcal{R} \times \prod_1^j X^{(r)}$, $0 \leq j \leq m$ ($\operatorname{Pr}_0(\mathcal{R}) = \operatorname{Conv}_0(\mathcal{R}) = \mathcal{R}$) are defined inductively as follows:

$$\operatorname{Pr}_1(\mathcal{R}) = \operatorname{Pr}_{\tau_1}(\mathcal{R}); \quad \operatorname{Pr}_j(\mathcal{R}) = \operatorname{Pr}_{\tau_j}(\operatorname{Pr}_{j-1}(\mathcal{R}))$$
$$\operatorname{Conv}_1(\mathcal{R}) = \operatorname{Conv}_{\tau_1}(\mathcal{R}); \quad \operatorname{Conv}_j(\mathcal{R}) = \operatorname{Conv}_{\tau_j}(\operatorname{Conv}_{j-1}(\mathcal{R}))$$

Evidently, $\operatorname{Conv}_j(\mathcal{R}) \subset \operatorname{Pr}_j(\mathcal{R})$, and there is a natural projection map onto the last factor $\rho_j \colon \operatorname{Pr}_j(\mathcal{R}) \to X^{(r)}$, $(a, z_1, \dots, z_j) \mapsto z_j$, $1 \leq j \leq m$.

Following Gromov [18], a relation $\rho \colon \mathcal{R} \to X^{(r)}$ is ample if for all U open in V and all codimension 1 tangent hyperplane fields $\tau \subset T(U)$,

$$\operatorname{Pr}_\tau(\mathcal{R}) = \operatorname{Conv}_\tau(\mathcal{R}). \tag{8.8}$$

Evidently (8.8) is equivalent to the property (8.7). More precisely, Gromov defines $\mathcal{R}$ to be ample if and only if the algebraic condition $\mathrm{Pr}_\infty \mathcal{R} = \mathrm{Conv}_\infty \mathcal{R}$ obtains. These "infinite" principal and convex hull extensions, not defined here, are essential to Gromov's algebraic formalism for studying ample relations. In the above notation, Gromov's ampleness condition is equivalent to: for all $m \geq 1$, $\mathrm{Conv}_m(\mathcal{R}) = \mathrm{Pr}_m(\mathcal{R})$. The equivalence (8.8) (the case $m = 1$) is Gromov's sufficient condition for ample relations, as is confirmed in Corollary (8.10) below. Principal extensions of ample relations are important because of their algebraic properties, developed below, which finesse the elaborate technicalities expressed in Proposition 8.5 for iterated convex hull extensions.

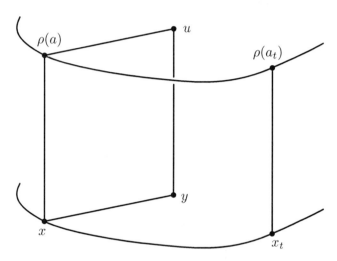

FIGURE 8.1

Let $\rho\colon \mathcal{R} \to X^{(r)}$ be continuous. The algebraic analysis of iterated principal and convex hull extensions depends on the following geometrical homeomorphism (an involution: $h \circ h = \mathrm{id}$).

$$h\colon \mathrm{Pr}_{\tau_2}(\mathrm{Pr}_{\tau_1}(\mathcal{R})) \to \mathrm{Pr}_{\tau_1}(\mathrm{Pr}_{\tau_2}(\mathcal{R})), \quad (a, u, y) \mapsto (a, x, y),$$

(reversing the indices 1,2) where the ordered points $\rho(a), u, y, x$ are the vertices of a parallelogram in the affine fiber $X_b^{(r)}$ ($b \in X^\perp$ is a base point), such that $\rho(a), u$, respectively u, y lie in a principal subspace associated to τ_1, respectively a principal subspace associated to τ_2. The restriction of h is an injection (same notation),

$$h\colon \mathrm{Pr}_{\tau_2}(\mathrm{Conv}_{\tau_1}(\mathcal{R})) \to \mathrm{Conv}_{\tau_1}(\mathrm{Pr}_{\tau_2}(\mathcal{R})). \tag{8.9}$$

To prove (8.9) we verify explicitly $\mathrm{im}\, h \subset \mathrm{Conv}_{\tau_1}(\mathrm{Pr}_{\tau_2}(\mathcal{R}))$. To this end let $(a, u, y) \in \mathrm{Pr}_{\tau_2}(\mathrm{Conv}_{\tau_1}(\mathcal{R}))$. There is a path $a_t \in \mathcal{R}_b$, $a_0 = a$, $t \in [0, 1]$, such that the path $\rho(a_t)$ surrounds u in the corresponding principal subspace associated to τ_1. Employing the translation $T\colon X_b^{(r)} \to X_b^{(r)}$, $T(z) = z + (y - u)$, observe that $T(u) = y$; $T(\rho(a)) = x$. The path $\rho(a_t)$ translates under T to a path x_t, $x_0 = x$, which surrounds y. Furthermore, for all $t \in [0,1]$, $\rho(a_t) - x_t = y - u$, i.e. for all $t \in [0,1]$, $\rho(a_t), x_t$ lie in a principal subspace associated to τ_2, from which it follows that $h((a, u, y)) = (a, x, y) \in \mathrm{Conv}_{\tau_1}(\mathrm{Pr}_{\tau_2}(\mathcal{R}))$.

Lemma 8.9. *Let $\mathcal{R}$ be an ample relation over $X^{(r)}$. Then $\mathrm{Pr}_1(\mathcal{R})$ is ample over $X^{(r)}$.*

Proof. Employing (8.8) we prove $\mathrm{Pr}_{\tau_2}(\mathrm{Pr}_{\tau_1}(\mathcal{R})) = \mathrm{Conv}_{\tau_2}(\mathrm{Pr}_{\tau_1}(\mathcal{R}))$. This follows formally from two applications of the above involution h:

$$\mathrm{Pr}_{\tau_2}(\mathrm{Pr}_{\tau_1}(\mathcal{R})) \xrightarrow{h} \mathrm{Pr}_{\tau_1}(\mathrm{Pr}_{\tau_2}(\mathcal{R})) = \mathrm{Pr}_{\tau_1}(\mathrm{Conv}_{\tau_2}(\mathcal{R}))$$

$$\xrightarrow{h} \mathrm{Conv}_{\tau_2}(\mathrm{Pr}_{\tau_1}(\mathcal{R})) \subset \mathrm{Pr}_{\tau_2}(\mathrm{Pr}_{\tau_1}(\mathcal{R})). \quad \square$$

Corollary 8.10. *Suppose $\mathcal{R}$ is an ample relation over $X^{(r)}$. Then $\mathrm{Pr}_j(\mathcal{R})$ is ample over $X^{(r)}$, and $\mathrm{Pr}_j(\mathcal{R}) = \mathrm{Conv}_j(\mathcal{R})$, $1 \le j \le m$.*

Proof. Employing the lemma, $\mathrm{Pr}_1(\mathcal{R}) = \mathrm{Pr}_{\tau_1}(\mathcal{R})$ is ample over $X^{(r)}$. Hence from (8.8),

$$\begin{aligned}
\mathrm{Pr}_2(\mathcal{R}) = \mathrm{Pr}_{\tau_2}(\mathrm{Pr}_1(\mathcal{R})) &= \mathrm{Conv}_{\tau_2}(\mathrm{Pr}_1(\mathcal{R})) \\
&= \mathrm{Conv}_{\tau_2}(\mathrm{Conv}_{\tau_1}(\mathcal{R})) = \mathrm{Conv}_2(\mathcal{R}).
\end{aligned}$$

Thus, if $\mathcal{R}$ is an ample relation over $X^{(r)}$, it follows that:

$$\mathrm{Pr}_2(\mathcal{R}) = \mathrm{Conv}_2(\mathcal{R}). \tag{8.10}$$

Suppose inductively on $j \ge 1$, $\mathrm{Pr}_j(\mathcal{R})$ is ample over $X^{(r)}$ and that $\mathrm{Pr}_j(\mathcal{R}) = \mathrm{Conv}_j(\mathcal{R})$. Thus (in abbreviated notation),

$$\begin{aligned}
\mathrm{Pr}_1(\mathrm{Pr}_{j+1}(\mathcal{R})) &= \mathrm{Pr}_2(\mathrm{Pr}_j(\mathcal{R})) \\
&= \mathrm{Conv}_2(\mathrm{Pr}_j(\mathcal{R})) \quad \text{by (8.10)} \\
&= \mathrm{Conv}_1(\mathrm{Pr}_1(\mathrm{Pr}_j(\mathcal{R}))) \quad \text{by (8.8)} \\
&= \mathrm{Conv}_1(\mathrm{Pr}_{j+1}(\mathcal{R})).
\end{aligned} \tag{8.11}$$

Employing (8.8), it follows from (8.11) that $\mathrm{Pr}_{j+1}(\mathcal{R})$ is ample. Similarly,

$$\begin{aligned}
\mathrm{Pr}_{j+1}(\mathcal{R}) &= \mathrm{Pr}_1(\mathrm{Pr}_j(\mathcal{R})) \\
&= \mathrm{Conv}_1(\mathrm{Pr}_j(\mathcal{R})) = \mathrm{Conv}_{j+1}(\mathcal{R}).
\end{aligned} \tag{8.12}$$

Thus (8.11), (8.12) prove the inductive step, and the corollary is proved. $\quad \square$

8.2.2. Recall that a formal solution to the relation $\rho\colon \mathcal{R} \to X^{(r)}$ consists of a pair $(\alpha, f) \in \Gamma(\mathcal{R}) \times \Gamma^r(X)$ such that $p^r_{r-1} \circ \rho \circ \alpha = j^{r-1}f \in \Gamma(X^{(r-1)})$ i.e. $\alpha \in \Gamma(\mathcal{R})$ is a continuous lift of the section $j^{r-1}f$ into $\mathcal{R}$. Recall also the space of short sections for the relation $\mathcal{R}$ (cf. §8.1.2).

Lemma 8.11. *Suppose $\mathcal{R}$ is ample over $X^{(r)}$. Let (α, f) be a formal solution to $\mathcal{R}$. Then locally over each chart U of V the section f is short i.e. there are smooth integrable codimension 1 tangent hyperplane fields $\tau_i \subset T(U)$, $1 \le i \le m$, and a holonomic section $\varphi \in \mathrm{Conv}_m(\mathcal{R}_U)$ (α_U, f_U denote the restrictions to U),*

$$\varphi = (\alpha_U, z_1, z_2, \ldots, z_{m-1}, j^r f_U) \in \mathrm{Conv}_m(\mathcal{R}),$$

such that the following relative property obtains:
Let $K = \{x \in U \mid \rho \circ \alpha(x) = j^r f(x)\}$. Then $z_i(x) = j^r f(x)$, $1 \le i \le m - 1$.

Proof. Employing VI Theorem 6.8 to the sections $\rho \circ \alpha_U, j^r f_U \in \Gamma(X_U^{(r)})$, there is a section $\varphi = (\alpha_U, z_1, \ldots z_{m-1}, j^r f_U) \in \mathrm{Pr}_m(\mathcal{R})$, $m = \dim \mathcal{H}(n,1)$, associated to smooth integrable codimension 1 tangent hyperplane fields $\tau_i \subset T(U)$, $\tau_i = \ker \ell_i$, where the monomials ℓ_i^r, $1 \le i \le m$, form a basis of $\mathcal{H}(n,1)$, and such that conclusion (1) obtains. Since $\mathcal{R}$ is ample over $X^{(r)}$, employing Corollary 8.10, $\mathrm{Pr}_m(\mathcal{R}) = \mathrm{Conv}_m(\mathcal{R})$, and the lemma is proved. $\qquad\square$

Theorem 8.12. $\mathbf{C^{r-1}}$-dense h-Principle. *Let $\rho\colon \mathcal{R} \to X^{(r)}$ be a microfibration and suppose $\mathcal{R}$ is ample over $X^{(r)}$. Let (α, f) be a formal solution of $\mathcal{R}$: $p^r_{r-1} \circ \rho \circ \alpha = j^{r-1}f \in \Gamma(X^{(r-1)})$. Let $\mathcal{N}$ be a neighbourhood of the image $j^{r-1}f(V)$ in $X^{(r-1)}$. There is a homotopy of formal solutions (α_t, f_t), $(\alpha_0, f_0) = (\alpha, f)$, $t \in [0,1]$, such that the following properties obtain:*

(i) *$\rho \circ \alpha_1 = j^r f_1$ i.e. the h-principle holds for $\alpha \in \Gamma(\mathcal{R})$.*

(ii) *(C^{r-1}-dense principle) : For each $t \in [0,1]$, the image $j^{r-1}f_t(V) \subset \mathcal{N}$.*

(iii) *Let $K = \{x \in V \mid \rho \circ \alpha(x) = j^r f(x)\}$. One can choose the homotopy (α_t, f_t) so that for all $t \in [0,1]$, $(\alpha_t, f_t) = (\alpha, f)$ on K (constant homotopy on K).*

Proof. Since $p^r\colon X^{(r)} \to X^{(r-1)}$ is a smooth bundle and $\rho\colon \mathcal{R} \to X^{(r)}$ is a microfibration, employing V Theorem 5.3, Corollary 5.5, up to a small homotopy of formal solutions (α_t, f_t), $t \in [0,1]$, one may assume: (i) $\rho \circ \alpha, f$ are smooth on $V \setminus K$; (ii) $\rho \circ \alpha = j^r f$ on $\mathfrak{Op}\, K$.

Following the general proof procedure of Chapter IV, let $(W_i)_{i \ge 1}$, $(U_i)_{i \ge 1}$ be locally finite coverings by closed charts of $\overline{V \setminus \mathfrak{Op}_1 K}$ where $\overline{\mathfrak{Op}_1 K} \subset \mathfrak{Op}\, K$, such that for all $i \ge 1$: (i) $W_i \subset \mathrm{int}\, U_i$; (ii) $U_i \equiv I^n$ ($n = \dim V$); (iii) $U_i \subset V \setminus K$. In particular the sections $\rho \circ \alpha \in \Gamma(X^{(r)})$, $f \in \Gamma(X)$ are smooth over each chart U_i, $i \ge 1$.

Furthermore one may assume that with respect to the smooth integrable codimension 1 tangent hyperplane fields $\tau_j \subset T(U_i)$, $1 \leq j \leq m$, (that are employed in Lemma 8.11), there is a sequence of closed charts Z_i^j in U_i, $1 \leq j \leq m$, such that:

(p_1) $Z_i^j \subset \text{int } Z_i^{j+1}$, $1 \leq j \leq m-1$; $W_i \subset \text{int } Z_i^1$; $Z_i^m \subset \text{int } U_i$.

(p_2) The chart $Z_i^j \equiv I^{n-1} \times [0,1]$ is a split manifold with respect to the integrable tangent hyperplane field $\tau_j \subset T(U_i)$, $1 \leq j \leq m$.

The proof procedure is an inductive process over successive charts U_i, $i \geq 1$, the main step of which is the following local extension lemma.

Local Extension Lemma 8.13. *Let (β, g) be a formal solution to $\mathcal{R}$. Let $U \equiv I^n$ be a closed chart on V, and let $A, W \subset V$ be closed sets, $W \subset \text{int } U$, such that g is smooth on U and such that the following properties obtain:*

(i) *There is a sequence of split manifold charts Z^j, $1 \leq j \leq m$, $W \subset \text{int } Z^1$; $Z^m \subset \text{int } U$, that satisfy (p_1), (p_2) above.*

(ii) *$\rho \circ \beta = j^r g$ on $\mathfrak{Op}\, A$ i.e. $\beta \in \Gamma(\mathcal{R})$ is holonomic on $\mathfrak{Op}\, A$.*

(iii) *The image $j^{r-1}g(V) \subset \mathcal{N}$ i.e. g is a C^{r-1}-approximation to f.*

Then there is a homotopy of formal solutions (β_t, g_t), $(\beta_0, g_0) = (\beta, g)$, $t \in [0,1]$, such that the following properties obtain:

(iv) *For all $t \in [0,1]$, $(\beta_t, g_t) = (\beta, g)$ on $\mathfrak{Op}\, A \cup \mathfrak{Op}\, \overline{V \setminus U}$.*

(v) *For all $t \in [0,1]$, the image $j^{r-1}g_t(V) \subset \mathcal{N}$.*

(vi) *$\rho \circ \beta_1 = j^r g_1$ on $\mathfrak{Op}\, A \cup \mathfrak{Op}\, W$. Thus the homotopy $\beta_t \in \Gamma(\mathcal{R})$ is constant on $\mathfrak{Op}\, A$ and it connects $\beta_0 = \beta$ which is holonomic on $\mathfrak{Op}\, A$ to β_1 which is holonomic on $\mathfrak{Op}\, A \cup \mathfrak{Op}\, W$.*

Proof. Employing the inclusion $\mathcal{R} \subset \text{Conv}_m(\mathcal{R})$ (cf. §8.1.1), and applying Lemma 8.11, there is a section,

$$\varphi = (\beta_U, z_1, z_2, \ldots, z_{m-1}, j^r g_U) = (\alpha, j^r g_U) \in \Gamma(\text{Conv}_m(\mathcal{R}_U)),$$

where $\alpha \in \Gamma(\text{Conv}_{m-1}(\mathcal{R}_U))$ and such that $\varphi = (\beta_U, j^r g_U, \ldots, j^r g_U)$ on $\mathfrak{Op}\, A \cap U$. Employing Lemma 8.2, the projection $\rho_{m-1} \colon \text{Conv}_{m-1}(\mathcal{R}_U) \to X_U^{(r)}$ is a microfibration. Since g is smooth on U, applying VII Local Extension Lemma 7.5 to the split manifold Z^m (with respect to the smooth, integrable, tangent hyperplane field τ_m), and to the closed set $Z^{m-1} \subset \text{int } Z^m$, there is a homotopy of holonomic sections $\varphi_t \in \Gamma(\text{Conv}_m(\mathcal{R}_U))$, $\varphi_0 = \varphi$,

$$\varphi_t = (\beta_U^t, z_1^t, \ldots, z_{m-1}^t, j^r g_U^t) = (\alpha_t, j^r g_U^t), \quad \alpha_t \in \Gamma(\text{Conv}_{m-1}(\mathcal{R}_U)),$$

such that the following properties are satisfied for all $t \in [0,1]$:

(vii) $\varphi_t = \varphi$ on $\mathfrak{Op}\, A \bigcup \mathfrak{Op}\,(\partial Z^m)$ (constant section); the map g_U^t is smooth on U. In particular, β_U^t, g_U^t extend to sections (same notation) $\beta_U^t \in \Gamma(\mathcal{R})$, respectively $g_U^t \in \Gamma^r(X)$, such that $\beta_U^t = \beta$, respectively $g_U^t = g$ on $V \setminus Z^m$.

(viii) The image $j^{r-1}g^t(V) \subset \mathcal{N}$.

(ix) $\rho_{m-1} \circ \alpha_1 (= z_{m-1}^1) = j^r g_U^1$ on $(\mathfrak{Op}\, A \cap U) \bigcup \mathfrak{Op}\, Z^{m-1}$.

Therefore the homotopy $\alpha_t \in \Gamma(\mathrm{Conv}_{m-1}(\mathcal{R}_U))$ is constant on $\mathfrak{Op}\, A \cap U$ and it connects $\alpha_0 = \alpha$ which is holonomic on $\mathfrak{Op}\, A \cap U$ to α_1 which is holonomic on $(\mathfrak{Op}\, A \cap U) \bigcup \mathfrak{Op}\, Z^{m-1}$. In particular, $\alpha_1 = (\beta_U^1, j^r g_U^1, \dots, j^r g_U^1)$ on $\mathfrak{Op}\, A \cap U$.

Employing (vii), (β_U^t, g_U^t), $t \in [0,1]$, is a homotopy rel $\overline{V \setminus U}$ of formal solutions to $\mathcal{R}$ such that $(\beta_U^0, g_U^0) = (\beta, g)$. Employing (ix), the restriction (same notation) $\alpha_1 \in \Gamma(\mathrm{Conv}_{m-1}(\mathcal{R}_{Z^{m-1}}))$ is holonomic: $\rho_{m-1}(\alpha_1) = j^r g_U^1$ on Z^{m-1}.

Since the section g_U^1 is smooth on U, there is an obvious downward induction beginning with the holonomic section $\varphi_1 \in \Gamma(\mathrm{Conv}_m(\mathcal{R}_U))$, such that applying VII Local Extension Lemma 7.5 inductively on j to the split manifolds Z^{m-j} (with respect to τ_{m-j}), $1 \leq j \leq m-1$, (at the jth level all the changes occur in the manifold Z^{m-j}) and then concatenating the resulting homotopies that are obtained inductively at each stage, one obtains a homotopy of formal solutions (γ^t, h^t) to the relation $\mathcal{R}$, $(\gamma^0, h^0) = (\beta, g)$, a homotopy of holonomic sections $H_t \in \Gamma(\mathrm{Conv}_m(\mathcal{R}_U))$,

$$H_t = (\gamma_U^t, y_1^t, \dots, y_{m-1}^t, j^r h_U^t) \quad t \in [0,1],$$

such that the following properties obtain for all $t \in [0,1]$:

(x) $H_0 = \varphi$; $H_t = \varphi$ on $\mathfrak{Op}\, A \bigcup \mathfrak{Op}\,(\partial Z^m)$ (constant section); the map h^t is smooth on U. In particular, all changes occur in $Z^m \subset U$. Thus $\gamma^t = \beta$, respectively $h^t = g$, on $\mathfrak{Op}\, A \bigcup \mathfrak{Op}\, \overline{V \setminus Z^m}$.

(xi) The restriction (same notation) $H_1 \in \Gamma(\mathcal{R}_{\mathfrak{Op}\, W})$ is holonomic: $\rho \circ \gamma^1 = j^r h^1$ on $\mathfrak{Op}\, W$. Explicitly,

$$H_1(x) = (\gamma^1(x), j^r h^1(x), \dots, j^r h^1(x)) \in \mathcal{R} \quad \text{for all } x \in \mathfrak{Op}\, W.$$

(xii) the image $j^{r-1}h^t(V) \subset \mathcal{N}$.

Employing (xi) the homotopy of formal solutions $(\gamma^t, j^r h^t)$ is such that $\rho \circ \gamma^1 = j^r h^1$ on $\mathfrak{Op}\, A \cup \mathfrak{Op}\, W$, and the approximation property (xii) is satisfied, which completes the proof of the local extension lemma. □

Returning to the proof of the theorem, employing the Local Extension Lemma 8.13 inductively to the pair of charts (W_i, U_i), $i \geq 1$, and to the closed set $\overline{\mathfrak{Op}_1 K}$, one obtains a sequence of homotopies of formal solutions (α_t^i, f_t^i) to $\mathcal{R}$, $t \in [0,1]$, such that the following properties are satisfied for all $i \geq 1$:

(1) $(\alpha_0^1, f_0^1) = (\alpha, f)$; $(\alpha_1^i, f_1^i) = (\alpha_0^{i+1}, f_0^{i+1})$ i.e. successive homotopies match up.

(2) For all $t \in [0,1]$, $(\alpha_t^i, f_t^i) = (\alpha_0^i, f_0^i)$ on $\overline{\mathfrak{Op}\, V} \setminus U_i$ i.e. the changes to α_t^i, f_t^i occur only in the chart U_i.

(3) For all $t \in [0,1]$, the image $j^{r-1} f_t^i(V) \subset \mathcal{N}$.

(4) For all $t \in [0,1]$, $(\alpha_t^i, f_t^i) = (\alpha, f)$ on $\overline{\mathfrak{Op}_1 K}$.

(5) $\rho \circ \alpha_1^i = j^r f_1^i$ on $\bigcup_{j=1}^i \mathfrak{Op}\, W_j$.

Concatenating the sequence of homotopies of formal solutions (α_t^i, f_t^i), one obtains a homotopy of formal solutions (α_t, f_t) of $\mathcal{R}$, $t \in [0,1]$, such that $(\alpha_0, f_0) = (\alpha, f)$; for all $t \in [0,1]$, the image $j^{r-1} f_t(V) \subset \mathcal{N}$; for all $t \in [0,1]$, $(\alpha_t, f_t) = (\alpha, f)$ on $\overline{\mathfrak{Op}_1 K}$; α_1 is holonomic on V:

$$\rho \circ \alpha_1 = j^r f_1 \text{ on } \overline{\mathfrak{Op}_1 K} \bigcup_{i=1}^{\infty} W_i = V,$$

hence the proof of the theorem is complete. $\qquad\square$

Theorem 8.14. C^i-dense h-principle. *Let $\rho \colon \mathcal{R} \to X^{(r)}$ be a microfibration. Fix $i \in \{0, \ldots, r-1\}$ and suppose the composed microfibration $p_s^r \circ \rho \colon \mathcal{R} \to X^{(s)}$ is ample for all s, $i + 1 \le s \le r$. Then $\mathcal{R}$ satisfies the strong C^i-dense h-principle (cf. I §1.3.2).*

Proof. Let $\alpha \in \Gamma(\mathcal{R})$ and let $f \in \Gamma^i(X)$ such that $j^i f = p_i^r \circ \rho \circ \alpha \in \Gamma(X^{(i)})$, the i-jet component of the section $\rho \circ \alpha \in \Gamma(X^{(r)})$. The pair (α, f) is not quite a formal solution to the relation $p_{i+1}^r \colon \mathcal{R} \to X^{(i+1)}$ since one requires also $f \in \Gamma^{i+1}(X)$ i.e. f is a C^{i+1}-section (cf. §8.1.2). We arrange this latter condition as follows. Since the projection map $p_{i+1}^r \circ \rho \colon \mathcal{R} \to X^{(i+1)}$ is a microfibration, employing V Remark 5.4 and the Approximation Theorem I 1.1, there is a homotopy $(\alpha_t, f_t) \in \Gamma(\mathcal{R}) \times \Gamma^i(X)$, $(\alpha_0, f_0) = (\alpha, f)$, $t \in [0,1]$, such that: (i) for all $t \in [0,1]$, $p_i^r \circ \alpha_t = j^i f_t$; (ii) $f_1 \in \Gamma^{i+1}(X)$ i.e. f_1 is a C^{i+1}-section. Consequently, up to a small homotopy of $\alpha \in \Gamma(\mathcal{R})$, one may assume that the pair (α, f) is a formal solution to the relation $p_{i+1}^r \circ \rho \colon \mathcal{R} \to X^{(i+1)}$. Since the microfibration $p_{i+1}^r \circ \rho \colon \mathcal{R} \to X^{(i+1)}$ is ample, employing Theorem 8.12, there is a homotopy of formal solutions (same notation) $(\alpha_t, f_t) \in \Gamma(\mathcal{R}) \times \Gamma^{i+1}(X)$, $(\alpha_0, f_0) = (\alpha, f)$, $t \in [0,1]$, such that α_1 is holonomic: $p_{i+1}^r \circ \rho \circ \alpha = j^{i+1} f_1$. Similarly, employing the microfibration $p_{i+2}^r \circ \rho \colon \mathcal{R} \to X^{(i+2)}$, there is a homotopy of formal solutions, holonomic at each stage up to jets of order $(i+1)$, that connects α_1 to a holonomic section $\alpha_2 \in \mathcal{R}$: $p_{i+2}^r \circ \rho \circ \alpha_2 = j^{i+2} f_2$, where $f_2 \in \Gamma^{i+2}(X)$. An obvious induction proves the theorem. $\qquad\square$

8.2.3. Weak Homotopy Equivalence. Recall §8.1.2 that a formal solution to a relation $\rho\colon \mathcal{R} \to X^{(r)}$ is a pair (α, f) where $\alpha \in \Gamma(\mathcal{R})$, $f \in \Gamma^r(X)$ such that $p^r_{r-1} \circ \rho \circ \alpha = j^{r-1}f \in \Gamma(X^{(r-1)})$. Let $\mathcal{F} = \mathcal{F}_\mathcal{R}$ denote the subspace of $\Gamma(\mathcal{R}) \times \Gamma(X^{(r)})$ consisting of all formal solutions of $\mathcal{R}$. Let $\mathcal{H} \subset \mathcal{F}$ be the subspace of holonomic solutions: a formal solution (α, f) is *holonomic* if in addition $\rho \circ \alpha = j^r f$. In case $\mathcal{R} \subset X^{(r)}$ and ρ is the inclusion map, then $(\alpha, f) \in \mathcal{H}$ if and only if $\alpha = j^r f$. The inclusion $i\colon \mathcal{H} \to \mathcal{F}$ is studied homotopically in this section.

The main work in immersion-theoretic topology over the past few decades has been to show that in cases of relations $\mathcal{R}$ of geometrical and topological interest, the inclusion $i\colon \mathcal{H} \to \mathcal{F}$ is a weak homotopy equivalence (cf. Hirsch [21], Smale [36], Phillips [31], Feit [12]). The computations of $\pi_i(\mathcal{F})$, $i \geq 0$, can be carried out in principle, and often in practice, by employing techniques from algebraic topology. The main result of this section is the following theorem.

Weak Homotopy Equivalence Theorem 8.15. *Let $\rho\colon \mathcal{R} \to X^{(r)}$ be a microfibration and suppose $\mathcal{R}$ is ample over $X^{(r)}$. Then the inclusion $i\colon \mathcal{H} \to \mathcal{F}$ induces a weak homotopy equivalence.*

The proof of this theorem follows from the C^{r-1}-dense h-Principle 8.12 applied to parameter spaces of spheres S^i, $i \geq 0$. In fact a suitable parametric version of Theorem 8.12 proves the weak homotopy equivalence theorem. We develop a somewhat different approach based on associated spaces of parametrized r-jets. Let Z be an auxiliary smooth compact manifold. Let $\mathrm{id} \times p\colon Z \times X \to Z \times V$ be the smooth bundle, $(z, y) \mapsto (z, p(y))$. Associated to this bundle are two jet space bundles over $Z \times V$ (s denotes the source map): $\mathrm{id} \times s\colon Z \times X^{(r)} \to Z \times V$; $s\colon (Z \times X)^{(r)} \to Z \times V$.

There is a natural projection $\pi\colon (Z \times X)^{(r)} \to Z \times X^{(r)}$ such that on r-jets, $\pi(j^r f(z, x)) = (z, j^r h_z(x)) \in Z \times X^{(r)}$, where $f = (\mathrm{id} \times h) \in \Gamma^r(Z \times X)$, $h \in C^0(Z, \Gamma^r(X))$. The projection π is compatible with restriction to $(r-1)$-jets: $\pi(j^{(r-1)} f(z, x)) = (z, j^{r-1} h_z(x))$. Furthermore, in local coordinates, the projection of derivatives $D^r f(z, x) \mapsto D^r_V h_z(x)$ (D^r_V denotes derivative in the V-variables) is linear in f. Hence there is an induced affine map on the fibers such that at the level of r-jets (the base points $m = j^{r-1} f(z, x)$; $\pi(m) = (z, w) \in Z \times X^{(r-1)}$),

$$\pi\colon (Z \times X)^{(r)}_m \to \{z\} \times X^{(r)}_w; \ j^r f(z, x) \mapsto (z, j^r_V h_z(x)) \in Z \times X^{(r)}. \quad (8.13)$$

Let $U \subset V$ be open and let $\tau \subset T(U)$ be a codimension 1 tangent hyperplane field on U. Associated to τ is the codimension 1 tangent hyperplane field $\tau_Z = T(Z) \times \tau \subset T(Z \times U)$. τ_Z is integrable if τ is integrable. Let $p^r_\perp\colon (Z \times X)^{(r)}_W \to (Z \times X)^\perp_W$ be the affine $\mathbf{R}^q$-bundle associated to τ_Z (restriction to $W = (Z \times U)$). Since local adapted coordinates for τ_x embed into local adapted coordinates for

$\tau_Z(z,x)$, employing (8.13), it follows that the projection π induces a fiberwise affine isomorphism of associated $\perp$-spaces:

$$
\begin{array}{ccc}
(Z \times X)_W^{(r)} & \xrightarrow{\ \pi\ } & Z \times X_U^{(r)} \\[4pt]
p_\perp^r \downarrow & & \downarrow \mathrm{id} \times p_\perp^r \\[4pt]
(Z \times X)_W^\perp & \xrightarrow[\ \pi\ ]{} & Z \times X_U^\perp
\end{array}
\tag{8.14}
$$

Lifting Properties. Let $f = (\mathrm{id} \times h) \in \Gamma^r(Z \times X)$, $h \in C^0(Z, \Gamma^r(X))$. Associated to f, h are the section spaces:

$$\Gamma_f = \{G \in \Gamma((Z \times X)^{(r)}) \mid p_{r-1}^r \circ G = j^{r-1} f\}.$$

$$\Gamma_h = \{g \in C^0(Z, \Gamma(X^{(r)})) \mid p_{r-1}^r \circ g_z = j^{r-1}(h_z) \in \Gamma(X^{(r-1)}) \text{ for all } z \in Z\}.$$

Continuous Lift Lemma 8.16. *There is a continuous lift* $\lambda \colon \Gamma_h \to \Gamma_f$ *such that the following properties are satisfied:*

(i) *for all* $g \in \Gamma_h$, $\pi \circ \lambda(g) = \mathrm{ev}\, g \in \Gamma(Z \times X^{(r)})$.

(ii) *Let* $g \in \Gamma_h$ *and let* $U \subset V$ *be open such that for all* $z \in U$, $g_z = j^r h_z$ *on* U. *Then for all* $(z,x) \in Z \times U$, $\lambda(g)(z,x) = j^r(\mathrm{id} \times h)(z,x) = j^r f(z,x)$.

(iii) *Let* $g^0, g^1 \in \Gamma_h$ *and let* $\tau \subset T(U)$ *a codimension 1 tangent hyperplane field on an open set* U *in* V *such that for all* $(z,x) \in W = Z \times U$, $g_z^1(x), g_z^2(x)$ *lie in a principal subspace of the bundle* $X_U^{(r)} \to X_U^\perp$ *associated to* τ. *Then for all* $(z,x) \in W$ $\lambda(g^1)(z,x), \lambda(g^2)(z,x)$ *lie in a principal subspace of the bundle* $(Z \times X)_W^{(r)} \to (Z \times X)_W^\perp$ *associated to* τ_Z *(cf. (8.14)).*

Proof. In local coordinates, there are two types of rth order components of r-jets in $(Z \times X)^{(r)}$: (i) rth order components that correspond to rth order derivatives in the V-variables; (ii) rth order components that correspond to rth order derivatives that include at least one derivative in the Z-variables. The lift λ is determined uniquely in local coordinates according to the following prescription on rth order components:

For each $g \in \Gamma_h$ there is a *unique* section $G = \lambda(g) \in \Gamma_f$ such that (the continuity of G follows from the continuity in local coordinates below for G):

(1) $\pi \circ G = \mathrm{ev}\, g \in \Gamma(Z \times X^{(r)})$ i.e. for all $(z,x) \in Z \times X$ the components of the r-jet $G(z,x)$ that include only derivatives in the V-variables are given by $g_z(x) \in X^{(r)}$.

(2) For all $(z,x) \in Z \times X$, in local coordinates, the ∂^α-component of the r-jet $G(z,x)$ is $\partial^\alpha f(z,x)$ for all rth order derivatives ∂^α, $|\alpha| = r$, that include at least one derivative in the Z-variables.

Properties (i), (ii) are satisfied by construction. Let $\lambda(g^i) = G^i$, $i \in \{1,2\}$. Since $\pi \circ G^i(z,x) = g_z^i(x) \in X^{(r)}$, $i \in \{1,2\}$, employing the affine isomorphism (8.14) induced by π, conclusion (iii) follows. $\qquad\square$

Let $\rho\colon \mathcal{R} \to X^{(r)}$ be a relation over $X^{(r)}$. Let $Z \times \mathcal{R} \to Z \times X^{(r)}$ be the pullback relation over $Z \times X^{(r)}$, and let $\rho_Z\colon \mathcal{R}_Z \to (Z \times X)^{(r)}$ be the corresponding relation over $(Z \times X)^{(r)}$ obtained by pulling back along the projection $\pi\colon (Z \times X)^{(r)} \to Z \times X^{(r)}$. Since the pullback of a microfibration is a microfibration, employing the fiberwise affine isomorphism (8.14), the following result obtains.

Lemma 8.17. *Let $\rho\colon \mathcal{R} \to X^{(r)}$ be a microfibration which is ample over $X^{(r)}$. Then $\rho_Z\colon \mathcal{R}_Z \to (Z \times X)^{(r)}$ is a microfibration which is ample over $(Z \times X)^{(r)}$ with respect to all tangent hyperplane fields τ_Z.*

The pullback $\mathcal{R}_Z = \pi^*(Z \times \mathcal{R})$ induces a continuous map on section spaces, $P\colon \Gamma(\mathcal{R}_Z) \to \Gamma(Z \times \mathcal{R}) \equiv C^0(Z, \Gamma(\mathcal{R}))$. In particular if $P(\beta) = \alpha$ then for all $(z,x) \in Z \times V$,

$$\pi \circ \rho_Z \circ \beta(z,x) = \rho \circ \alpha_z(x) \in X^{(r)}. \qquad (8.15)$$

Let $Q\colon \Gamma(Z \times X) \to C^0(Z, \Gamma(X))$ be the canonical homeomorphism. Explicitly, $Q(g) \in C^0(Z, \Gamma(X))$ is the map $z \mapsto h_z$, where $g = (\mathrm{id} \times h)$.

Let $\varphi = P \times Q\colon \Gamma(\mathcal{R}_Z) \times \Gamma(Z \times X) \to C^0(Z, \Gamma(\mathcal{R})) \times C^0(Z, \Gamma(X))$.

Let $\mathcal{F}_Z$ denote the space of formal solutions of the relation $\mathcal{R}_Z$. $\mathcal{H}_Z \subset \mathcal{F}_Z$ denotes the subspace of holonomic solutions. Thus $(\beta, g) \in \mathcal{F}_Z$ if and only if $\beta \in \Gamma(\mathcal{R}_Z)$, $g \in \Gamma^r(Z \times X)$, such that on the level of $(r-1)$-jets,

$$p_{r-1}^r \circ \rho_Z \circ \beta = j^{r-1}g \in \Gamma((Z \times X)^{(r-1)}). \qquad (8.16)$$

Similarly, $(\beta, g) \in \Gamma(\mathcal{H}_Z)$ if and only if $\rho_Z \circ \beta = j^r g$. The relation between $\mathcal{F}_Z$, $\mathcal{F}$ is as follows.

Lemma 8.18. $\varphi\colon \mathcal{F}_Z \to C^0(Z, \mathcal{F})$, *respectively* $\varphi\colon \mathcal{H}_Z \to C^0(Z, \mathcal{H})$.

Proof. Let $(\beta, g) \in \mathcal{F}_Z$. Thus $\varphi(\beta, g) = (P(\beta), Q(g))$. Let $P(\beta) = \alpha \in C^0(Z, \Gamma(\mathcal{R}))$, and let $g = (\mathrm{id} \times h)$. Employing (8.15), (8.16), and the projection map $\pi\colon (Z \times X)^{(r-1)} \to Z \times X^{(r-1)}$, one calculates: for each $(z,x) \in Z \times V$ (one identifies $\Gamma(Z \times X^{(r-1)}) \equiv C^0(Z, X^{(r-1)})$),

$$\pi \circ p_{r-1}^r \circ \rho_Z(\beta)(z,x) = \pi \circ j^{r-1}g(z,x) = j^{r-1}h_z(x) \in X^{(r-1)}$$
$$= p_{r-1}^r \circ \rho(\alpha_z)(x).$$

Hence $p_{r-1}^r \circ \rho(\alpha_z) = j^{r-1}h_z = j^{r-1}(Q(g)(z))$ i.e. $(P(\beta)(z), h_z) \in \mathcal{F}$, from which it follows that $(P(\beta), Q(g)) \in C^0(Z, \mathcal{F})$ i.e. $\varphi\colon \mathcal{F}_Z \to C^0(Z, \mathcal{F})$. A similar argument on the level of r-jets proves that $\varphi\colon \mathcal{H}_Z \to C^0(Z, \mathcal{H})$. $\qquad\square$

Returning to the proof of the weak homotopy equivalence theorem, let $A\colon Z \to \mathcal{F}$ be continuous, $A(z) = (\alpha_z, f_z) \in \Gamma(\mathcal{R}) \times \Gamma^r(X)$. Thus the section f_z is C^r on V but only C^0 in $z \in Z$. We perturb the map A as follows to obtain smoothness also in the Z-variables.

Lemma 8.19. *Let $A\colon Z \to \mathcal{F}$, respectively $A\colon Z \to \mathcal{H}$, $A(z) = (\alpha_z, f_z)$, be continuous. Up to a small homotopy of the map A, one may assume that the section $g\colon Z \times V \to Z \times X$, $(z, x) \mapsto (z, f_z(x))$ is smooth. Furthermore, let $K \subset Z \times V$ be closed such that the restriction map (same notation) $g\colon \mathfrak{Op}\,K \to Z \times X$ is of class C^r, $r \geq 0$. Then one can choose the homotopy A_t, $A_0 = A$, $t \in [0,1]$, to be constant over (a smaller) $\mathfrak{Op}_1 K$ and such that the section g is smooth on $\overline{(Z \times V) \setminus \mathfrak{Op}\,K}$.*

Proof. The lemma follows from smooth approximation theory, I Theorem 1.1, and from the lifting properties V Theorem 5.3 applied to the microfibration $\mathrm{id} \times (p_{r-1}^r \circ \rho)\colon Z \times \mathcal{R} \to Z \times X^{(r-1)}$ in the case of the map $A\colon Z \to \mathcal{F}$, respectively the microfibration $\mathrm{id} \times \rho\colon Z \times \mathcal{R} \to Z \times X^{(r)}$ in the case of the map $A\colon Z \to \mathcal{H}$. $\square$

Proposition 8.20 Parametric h-Principle. *Let $A\colon Z \to \mathcal{F}$, $A_z = (\alpha_z, f_z)$, be continuous. Let $K \subset Z \times V$ be closed such that for all $(z, x) \in \mathfrak{Op}\,K$, $\rho \circ \alpha_z(x) = j^r f_z(x)$. There is a homotopy, $H\colon [0,1] \times Z \to \mathcal{F}$, $H_0 = A$, such that $H_1\colon Z \to \mathcal{H}$ and $H_t = A$ over $\mathfrak{Op}\,K$ (constant homotopy).*

Proof. Employing Lemma 8.19, up to a small homotopy of A, one may assume that the section $g\colon Z \times V \to Z \times X$, $(z, x) \mapsto (z, f_z(x))$ is smooth on $\overline{(Z \times V) \setminus \mathfrak{Op}\,K}$. In particular $j^r g \in \Gamma((Z \times X)^{(r)})$, and $Q(g) = f \in C^0(Z, \Gamma(X))$.

We lift $\alpha\colon Z \to \Gamma(\mathcal{R})$ back to $\beta \in \Gamma(\mathcal{R}_Z)$ such that $(\beta, g) \in \mathcal{F}_Z$. To this end, let $\rho \circ \alpha = J\colon Z \to \Gamma(X^{(r)})$. Employing Lemma 8.16 the lift $\lambda(J) = G \in \Gamma((Z \times X)^{(r)})$ satisfies the properties:

(i) $p_{r-1}^r \circ G = j^{r-1}g$.

(ii) $G = j^r g$ on $\mathfrak{Op}\,K$.

Let $\beta \in \Gamma(\mathcal{R}_Z)$ be the restriction of the pullback $\pi^*(\mathrm{id} \times \alpha)$ to the image $G(Z \times V) \subset (Z \times X)^{(r)}$. Employing (i), $p_{r-1}^r \circ \rho_Z \circ \beta = j^{r-1}g$ i.e. $(\beta, g) \in \mathcal{F}_Z$. Since $Q(g) = f$ it follows that $\varphi((\beta, g)) = (\alpha, f)$. Furthermore, employing (ii), $(\beta, g) \in \mathcal{H}_Z$ over $\mathfrak{Op}\,K$.

Employing Lemma 8.17, the relation $\mathcal{R}_Z$ is ample with respect to all codimension 1 tangent hyperplane fields τ_Z, and $\rho_Z\colon \mathcal{R}_Z \to (Z \times X)^{(r)}$ is a microfibration. Applying Theorem 8.12 to the formal solution $(\beta, g) \in \mathcal{F}_Z$ there is a homotopy of formal solutions $(\beta_t, g_t) \in \mathcal{F}_Z$, $(\beta_0, g_0) = (\beta, g)$, $t \in [0,1]$, such that $(\beta_1, g_1) \in \mathcal{H}_Z$, and $(\beta_t, g_t) = (\beta, g)$ on $\mathfrak{Op}\,K$. Note that this application of Theorem 8.12 requires the Local Extension Lemma 8.13 on charts only with respect to a sequence of integrable hyperplane fields $(\tau_i)_Z$, $1 \leq i \leq m = \dim \mathcal{H}(n,1)$, along which, employing Lemma 8.17, $\mathcal{R}_Z$ is ample.

Let $H: [0,1] \to C^0(Z, \mathcal{F})$ be the homotopy $H_t = \varphi((\beta_t, g_t))$, $t \in [0,1]$. Thus $H_0 = A$, $H_1 \in C^0(Z, \mathcal{H})$ and $H_t = \varphi((\beta, g)) = A$ over $\mathfrak{Op}\, K$ (constant homotopy), which completes the proof of the proposition. $\qquad\square$

Remark 8.21. The homotopy $(\beta_t, g_t) \in \mathcal{F}_Z$ constructed above satisfies the following approximation property not required in the conclusion. Let $\mathcal{N}$ be a neighbourhood of the image $j^r g(Z \times V)$ in $(Z \times V)^{(r-1)}$. Employing Theorem 8.12(ii), for all $t \in [0,1]$, the image $j^r g_t(Z \times V) \subset \mathcal{N}$.

To complete the proof of Theorem 8.15, employing the Parametric h-Principle, Proposition 8.20, the proof naturally splits into two parts ($\ell = (\alpha_0, f_0) \in \mathcal{H}$ is a base point):

  (1) The induced map $i_*\colon \pi_j(\mathcal{H}, \ell) \to \pi_j(\mathcal{F}, \ell)$ is surjective, for all $j \geq 0$.
  (2) The induced map $i_*\colon \pi_j(\mathcal{H}, \ell) \to \pi_j(\mathcal{F}, \ell)$ is injective, for all $j \geq 0$.

Surjectivity. Let $Z = S^j$ and let $A\colon (Z, z_0) \to (\mathcal{F}, \ell)$ be continuous, $A = (\alpha, f)$, where $\alpha\colon Z \to \Gamma(\mathcal{R})$, $f\colon Z \to \Gamma^r(X)$. Up to a small homotopy of A one may assume A is constant on $\mathfrak{Op}\, z_0$: $A(z) = \ell$ for all $z \in \mathfrak{Op}\, z_0$. Subjectivity then follows from the Parametric h-Principle, Proposition 8.20, applied to the closed manifold $Z = S^j$ and the closed subset $\{z_0\} \times V$.

Injectivity. Let $W = [0,1] \times Z$, $Z = S^j$, and let $A_0, A_1\colon (Z, z_0) \to (\mathcal{H}, \ell)$ be continuous such that there is a homotopy rel z_0, $B\colon W \to \mathcal{F}$, $B_i = A_i$, $i \in \{0,1\}$. Let $B = (\alpha, f)$, where $\alpha \in C^0(W, \Gamma(\mathcal{R}))$, $f \in C^0(W, \Gamma^r(X))$. In particular, for each $z \in Z$, (α_z^t, f_z^t), $t \in [0,1]$, is a homotopy of formal solutions to the relation $\mathcal{R}$. Up to a small homotopy of B one may assume that: (i) $B = \ell$ on $\mathfrak{Op}\, L$, where $L = [0,1] \times \{z_0\} \subset W$ (constant homotopy); (ii) $B = A_0$ on $Z \times [0, \epsilon]$, respectively $B = A_1$ on $Z \times [1 - \epsilon, 1]$ for a small $\epsilon > 0$. Injectivity then follows from Proposition 8.20 applied to the closed manifold W and the closed subset $([0,1] \times \{z_0\} \cup \partial W) \times V$. $\qquad\square$

Remark 8.22. The proof of injectivity yields additional information: the maps $A_i\colon S^j \to \mathcal{F}$, $i = 0, 1$ are homotopic by a homotopy $B\colon [0,1] \times S^j \to \mathcal{F}$, which itself is homotopic to $C\colon [0,1] \times S^j \to \mathcal{H}$, rel $L \cup (S^j \times \{0,1\})$. The relative homotopy from B to C, though not required, is a consequence of the general theory.

  To conclude this section we prove a generalized weak homotopy equivalence theorem that respects holonomicity up to jets of a fixed lower order i, $i \in \{0, \ldots, r-1\}$. The case $i = 0$ includes the classical results in Chapter IV. Let $\rho\colon \mathcal{R} \to X^{(r)}$ be continuous. Define,

$$\mathcal{H}_i = \{(\alpha, f) \in \Gamma(\mathcal{R}) \times \Gamma^i(X) \mid p_i^r \circ \rho \circ \alpha = j^i f\}.$$

In particular, $\mathcal{H}_r = \mathcal{H}$, the space of holonomic solutions to the relation $\mathcal{R}$ over $X^{(r)}$; $\mathcal{H}_k \subset \mathcal{H}_i$ for all $i < k$.

Theorem 8.23. *Let $\rho \colon \mathcal{R} \to X^{(r)}$ be a microfibration. Let $i \in \{0, \ldots, r-1\}$ and suppose the composed microfibration $p_s^r \circ \rho \colon \mathcal{R} \to X^{(s)}$ is ample for all s, $i+1 \leq s \leq r$. Then the inclusion $\mathcal{H} = \mathcal{H}_r \subset \mathcal{H}_i$ induces a weak homotopy equivalence.*

Proof. Let $\mathcal{F}_{i+1}$ be the space of formal solutions associated to the ample microfibration $p_{i+1}^r \circ \rho \colon \mathcal{R} \to X^{(i+1)}$. Thus $\mathcal{H}_{i+1} \subset \mathcal{F}_{i+1}$, the subspace of holonomic solutions, and $\mathcal{F}_{i+1} \subset \mathcal{H}_i$ is a proper subset i.e. the map f of the pair $(\alpha, f) \in \mathcal{H}_i$ is of class C^i rather than of class $i+1$ (cf. §8.2.3).

To prove the theorem, let Z be an auxiliary compact smooth manifold of parameters. Let $A \in C^0(Z, \mathcal{H}_i)$, $A(z) = (\alpha_z, f_z)$. Applying Lemma 8.19, up to a small homotopy, one may assume that for each $z \in Z$, $f_z \in \Gamma^{i+1}(X)$ i.e. $A \in C^0(Z, \mathcal{F}_{i+1})$. Furthermore, this homotopy is rel K in case $K \subset Z \times V$ is closed, and over $\mathfrak{Op}\, K$, the map $(z, x) \mapsto (z, f_z(x)) \in Z \times X$ is of class C^r and satisfies $p_i^r \circ \rho \circ \alpha_z(x) = j^r f_z(x)$.

Applying Proposition 8.20, there is a homotopy rel K, $A_t \in C^0(Z, \mathcal{F}_{i+1})$, $A_0 = A$, $t \in [0, 1]$, such that $A_1 \in C^0(Z, \mathcal{H}_{i+1})$. Thus $A_1 = (\alpha^1, f^1)$ satisfies the property that for all $z \in Z$,

$$p_{i+1}^r \circ \rho \circ \alpha_z^1 = j^{i+1} f_z^1 \in \Gamma(X^{(i+1)}).$$

An obvious induction on j, $i \leq j \leq r-1$, proves A is homotopic rel K to $B \in C^0(Z, \mathcal{H})$ ($\mathcal{H} = \mathcal{H}_r$). For all $k \geq 0$, setting $Z = S^k$ (for surjectivity), $Z = S^k \times [0, 1]$ (for injectivity), the theorem is proved. $\square$

In case $i = 0$ then $(\alpha, f) \in \mathcal{H}_0$ if and only if f is the 0-jet component of $\rho \circ \alpha \in \Gamma(X^{(r)})$. We therefore identify $\mathcal{H}_0 \equiv \Gamma(\mathcal{R})$, $(\alpha, f) \mapsto \alpha$, to obtain the following special case for $i = 0$.

Corollary 8.24. *Suppose the composed microfibration $p_s^r \circ \rho \colon \mathcal{R} \to X^{(s)}$ is ample for all $s \in \{1, \ldots, r\}$. Then the $\mathcal{H} = \mathcal{H}_r \to \Gamma(\mathcal{R})$, $(\alpha, f) \mapsto \alpha$, induces a weak homotopy equivalence.*

In case $\mathcal{R} \subset X^{(r)}$ we identify $\mathcal{H} = \Gamma_{\mathcal{R}}(X) = \{h \in \Gamma^r(X) \mid j^r h \in \Gamma(\mathcal{R})\}$. Assuming $\mathcal{R} \subset X^{(r)}$ is open, under the ampleness hypothesis of Corollary 8.24, the map $\Gamma_{\mathcal{R}}(X) \to \Gamma(\mathcal{R})$, $h \mapsto j^r h$, induces a weak homotopy equivalence, which reproves the classical Weak Homotopy Equivalence Theorem 4.7 studied in Chapter IV in case $r = 1$. However in case $i \geq 1$, Theorem 8.23 (cf. also the approximation property in Remark 8.21) lies beyond the scope of classical immersion-theoretic topology.

§3. Examples

8.3.1. There are two categories of applications of the main results of this chapter on the h-principle to solving, i.e. constructing holonomic sections, of relations over jet spaces. The first category of applications concerns the solutions of closed relations i.e. systems of PDEs. The results of this chapter apply only to prove the existence of approximate solutions to systems of PDEs which satisfy certain convexity properties and for which the solutions satisfy a density property on all lower order derivatives. This type of example is illustrated by Theorem 8.6 which is an application of the theory of short maps to the construction of approximate solutions to certain systems of PDEs. In Chapter IX this application of the h-principle is extended to include a convergence process for a sequence of approximate solutions in order to obtain, in the limit, C^r-solutions to rth order systems of PDEs which satisfy certain local convexity, "nowhere flatness", conditions. We defer to Chapter IX the development of this category of applications of the h-principle.

The second category of applications, to which we now turn, concerns the applications of the h-principle to solving open relations in jet spaces which are of interest in topology and geometry. Let V, W be smooth manifolds, $\dim V = n$, $\dim W = q$. Let $X = V \times W \to V$ be the product bundle, fiber W. In particular, $\Gamma^r(X) \equiv C^r(V, W)$, $r \geq 0$.

Let $\alpha \in \Gamma(\mathcal{R})$ be a continuous section of a relation $\rho \colon \mathcal{R} \to X^{(r)}$. Recall, I §1.3.1, that local coordinates in $X^{(r)}$ for the section $\rho \circ \alpha \in \Gamma(X^{(r)})$ are of the form ($f \in \Gamma(X)$ is the 0-jet component of $\rho \circ \alpha$; U is a chart on V): for all $x \in U$,

$$\rho \circ \alpha(x) = \left(x, f(x), (\varphi_\beta(x))_{|\beta| \leq r}\right),$$

where $\varphi_\beta \in C^0(U, \mathbf{R}^q)$ for all multi-indices β, $|\beta| \leq r$. The coordinates $(\varphi_\beta(x) \in \mathbf{R}^q)_{|\beta|=p}$ represent a polynomial map in $\mathcal{H}_p(n, q)$, $1 \leq p \leq r$.

8.3.2. Free Maps. Let $\mathcal{R}^r \subset X^{(r)}$ be the (generalized) free map relation in $X^{(r)}$: In local coordinates on $X^{(r)}$ as above, $\alpha(x) \in \mathcal{R}^r$ if and only if all of the vectors $\varphi_\beta \in \mathbf{R}^q$, $1 \leq |\beta| \leq r$, are linearly independent in $\mathbf{R}^q$. The free map relation is non-vacuous if and only if $q \geq D_r = d_1 + \cdots + d_r$, where $d_p = \dim \mathcal{H}_p(n, 1)$, $1 \leq p \leq r$. Since linear independence is an open condition it follows that $\mathcal{R}^r$ is open in $X^{(r)}$. Employing I (1.7), a section $\alpha \in \Gamma(\mathcal{R}^r)$ is holonomic i.e. $\alpha = j^r f$, if and only if in local coordinates on all charts U of the manifold V ($f(U) \subset \mathbf{R}^q$), for all $x \in U$ the vectors,

$$\frac{\partial^r f}{\partial u_{i_1} \ldots \partial u_{i_p}}(x) \in \mathbf{R}^q, \quad 1 \leq i_1 \leq i_2 \leq \cdots \leq i_p \leq n, \quad 1 \leq p \leq r,$$

are linearly independent. A map $f \in C^r(V, W)$ is *free* if $j^r f \in \Gamma(\mathcal{R}^r)$. In case $r = 2$ then (cf. I §1.3.1) $\alpha = j^2 f \in \Gamma(\mathcal{R}^2)$ if and only if over all charts $U \subset V$, for all $x \in U$ the vectors,

$$\frac{\partial f}{\partial u_i}(x), \frac{\partial^2 f}{\partial u_j, \partial u_k}(x), \quad 1 \le i \le n, 1 \le j \le k \le n,$$

are linearly independent, which is precisely the classical free map condition on a map $f \in C^2(V, W)$ ($q \ge n + n(n+1)/2$). In case $r = 1$ then $\mathcal{R}^1 \subset X^{(1)}$ is the immersion relation.

The projection map $p_r^s \colon X^{(s)} \to X^{(r)}$, $s \ge r$, induced by the restriction of jets $[f]_s \mapsto [f]_r$ (cf. I §1.2.1), induces a natural projection map $\rho = p_r^s \colon \mathcal{R}^s \to \mathcal{R}^r$. In local coordinates this projection map is just the restriction (pointwise) to a subset of linearly independent vectors. Since $\mathcal{R}^s$ is open in $X^{(s)}$, employing the bundle $p_r^s \colon X^{(s)} \to X^{(r)}$, it follows that the projection $\rho \colon \mathcal{R}^s \to X^{(r)}$ is a microfibration.

Lemma 8.25. $\mathcal{R}^r \subset X^{(r)}$ *is ample if and only if* $q > D_r$ *(known as the extra dimensional case).*

Proof. Let $x \in V$ and let $\tau \subset T(\mathfrak{Op}\, x)$ be a codimension 1 tangent hyperplane field. Associated to τ is the affine $\mathbf{R}^q$-bundle $p_\perp^r \colon X_U^{(r)} \to X_U^\perp$ ($U = \mathfrak{Op}\, x$). We prove that for all $(b, a) \in X_U^\perp \times \mathcal{R}_b^r$ (cf. VIII §8.2.1),

$$X_b^{(r)} = \mathrm{Conv}(\mathcal{R}_b^r, a),$$

if and only if $q > D_r$. To this end let $b = j^\perp f(x) \in X_U^\perp$. With respect to a basis $v = (v_1, \dots, v_n)$ of $T_x(V)$ which is adapted to τ (cf. VI §6.1.1),

$$j^\perp f(x) = (j^{r-1} f(x), \partial_v^\alpha f(x)), \tag{8.17}$$

where $|\alpha| = r$ and $\partial_v^\alpha \neq \partial_{v_n}^r$ i.e. only the pure derivative $\partial_{v_n}^r$ is excluded. Since $a \in \mathcal{R}_b^r$ i.e. $a = (b, y) \in X_b^{(r)} = \{b\} \times \mathbf{R}^q$ satisfies the free map condition, it follows that all the derivative vectors, $\partial_v^\alpha f(x) \in \mathbf{R}^q$, $1 \le |\alpha| \le r$, $\partial_v^\alpha \neq \partial_{v_n}^r$ are linearly independent and span a subspace L of $\mathbf{R}^q$, $\dim L = D_r - 1$. Consequently the fiber $\mathcal{R}_b^r$ is the complement of the subspace L in $X_b^{(r)} \equiv \mathbf{R}^q$. In case $q = D_r$ the subspace L has codimension 1; hence the complement $\mathcal{R}_b^r$ consists of 2 half-spaces, each of whose convex hulls is itself i.e. condition (8.17) is not satisfied and $\mathcal{R}^r$ is not ample in the equidimensional case $q = D_r$. In case $q > D_r$, then the subspace L has codimension 2, from which it follows that the complement $\mathcal{R}_b^r$ is path connected and whose convex hull is $X_b^{(r)}$ i.e. condition (8.17) is satisfied and hence $\mathcal{R}$ is ample. $\qquad\square$

Analogous arguments prove the following extension of Lemma 8.25 to the microfibration $\rho\colon \mathcal{R}^s \to X^{(r)}$. The extra dimensional case is not required in case $s > r$.

Lemma 8.26. *Let $s > r \geq 1$ be integers. The projection map $\rho\colon \mathcal{R}^s \to X^{(r)}$ is ample for all $q \geq D_s$.*

Proof. Briefly, in the above notation, let $a \in \mathcal{R}_b^s$ and $c = (b, y) \in X_b^{(r)}$ where $y \in \mathbf{R}^q \setminus L$. There exists an extension of c to an element $a' \in \mathcal{R}_b^s$ i.e. $\rho(a') = c$ (restriction to a subset of independent vectors). Since $s > r$, up to a change in sign of the last vector in a', one easily proves that a', a lies in the same path component of $\mathcal{R}_b^s$. Consequently $X_b^{(r)} = \operatorname{Conv}(\mathcal{R}_b^s, a)$, which proves that $\mathcal{R}^s$ is ample over $X^{(r)}$ in the case $s > r$. $\qquad\square$

Theorem 8.27. *Let $\mathcal{R}^s \subset X^{(s)}$ be the free map relation, $s \geq 1$. Suppose $q > D_s$.*

(i) *The inclusion $i\colon \mathcal{R}^s \to X^{(s)}$ satisfies the strong C^r-dense h-principle for all r, $0 \leq r \leq s - 1$.*

(ii) *The inclusion map $i\colon \mathcal{H} \to \mathcal{F}$ (cf. §8.2.3) induces a weak homotopy equivalence.*

(iii) *Let $\mathcal{H}_r = \{(\alpha, f) \in \Gamma(\mathcal{R}^s) \times \Gamma^r(X) \mid p_r^s \circ \alpha = j^r f\}$. For all $s \geq r \geq 0$, the inclusion $\mathcal{H} = \mathcal{H}_s \to \mathcal{H}_r$ induces a weak homotopy equivalence.*

Proof. The projection map $\rho\colon \mathcal{R}^s \to X^{(r)}$ is a microfibration and, employing Lemmas 8.25, 8.26, the relation $\mathcal{R}^s$ is ample over $X^{(r)}$ for all r, $1 \leq r < s$. The theorem follows from Theorems 8.14, 8.15, 8.23. $\qquad\square$

As a particular example in case $r = 1$, $s = 2$, let $\alpha \in \Gamma(\mathcal{R}^2)$ such that $p_1^2 \circ \alpha = j^1 f$, where $f \in C^1(V, W)$ is an immersion i.e. $j^1 f$ lifts to α which satisfies the classical free map condition. If $q > n + n(n+1)/2$ it follows from Theorem 8.27(i) that one can approximate f in the fine C^1-topology by a free map $g \in C^2(V, W)$. In fact this application of Theorem 8.14 gives more precise information. Let $\mathcal{N}$ be a neighbourhood of the image $j^1 f(V)$ in $X^{(1)}$. There is a homotopy $\alpha_t \in \Gamma(\mathcal{R}^2)$, $t \in [0, 1]$, $\alpha_0 = \alpha$, such that the projection $p_0^1 \circ \alpha_t = j^1 g_t$, where $g_t \in C^1(V, W)$, $t \in [0, 1]$, is a regular homotopy of immersions, $g_0 = f$, g_1 is a free map, and for all $t \in [0, 1]$, the image $j^1 g_t(V) \subset \mathcal{N}$. As explained in I §1.3.2, this sort of result in the fine C^1-topology cannot be proved by the covering homotopy methods due originally to Smale [34] and generalized extensively in Gromov [16].

8.3.3. Submersions of Open Manifolds. Let V, W be smooth manifolds such that $\dim V = n$, $\dim W = q$, $n \geq q$. Let $X = V \times W \to V$ be the product bundle, fiber W. Let $\mathcal{R} \subset X^{(1)}$ be the submersion relation defined by germs of C^1-maps

$h\colon V \to W$ of maximal rank q. As explained in I §1.3.1, a section $\alpha \in \Gamma(\mathcal{R})$, whose 0-jet component is $g \in C^0(V, W)$, corresponds to a bundle epimorphism $F\colon T(V) \to T(W)$ which covers g. Since maximal rank is an open condition it follows that $\mathcal{R}$ is open in $X^{(1)}$.

However $\mathcal{R}$ is not ample in $X^{(1)}$. To see this, let $x \in V$ and let $\tau \subset T(\mathfrak{Op}\, x)$ be a codimension 1 tangent hyperplane field. Associated to τ is the affine $\mathbf{R}^q$-bundle $p_\perp^1\colon X_U^{(1)} \to X_U^\perp$ $(U = \mathfrak{Op}\, x)$. We show that for all $b = j^\perp f(x) \in X_U^\perp$, $X_b^{(1)} \neq \mathrm{Conv}(\mathcal{R}_b, a)$.

To this end let $v = (v_1, \ldots, v_n)$ be a basis of $T_x(V)$ which is adapted to τ. In local coordinates let $a = (j^\perp f(x), y) \in \mathcal{R}_b$: the vectors $\partial_{v_i} f(x), y \in \mathbf{R}^q$, $1 \leq i \leq n - 1$, span $\mathbf{R}^q$. Let L be the subspace of $\mathbf{R}^q$ spanned by the derivative vectors $\partial_{v_i} f(x)$, $1 \leq i \leq n - 1$. Thus $\dim L \geq q - 1$. In case $L = \mathbf{R}^q$ then $X_b^{(1)} = \mathcal{R}_b$. However in case L has codimension 1 then $\mathcal{R}_b = \mathbf{R}^q \setminus L$ consists of 2 half spaces, each of whose convex hulls is itself i.e. $X_b^{(1)} \neq \mathrm{Conv}(\mathcal{R}_b, a)$, from which it follows that $\mathcal{R}$ is not ample. Simple examples show that in general the h-principle fails for the submersion relation. However, in case the base manifold V is open then it is a classical result in immersion-theoretic topology, Phillips [31], that the submersion relation $\mathcal{R}$ satisfies the weak homotopy equivalence property which, in our terminology, is formulated as follows. Recall that $\mathcal{F}, \mathcal{H}$ denote the space of formal solutions, respectively the subspace of holonomic solutions, of the relation $\mathcal{R} \subset X^{(1)}$. Classically, let $\mathrm{Sub}(V, W)$ denote the subspace of submersions in $C^1(V, W)$. We identify $\mathrm{Sub}(V, W) \equiv \mathcal{H}$ via the homeomorphism $h \mapsto (j^1 h, h)$.

Submersion Theorem 8.28 (Phillips). *Let V, W be smooth manifolds such that $\dim V = n$, $\dim W = q$, $n \geq q$. Let $\mathcal{R} \subset X^{(1)}$ be the submersion relation. If V is an open manifold then the map $i\colon \mathcal{H} = \mathrm{Sub}(V, W) \to \Gamma(\mathcal{R})$, $h \mapsto j^1 h$, induces a weak homotopy equivalence.*

Consequently, in the case of the (non-ample) submersion relation $\mathcal{R} \subset X^{(1)}$, Theorem 8.28 generalizes the weak homotopy equivalence result Corollary 8.24 if the domain V is an open manifold.

In what follows Theorem 8.28 is proved by convex integration. The point here is that the open manifold V admits of a handle decomposition by handles of index $\leq n - 1$ i.e. handles of the form $(D^{n-k}, S^{n-k-1}) \times [-1, 1]^k$, $1 \leq k \leq n$, whose cores D^{n-k} are not top dimensional. The existence of these extra "normal directions" in the factor $[-1, 1]^k$ is essential to the proof of the theorem, Phillips [31], which is based on a covering homotopy construction. The proof below replaces the covering homotopy construction over these "thickened" cores by convex integration with respect to a certain ample microfibration $\rho\colon \mathcal{S} \to Y^{(1)}$, where $Y = D^{n-k} \times W \to D^{n-k}$ is the product bundle, fiber W, over the core D^{n-k}, $k \geq 1$.

Proof of Theorem 8.28. We consider the following general data: Let $V = V' \times \mathbf{R}^k$ (a split manifold), dim $V' = n - k$, $k \geq 1$. We identify $V' = V' \times \{0\} \subset V$. Let $X = V \times W \to V$ be the product bundle, fiber W; let also $Y = V' \times W \to V'$ be the product bundle, fiber W over the submanifold V'. Let $\mathcal{R} \subset X^{(1)}$ be the submersion relation. Let $\mathcal{S} \subset \mathcal{R}$ be the subspace defined by 1-jets whose source is in the submanifold V'. $\mathcal{S}$ is a submanifold of codimension k of the open manifold $\mathcal{R}$.

Let $\rho \colon \mathcal{S} \to Y^{(1)}$ be the projection map induced by the restriction of germs of C^1-functions to the submanifold V': $\rho(j^1 h(x)) = j^1(h|V')(x)$, $x \in V'$. In local coordinates that respect the splitting $V = V' \times \mathbf{R}^q$, let $a = (x, y, \varphi_1, \ldots, \varphi_n) \in \mathcal{S}$. Then $\rho(a) = (x, y, \varphi_1, \ldots, \varphi_{n-k}) \in Y^{(1)}$.

Algebraically, in these local coordinates, $\rho(x, y, A) = (x, y, B)$ where A is a $q \times n$ matrix and B is the $q \times (n - k)$ matrix obtained from A by deleting the last k columns. Since $\mathcal{R}$ is the submersion relation then the above matrix A has maximal rank q, from which it follows that the image $\rho(\mathcal{S}) \subset Y^{(1)}$ consists of 1-jets of germs of C^1-functions of rank $\geq q - k$.

Lemma 8.29. *The projection map $\rho \colon \mathcal{S} \to Y^{(1)}$ is a microfibration and $\mathcal{S}$ is ample over $Y^{(1)}$.*

Proof. The point of the lemma is that although the relation $\mathcal{R} \subset X^{(1)}$ is not ample, the above splitting of the manifold V ensures that the corresponding "over" relation $\rho \colon \mathcal{S} \to Y^{(1)}$ is ample.

Let $\mathcal{L}_r(n, q)$ denote the linear space of linear maps of rank $\geq r$ from $\mathbf{R}^n$ to $\mathbf{R}^q$. Let $\pi \colon \mathcal{L}_q(n, q) \to \mathcal{L}_{q-k}(n - k, q)$ be the linear map obtained by restriction to the subspace $\mathbf{R}^{n-k} \subset \mathbf{R}^n$. In the above local coordinates for $\rho \colon \mathcal{S} \to Y^{(1)}$, $\rho(x, y, A) = (x, y, B)$, $B = \pi(A)$. Since $d\pi = \pi$ is onto it follows that the map ρ is a submersion hence a microfibration.

We prove that $\mathcal{S}$ is ample over $Y^{(1)}$. Let $x \in V'$ and let $\tau \subset T(\mathfrak{Op}\, x)$ be a codimension 1 tangent hyperplane field. Associated to τ is the affine $\mathbf{R}^q$-bundle $p_{\perp}^r \colon Y_U^{(1)} \to Y_U^{\perp}$ ($U = \mathfrak{Op}\, x$). We prove that for all $(b, a) \in Y_U^{\perp} \times \mathcal{S}_b$ (cf. VII 8.2.1),

$$Y_b^{(1)} = \mathrm{Conv}(\mathcal{S}_b, a). \tag{8.18}$$

Let $b = j^{\perp} h(x) = (x, h(x), \partial_1 h(x) \ldots, \partial_{n-k-1} h(x)) \in Y^{\perp}$ in the above split local coordinates. Thus $a = (b, \varphi_1, \ldots \varphi_{k+1}) \in \mathcal{S}_b$ i.e. the span of the vectors $\partial_j h(x), \varphi_i \in \mathbf{R}^q$, $1 \leq j \leq n - k - 1$, $1 \leq i \leq k + 1$, is $\mathbf{R}^q$.

To prove (8.18) fix $(b, z) \in Y_h^{(1)}$ and let $L \subset \mathbf{R}^q$ be the subspace spanned by the $n - 2$ vectors obtained from $a \in \mathcal{S}_b$ by deleting φ_1, φ_{k+1}:

$$L = \mathrm{span}\, \{\partial_j h(x), \varphi_2, \ldots, \varphi_k\}, \quad 1 \leq j \leq n - k - 1.$$

Thus dim $L \geq q - 2$. We consider the three cases, $q - 2 \leq \dim L \leq q$.

(i) $\dim L = q$. Let $z^t \in \mathbf{R}^q$, $t \in [0,1]$, be a path that connects $z^0 = \varphi_1$ to $z^1 = z$. Define $a^t = (b, z^t, \varphi_2, \dots, \varphi_{k+1})$. Then a^t is a path in $\mathcal{S}_b$, $a^0 = a$, such that $\rho(a^1) = (b, z) \in Y_b^{(1)}$, which proves (8.18).

(ii) $\dim L = q - 1$. Let (z^t, φ^t), $t \in [0,1]$, be a pair of paths in $\mathbf{R}^q$ such that $(z^0, \varphi^0) = (\varphi_1, \varphi_{k+1})$, $z^1 = z$, and for all $t \in (0,1]$ the subspace spanned by L, φ^t is $\mathbf{R}^q$. Define $a^t = (b, z^t, \varphi_2, \dots, \varphi^t)$. Then a^t is a path in $\mathcal{S}_b$, $a^0 = a$, such that $\rho(a^1) = (b, z) \in Y_b^{(1)}$, which proves (8.18).

(iii) $\dim L = q - 2$. Let $w \in \mathbf{R}^q \setminus L$. Since $\dim L^\perp = 2$ it follows that there is a pair of paths in $\mathbf{R}^q \setminus L$, (w^t, φ^t), $t \in [0,1]$, $(w^0, \varphi^0) = (\varphi_1, \varphi_{k+1})$, such that $w^1 = w$ and for all $t \in [0,1]$ the subspace spanned by L, w^t, φ^t is $\mathbf{R}^q$. Define $a^t = (b, w^t, \varphi_2, \dots, \varphi^t)$. Then a^t is a path in $\mathcal{S}_b$, $a^0 = a$, such that $\rho(a^1) = (b, w)$, which proves (8.18) and the proof of the lemma is complete. $\square$

In what follows, $\mathcal{F}_{\mathcal{S}}$, $\mathcal{H}_{\mathcal{S}}$, denote the space of formal solutions, respectively the subspace of holonomic solutions, of the relation $\rho \colon \mathcal{S} \to Y^{(1)}$ (cf. §8.2.3). In particular, $(\alpha, h) \in \mathcal{H}_{\mathcal{S}}$ if and only if $h \in C^1(V', W)$ and the section $\alpha \in \Gamma(\mathcal{S})$ satisfies $\rho \circ \alpha = j^1 h$: in split local coordinates, for all $x \in V'$,

$$\alpha(x) = (x, h(x), \partial_1 h(x), \dots, \partial_{n-k} h(x), \varphi_1(x), \dots, \varphi_k(x)),$$

where the n vectors $\partial_j h(x), \varphi_r(x) \in \mathbf{R}^q$, $1 \le j \le n - k$, $1 \le r \le k$, span $\mathbf{R}^q$. Suppose V' is compact. Applying the exponential map on $T(W)$ to the vectors $\varphi_r(x)$ (along the image $h(V') \subset W$), $1 \le r \le k$, for suitably small $\epsilon > 0$ one constructs a map,

$$\mu \colon \Gamma(\mathcal{H}_{\mathcal{S}}) \to \mathrm{Sub}((V' \times [-\epsilon, \epsilon]^k), W). \tag{8.19}$$

such that the submersion $\mu((\alpha, h))$ extends h on V'. Conversely, if $g \in \mathrm{Sub}(\mathfrak{O}\mathfrak{p}V', W)$ then the restriction of the tangent map dg to the zero-section V' induces a section $(dg|V', g|V') \in \Gamma(\mathcal{H}_{\mathcal{S}})$. Throughout, the map μ is treated as an equivalence (technically a weak homotopy equivalence).

Applying the Parametric h-Principle, Proposition 8.20, to the ample relation $\mathcal{S}$ over $Y^{(1)}$ one obtains the following result.

Corollary 8.30. *Let $A \colon Z \to \mathcal{F}_{\mathcal{S}}$ be continuous, $A(z) = (\alpha_z, f_z) \in \Gamma(\mathcal{S}) \times \Gamma^r(Y)$, where Z is a compact smooth manifold (a space of parameters). Let $K \subset Z \times V'$ be closed such that for all $(z, x) \in \mathfrak{O}\mathfrak{p}\, K$, $\rho \circ \alpha_z(x) = j^1 f_z(x)$ (i.e. over $\mathfrak{O}\mathfrak{p}\, K$, A is holonomic).*

There is a homotopy $H \colon Z \times [0,1] \to \mathcal{F}_{\mathcal{S}}$, $H_0 = A$, $H_t = A$ over $\mathfrak{O}\mathfrak{p}\, K$ (constant homotopy), such that $H_1 \colon Z \to \mathcal{H}_{\mathcal{S}}$.

Returning to the proof of Theorem 8.28 (cf. Haefliger [20]), there is a smooth proper Morse function $f\colon V \to [0, \infty)$ which has no critical point of maximal index n. At each critical point $w \in V$ there is a $\delta > 0$ such that $c = f(w)$ is the only critical level in the interval $[c - \delta, c + \delta]$, For each $t \geq 0$ let $V^t = \{x \in V \mid f(x) \leq t\}$. Then the manifold $V^{c+\delta}$ is obtained from $V^{c-\delta}$ by adding a handle $(D^{n-k}, S^{n-k-1}) \times [-1, 1]^k$, where $k \geq 1$ is the index of the critical point $w \in V$.

The proof proceeds inductively on the set of critical points $\{w_i\}$ which are ordered so that the critical values $c_i = f(w_i)$ are increasing. For notational convenience, for each $\epsilon \in (0, 1]$ let,

$$A_\epsilon = D^{n-k} \times [-\epsilon, \epsilon]^k, \quad B_\epsilon = (\mathfrak{Op}\, \partial D^{n-k}) \times [-\epsilon, \epsilon]^k \subset D^{n-k} \times [-1, 1]^k.$$

The main step in the proof of Theorem 8.28 is to prove that the restriction map, $\rho : \mathrm{Sub}(A_1, W) \to \mathrm{Sub}(B_1, W)$, is a Serre fibration. To this end, let Z be an auxiliary compact smooth manifold and consider the following commutative diagram of continuous maps ($A = A_1$; $B = B_1$),

$$
\begin{array}{ccc}
Z \times \{0\} & \xrightarrow{\;f\;} & \mathrm{Sub}(A, W) \\
\Big\downarrow{\scriptstyle i} & & \Big\downarrow{\scriptstyle \rho} \\
Z \times [0, 1] & \xrightarrow[\;F\;]{} & \mathrm{Sub}(B, W)
\end{array}
\qquad (8.20)
$$

The problem is to construct a continuous lift $G\colon Z \times [0, 1] \to \mathrm{Sub}(A, W)$ such that: (i) $G_0 = f$; (ii) $\rho \circ G = F$. In fact it is sufficient to construct a continuous lift for a suitably small $\epsilon > 0$,

$$G'\colon Z \times [0, 1] \to \mathrm{Sub}(A_\epsilon, W),$$

such that $G'_0 = f\colon Z \times \{0\} \to \mathrm{Sub}(A_\epsilon, W)$, and $\rho \circ G' = F\colon Z \times [0, 1] \to \mathrm{Sub}(B_\epsilon, W)$. Indeed, the submersion relation is stable in the sense of Gromov [16], (cf. also Haefliger [20]). Consequently, employing a homotopy of diffeomorphisms which connects the identity map to a diffeomorphism of $D^{n-k} \times [-1, 1]^k$ into a suitably small neighbourhood of $D^{n-k} \times \{0\} \cup (\partial D^{n-k} \times [-1, 1]^k)$, the lift G' extends to a lift G as above which satisfies the Serre fibration property. Furthermore we assume that the evaluation maps $\mathrm{ev}\, f\colon Z \times A \to W$, $\mathrm{ev}\, F\colon Z \times [0, 1] \times B \to W$ are of class C^1.

Setting $V' = D^{n-k} \times \{0\}$ (the zero-section), $Y = V' \times W \to V'$, recall the ample microfibration $\rho\colon \mathcal{S} \to Y^{(1)}$ defined above. Corollary 8.30 proves the following lemma.

Lemma 8.31. Let $P = Z \times [0, 1]$ and let $\sigma\colon P \to \mathcal{F}_\mathcal{S}$, $\sigma(y) = (\alpha_y, h_y)$, be continuous such that $\rho \circ \alpha_y(x) = \jmath^1 h_y(x)$ for all $(y, x) \in \mathfrak{Op}\, K$ where,

$$K = (Z \times \{0\} \times V') \cup (P \times \partial V') \subset P \times V'.$$

(Thus over $\mathfrak{Op}\, K$, σ is holonomic.) There is a homotopy rel$\mathfrak{Op}\, K$, $H\colon P \times [0, 1] \to \mathcal{F}_\mathcal{S}$, such that $H_0 = \sigma$, $H_1\colon P \to \mathcal{H}_\mathcal{S}$.

Employing the equivalence $\mu\colon \mathcal{H}_{\mathcal{S}} \to \mathrm{Sub}(A_\epsilon, W)$ (cf. (8.19)), we construct the map σ of the lemma as follows. Over K the map σ is given by the data of diagram (8.20) restricted to A_ϵ. Explicitly, over $Z \times \{0\} \times V'$, σ is defined by the map f; over $P \times \partial V'$, σ is defined by the map F. This data extends to $\mathfrak{Op}\, K$. The extension of this data on $\mathfrak{Op}\, K$ to $\sigma\colon P \to \mathcal{F}_{\mathcal{S}}$ follows from bundle considerations applied to a strong deformation retract on $P \times V'$ to the subspace K. The map $G = \mu \circ H_1\colon P \to \mathrm{Sub}(A_\epsilon, W)$ solves the Serre fibration lifting problem of diagram (8.20) at each of the critical points. The proof of Theorem 8.28 is completed by a taking the limit in a tower of weak homotopy equivalences as detailed in Gromov [16], Haefliger [20], Phillips [31].   □

Remark 8.32. In case $n = q$ then Theorem 8.28 reduces to the classical classification of immersions theorem in the (non-ample) equidimensional case, provided the source manifold V is open. Also, in case V is open, the techniques employed in the proof of Theorem 8.28 easily adapt to prove the Weak Homotopy Equivalence Theorem 8.27(ii) in the (non-ample) equidimensional case $q = D_s$ for the free map relation $\mathcal{R}^s \subset X^{(s)}$. It is not clear to what extent Convex Integration theory can be employed to replace the covering homotopy method of Gromov [16] for proving classification theorems for open relations $\mathcal{R} \subset X^{(r)}$ that may not be ample, in case the base manifold V is open. For example, Gromov [16] proves the analogue of IV Theorem 4.13 for the classification of symplectic 2-forms on open, even dimensional smooth manifolds V (the corresponding non-degeneracy relation on 2-forms is not ample). No proof of this result by Convex Integration theory is known.

§4. Relative h-Principles

8.4.1. Let $\rho\colon \mathcal{R} \to \mathcal{R}_0 \subset X^{(r)}$ be a relation over the subspace $\mathcal{R}_0$ in $X^{(r)}$. Recall that elements in the extensions $\mathrm{Pr}_n(\mathcal{R})$, $\mathrm{Conv}_n(\mathcal{R})$, $n \geq 1$, consist of piecewise principal paths in $X^{(r)}$, respectively (cf. Proposition 8.5) certain families of surrounding paths in successive principal subspaces in $X^{(r)}$. For certain problems it is reasonable to ask whether these piecewise principal paths and families of surrounding paths can be chosen to lie in the subspace $\mathcal{R}_0$ i.e. that all constructions be carried out relative to the subspace $\mathcal{R}_0$. For example, $\mathcal{R}_0 \subset X^{(1)}$ is the immersion relation and $\mathcal{R} \subset \mathcal{R}_0$ represents immersions which satisfy some additional property, for example totally real embeddings (studied below). However, in general the affine geometry of convex hull extensions imposes strong restrictions on the subspace $\mathcal{R}_0$ in order to carry out the above programme of constructions relative to $\mathcal{R}_0$.

Define $\mathrm{Pr}_1(\mathcal{R}|\mathcal{R}_0) \subset \mathrm{Pr}_1(\mathcal{R})$ to be the subspace of those pairs $(a, x) \in \mathrm{Pr}_1(\mathcal{R})$ for which $(1-t)\rho(a) + tx \in \mathcal{R}_0$ for all $t \in [0,1]$ i.e. the segment joining

$\rho(a), x$ is contained in $\mathcal{R}_0$. Inductively we define,

$$\mathrm{Pr}_n(\mathcal{R}|\mathcal{R}_0) = \mathrm{Pr}_1(\mathrm{Pr}_{n-1}(\mathcal{R}|\mathcal{R}_0)|\mathcal{R}_0), \ n \geq 2.$$

Thus $y = (a, x_1, \ldots, x_n) \in \mathrm{Pr}_n(\mathcal{R}|\mathcal{R}_0)$, $n \geq 2$, if and only if $y \in \mathrm{Pr}_n(\mathcal{R})$ and each of the segments joining $\rho(a), x_1$, respectively x_{j-1}, x_j, lies in the subspace $\mathcal{R}_0$, $2 \leq j \leq n$. Let $\mathrm{Conv}_1(\mathcal{R}|\mathcal{R}_0) = \mathrm{Conv}_1(\mathcal{R}) \cap \mathrm{Pr}_1(\mathcal{R}|\mathcal{R}_0)$; define inductively,

$$\mathrm{Conv}_n(\mathcal{R}|\mathcal{R}_0) = \mathrm{Conv}_1(\mathrm{Conv}_{n-1}(\mathcal{R}|\mathcal{R}_0)|\mathcal{R}_0), \ n \geq 2.$$

In particular, $\mathrm{Conv}_n(\mathcal{R}|\mathcal{R}_0) \subset \mathrm{Pr}_n(\mathcal{R}|\mathcal{R}_0)$ for all $n \geq 1$. There is the inclusion map $i \colon \mathcal{R} \to \mathrm{Conv}_n(\mathcal{R}|\mathcal{R}_0)$, $a \mapsto (a, \rho(a), \ldots, \rho(a))$. The projection $\rho \colon \mathrm{Pr}_n(\mathcal{R})$ $\to X^{(r)}$, $(a, x_1, \ldots, x_n) \mapsto x_n$ restricts to a projection map (same notation) $\rho \colon$ $\mathrm{Conv}_n(\mathcal{R}|\mathcal{R}_0) \to X^{(r)}$. In case $\mathcal{R}_0 = X^{(r)}$ then for all $n \geq 1$,

$$\mathrm{Pr}_n(\mathcal{R}|\mathcal{R}_0) = \mathrm{Pr}_n(\mathcal{R}); \quad \mathrm{Conv}_n(\mathcal{R}|\mathcal{R}_0) = \mathrm{Conv}_n(\mathcal{R}).$$

Definition 8.33. Let $\mathcal{R}_0 \subset X^{(r)}$ be open. A relation $\rho \colon \mathcal{R} \to \mathcal{R}_0 \subset X^{(r)}$ is *relatively ample* with respect $\mathcal{R}_0$ if for all $n \geq 1$,

$$\mathrm{Conv}_n(\mathcal{R}|\mathcal{R}_0) = \mathrm{Pr}_n(\mathcal{R}|\mathcal{R}_0)$$

In case $\mathcal{R}_0 = X^{(r)}$ then $\mathcal{R}$ is relatively ample with respect to $X^{(r)}$ if and only if $\mathcal{R}$ is ample.

Lemma 8.34. *let $\mathcal{R}_0 \subset X^{(r)}$ be open. Let $\rho \colon \mathcal{R} \to \mathcal{R}_0$ be a microfibration. For all $n \geq 1$ the projection map $\rho \colon \mathrm{Conv}_n(\mathcal{R}|\mathcal{R}_0) \to \mathcal{R}_0$ is a microfibration.*

Proof. The proof is by induction on n. The case $n = 1$ follows the proof of the corresponding microfibration property VII Lemma 7.1 in the case $\mathcal{R}_0 = X^{(r)}$ i.e. since $\mathcal{R}_0$ is open the proof easily relativizes to $\mathcal{R}_0$. Details are left to the reader. □

Definition 8.35. A relation $\rho \colon \mathcal{R} \to \mathcal{R}_0 \subset X^{(r)}$ is *relatively dense* in an open set $\mathcal{R}_0 \subset X^{(r)}$ if for all open sets $U \subset \mathcal{R}_0$ and all principal subspaces R such that $R \cap U \neq \emptyset$ the following property obtains:

$$\mathrm{Conv}_1(\rho^{-1}(U)|U) = \mathrm{Pr}_1(\rho^{-1}(U)|U). \tag{8.21}$$

For example, let $(a, x) \in \mathrm{Pr}_1(\rho^{-1}(U)|U)$ with respect to a principal subspace R. Employing (8.21) there is a path $a_t \in \rho^{-1}(U)$, $a_0 = a$, $t \in [0, 1]$, such that the path $\rho(a_t)$ lies in $R \cap U$ and surrounds x. Let J be the segment in $R \cap U$

that joins $\rho(a), x$. Applying (8.21) to the open sets of the form $\mathfrak{Op}\, J \subset \mathcal{R}_0$, one concludes that $J \subset \overline{\rho(\mathcal{R})}$.

Gromov [18] introduced (8.21) as the main criterion for a relation $\mathcal{R}$ to be relatively ample with respect to $\mathcal{R}_0$. This is proved as Theorem 8.40 below. The main construction of relatively dense relations is as follows. Let $\Sigma \subset X^{(r)}$ be a closed stratified subset such that the intersection $\Sigma \cap R$ is either equal to R or has codimension ≥ 2 in R for all principal subspaces $R \subset X^{(r)}$. Consequently for all open sets $U \subset X^{(r)}$,

$$\mathrm{Conv}_1((U \setminus \Sigma)|U) = \mathrm{Pr}_1((U \setminus \Sigma)|U). \qquad (8.22)$$

It follows that the inclusion $i\colon \mathcal{R} = \mathcal{R}_0 \backslash \Sigma \to \mathcal{R}_0$ satisfies (8.21) i.e. $\mathcal{R}$ is relatively dense in $\mathcal{R}_0$ for all open subsets $\mathcal{R}_0$ in $X^{(r)}$. For example, let $X = V \times W \to V$ be the product bundle, $\dim V = n$, $\dim W = q$. Let $\mathcal{R}^k$ be the k-mersion relation in $X^{(1)}$. Then $\mathcal{R}^k$ is open, and in case $k < q$, $\mathcal{R}^k$ is the complement in $X^{(1)}$ of a closed stratified subset Σ that has the codimension ≥ 2 property above. Thus for all open sets $\mathcal{R}_0 \subset X^{(1)}$ the relation $\mathcal{R} = \mathcal{R}_0 \setminus \Sigma$ is relatively dense by inclusion into $\mathcal{R}_0$ ($\mathcal{R}$ represents those germs of C^1 maps in $\mathcal{R}_0$ that have rank $\geq k$).

Theorem 8.36. Relative h-Stability. *Let $\mathcal{R}_0 \subset X^{(r)}$ be open, and let $\rho\colon \mathcal{R} \to \mathcal{R}_0$ be a microfibration that is relatively dense with respect to $\mathcal{R}_0$. Let $\varphi \in \Gamma(\mathrm{Conv}_\tau(\mathcal{R}|\mathcal{R}_0))$ be holonomic, $\varphi = (\alpha, j^r f)$. Let also $\mathcal{N}$ be a neighbourhood of the image $j^{r-1} f(V)$ in $X^{(r-1)}$.*

There is a homotopy of holonomic sections, $H\colon [0,1] \to \Gamma(\mathrm{Conv}_\tau(\mathcal{R}|\mathcal{R}_0))$, $H_t = (\alpha_t, j^r f_t)$, $H_0 = \varphi$, $t \in [0,1]$, such that the following properties are satisfied:

(i) *$H_1 = (\alpha_1, j^r f_1) \in \Gamma(\mathcal{R})$ i.e. $\rho \circ \alpha_1 = j^r f_1 \in \Gamma(\mathcal{R})$.*

(ii) *For all $t \in [0,1]$, the image $j^{r-1} g_t(V) \subset \mathcal{N}$.*

(iii) *Let $K \subset V$ be closed such that $\varphi \in \Gamma(\mathcal{R})$ on $\mathfrak{Op}\, K$ i.e. for all $x \in \mathfrak{Op}\, K$, $\rho \circ \alpha(x) = j^r f(x)$. Then $H_t = \varphi$ (constant homotopy) on (a smaller) $\mathfrak{Op}_1 K$.*

Proof. The proof follows that of VII Theorem 7.2, subject to trivial modifications relative to the open set $\mathcal{R}_0$. In particular, the relative density property implies the existence of a C-structure (h, H) with respect to $j^r f_0, \alpha$ such that for all $(x, t) \in V \times [0,1]$, $(h_t(x), j^r f_0(x)) \in \mathrm{Pr}_\tau(\mathcal{R}|\mathcal{R}_0)$. Furthermore since $\mathcal{R}_0$ is open, the extension of C-structures required by Lemma 7.10 also takes place in $\mathrm{Pr}_\tau(\mathcal{R}|\mathcal{R}_0)$. Details are left to the reader. $\qquad\square$

Let $\mathcal{P}_\infty \equiv \mathcal{P}_\infty(\mathcal{R}|\mathcal{R}_0)$ consist of all pairs $(a, y(t)) \in \mathcal{R} \times C^0([0,1], \mathcal{R}_0)$, such that $y(0) = \rho(a)$ and $y(t)$ is a path in the fiber $\mathcal{R}_0 \cap X_b^{(r)}$, where the base point $b = p_{r-1}^r \circ \rho(a) \in X^{(r-1)}$. There is an inclusion $i\colon \mathcal{R} \to \mathcal{P}_\infty$, $a \mapsto (a, C_a)$, where C_a is the constant path at $\rho(a) \in \mathcal{R}_0$.

Let $\rho\colon \mathcal{P}_\infty \to \mathcal{R}_0$ be the projection map $\rho(a, y(t)) = y(1)$. Thus a section $\varphi \in \Gamma(\mathcal{P}_\infty)$ consists of a pair of sections $\varphi = (\alpha, h)$ where $\alpha \in \Gamma(\mathcal{R})$ and the path $h \in C^0([0,1], \Gamma(\mathcal{R}_0))$ is a fiberwise homotopy in $\mathcal{R}_0$ that connects the sections $\rho \circ \alpha, h(1) \in \Gamma(\mathcal{R}_0)$. The section φ is holonomic if there is a C^r-section $g \in \Gamma(X)$ such that $h(1) = j^r g \in \Gamma(\mathcal{R}_0)$. The section $\varphi \in \Gamma(\mathcal{R})$ if $h_t = \rho \circ \alpha$ (constant path).

In case $\rho\colon \mathcal{R} \to \mathcal{R}_0$ is relatively ample there is an interesting connection to the theory of short sections in §8.1.2. This connection is useful also for the proof of Theorem 8.37 below. Let $\varphi = (\alpha, h) \in \Gamma(\mathcal{P}_\infty)$ be holonomic: $h(1) = j^r f \in \Gamma(X^{(r)})$. In particular the pair (α, f) is a formal solution to the relation $\mathcal{R}$. Over a chart $U \subset V$, employing VI Lemma 6.9 and subsequent remarks (or globally over V employing VI Proposition 6.12) one approximates $h(t)|U$ rel $\{0, 1\}$, by a sufficiently close piecewise principal homotopy in $\mathcal{R}_0$ i.e. one may assume that, for m sufficiently large, over U the homotopy $h(t)|U$ induces a holonomic section (same notation) $h \in \Gamma_U(\mathrm{Pr}_m(\mathcal{R}|\mathcal{R}_0))$. Since $\mathcal{R}$ is relatively ample over $\mathcal{R}_0$, it follows that $h \in \Gamma_U(\mathrm{Conv}_m(\mathcal{R}|\mathcal{R}_0))$; furthermore, $\rho \circ h = j^r f|U$ i.e. h is holonomic. In particular, the formal solution (α, f) is short with respect to the relation $\mathcal{R}$. In this way, the theory of relatively ample relations falls within the purview of the theory of short sections.

Theorem 8.37. (Relative C^{r-1}-Dense h-Principle.) *Let $\mathcal{R}_0$ be open in $X^{(r)}$. Let $\rho\colon \mathcal{R} \to \mathcal{R}_0$ be a microfibration that is relatively ample in $\mathcal{R}_0$. Let $\varphi = (\alpha, h) \in \Gamma(\mathcal{P}_\infty)$ be holonomic: $h(1) = j^r g \in \Gamma(\mathcal{R}_0)$. Let also $\mathcal{N}$ be a neighbourhood of the image $j^{r-1} g(V)$ in $X^{(r-1)}$.*

Then there is a homotopy of holonomic sections $\varphi_t = (\alpha_t, h_t) \in \Gamma(\mathcal{P}_\infty)$, $\varphi_0 = \varphi$, $h_t(1) = j^r g_t$, $t \in [0,1]$, such that the following properties obtain:

(i) *$\varphi_1 \in \Gamma(\mathcal{R})$ i.e. $h_1(s) = \rho \circ \alpha_1 = j^r g_1$ (constant path). Thus the homotopy $\alpha_t \in \Gamma(\mathcal{R})$, $t \in [0,1]$, solves the h-principle with respect to $\alpha \in \Gamma(\mathcal{R})$.*

(ii) *For all $t \in [0,1]$, the image $j^{r-1} g_t(V) \subset \mathcal{N}$.*

(iii) *(Relative Theorem) : Let $K \subset V$ be closed such that $\varphi \in \Gamma(\mathcal{R})$ on $\mathfrak{Op}\, K$ i.e. for all $x \in \mathfrak{Op}\, K$, $\rho \circ \alpha(x) = j^r g(x)$; $h_t(x) = j^r g(x)$ (constant path). Then $\varphi_t = \varphi$ on (a smaller) $\mathfrak{Op}_1(K)$.*

(iv) *In particular, $h_t(1) = j^r g_t \in \Gamma(\mathcal{R}_0)$, $t \in [0,1]$, is a homotopy of solutions to the relation $\mathcal{R}_0$ that connects $j^r g$ to $j^r g_1 = \rho \circ \alpha_1$.*

Proof. Note that, employing (i), (iv), (α_t, g_t) is a homotopy of short formal solutions to the relation $\mathcal{R}$, $(\alpha_0, g_0) = (\alpha, g)$, such that (α_1, g_1) solves $\mathcal{R}$: $\rho \circ \alpha_1 = j^r g_1$.

We follow the proof of Theorem 8.12 (C^{r-1}-Dense h-Principle), subject to the following modifications relative to the open set $\mathcal{R}_0 \subset X^{(r)}$. Over a chart $U =$

$\mathfrak{Op}\,x$ on V let $\psi = (\beta, \ell) \in \Gamma_U(\mathcal{P}_\infty)$ be holonomic. Thus the path ℓ_t is a fiberwise homotopy in $\mathcal{R}_0$ that connects the sections $\rho \circ \beta, \ell(1) = j^r f \in \Gamma_U(\mathcal{R}_0)$.

As explained above, employing VI Lemma 6.9 and subsequent remarks, one approximates $\ell(t)$ rel $\{0,1\}$, by a sufficiently close piecewise principal homotopy in $\mathcal{R}_0$. Since $\mathcal{R}$ is relatively ample over $\mathcal{R}_0$, the homotopy $\ell(t)$ induces a holonomic section (same notation) $\ell \in \Gamma_U(\mathrm{Conv}_m(\mathcal{R}|\mathcal{R}_0))$. One then formally follows the proof of the Local Extension Lemma 8.13, employing the Relative h-Stability Theorem 8.36 above to perform inductively the m successive convex integrations over the chart U. Consequently one obtains a homotopy of holonomic sections $\psi_t \in \Gamma_U(\mathrm{Conv}_m(\mathcal{R}|\mathcal{R}_0))$, $\psi_0 = \psi$, $\psi_1 \in \Gamma_U(\mathcal{R})$. Employing the equivalence between elements of $\mathrm{Pr}_m(\mathcal{R}_0)$ and piecewise principal paths in $\mathcal{R}_0$ with m-segments, the homotopy ψ_t induces a homotopy of holonomic sections (same notation) $\psi_t \in \Gamma_U(\mathcal{P}_\infty)$ that connects ψ to the holonomic section $\psi_1 \in \Gamma_U(\mathcal{R})$. Proceeding inductively over charts of V as in the proof of Theorem 8.12, the theorem is proved. Details are left to the reader. □

Corollary 8.38. *If $\mathcal{R}_0$ satisfies the C^{r-1}-dense h-principle, then $\mathcal{R}$ also satisfies the C^{r-1}-dense h-principle.*

Proof. Let (α, f) be a formal solution to $\mathcal{R}$: the sections $\alpha \in \Gamma(\mathcal{R})$, $f \in \Gamma^r(X)$ satisfy $p_{r-1}^r \circ \rho \circ \alpha = j^{r-1}f \in \Gamma(X^{(r-1)})$. Let $\mathcal{N}$ be a neighbourhood of the image $j^{r-1}f(V) \subset X^{(r-1)}$. By hypothesis on $\mathcal{R}_0$, there is a homotopy of formal solutions (α_t, f_t) to the relation $\mathcal{R}_0$, $(\alpha_0, f_0) = (\rho \circ \alpha, f)$, such that:

(i) $(\alpha_1, f_1) \in \Gamma(\mathcal{R}_0)$ i.e. $\alpha_1 = j^r f_1 \in \Gamma(\mathcal{R}_0)$.

(ii) for all $t \in [0,1]$, the image $j^{r-1}f_t(V) \subset \mathcal{N}$.

Thus the path $\alpha_t \in \Gamma(\mathcal{R}_0)$ connects $\alpha_0 = \rho \circ \alpha \in \Gamma(\mathcal{R}_0)$ to the holonomic section $\alpha_1 = j^r f_1 \in \Gamma(\mathcal{R}_0)$. Consequently $(\alpha, \alpha_t) \in \Gamma(\mathcal{P}_\infty)$ is a holonomic section, to which Theorem 8.38 applies, and the corollary is proved. □

8.4.2. The Affine Geometry of $\mathrm{Conv}_n(\mathcal{R}|\mathcal{R}_0)$.

The following affine geometry is useful for studying iterated convex hull extensions of a relation $\rho\colon \mathcal{R} \to X^{(r)}$. Let $y = (a, x_1, \ldots, x_m) \in \mathrm{Pr}_m(\mathcal{R})$. Thus $\rho(a), x_j$, $1 \le j \le m$ lie in an affine fiber $X_b^{(r)}$ of the bundle $p_{r-1}^r\colon X^{(r)} \to X^{(r-1)}$, $b \in X^{(r-1)}$. Associated to y is the (possibly degenerate) parallelopiped $P_m \subset X_b^{(r)}$ with vertex $\rho(a)$, defined inductively as follows. P_1 is the line segment that joins $\rho(a)$ to x_1. Assuming P_{k-1} is defined, let $P_k = \bigcup_{t \in [0,1]} T_t(P_{k-1})$, the union of the translates of P_{k-1} in the direction of $t(x_k - x_{k-1})$, $t \in [0,1]$, where,

$$T_t(z) = z + t(x_k - x_{k-1}), \ 2 \le k \le m.$$

In particular the edges of P_m at the vertex $\rho(a)$ are parallel respectively to the successive principal subspaces R_j associated to $y \in \mathrm{Pr}_m(\mathcal{R})$, $1 \le j \le m$. The parallelopiped P_m depends continuously on $y \in \mathrm{Pr}_m(\mathcal{R})$.

Lemma 8.39. *Let $\mathcal{R}_0 \subset X^{(r)}$ be open and let $\rho\colon \mathcal{R} \to \mathcal{R}_0$ be a relation which is relatively dense in $\mathcal{R}_0$. Let $y \in \mathrm{Pr}_m(\mathcal{R}|\mathcal{R}_0)$ and let P_m be the parallelopiped associated (as above) to y. Let also W be a neighbourhood of P_m in $X^{(r)}$. Then the following properties obtain:*

  (i) $y \in \mathrm{Conv}_m(\rho^{-1}(W)|W)$. *In particular,* $y \in \mathrm{Conv}_m(\mathcal{R}|\mathcal{R}_0)$ *if $P_m \subset \mathcal{R}_0$.*

  (ii) $P_m \subset \overline{\rho(\mathcal{R})}$.

Proof. In case $m = 1$ the conclusions (i), (ii) follow from (8.21) and the subsequent discussion. Inductively, suppose the lemma is true for all $m \leq k$. Let $y = (a, x_1, \ldots, x_{k+1}) \in \mathrm{Pr}_{k+1}(\mathcal{R}|\mathcal{R}_0)$; R_{k+1} is the principal subspace associated to y with respect to x_{k+1}, x_k. Let $P_k \subset X_b^{(r)}$ be the parallelopiped associated to $(a, x_1, \ldots, x_k) \in \mathrm{Pr}_k(\mathcal{R})$. Thus $P_{k+1} = \bigcup_{t \in [0,1]} T_t(P_k)$ where $T_t\colon X_b^{(r)} \to X_b^{(r)}$ is the one-parameter family of translations $T_t(z) = z + t(x_{k+1} - x_k)$, $t \in \mathbf{R}$. By hypothesis $P_{k+1} \subset W$. In particular, for all $t \in [0,1]$ the cross-sections $T_t(P_k) \subset W$. Let $U \subset W$ be a neighbourhood of the segment J that joins $\rho(a)$ to $u = T_1(\rho(a))$. Note that $\rho(a), x_k, x_{k+1}, u$ form the vertices of a parallelogram in $X_b^{(r)}$. Employing the translation $g(z) = z + \rho(a) - x_k$, let $R = g(R_{k+1})$, the translate of the principal subspace R_{k+1}. Thus $u = g(x_{k+1}) \in R$, $\rho(a) = g(x_k) \in R$.

Applying (8.21), there is a path $a_t \in \mathcal{R}$, $a_0 = a$, $t \in [0,1]$, such that the path $\rho(a_t) \in R \cap U$ surrounds u. Employing the translations $S_t\colon X_b^{(r)} \to X_b^{(r)}$, $S_t(z) = z + \rho(a_t) - \rho(a)$, $t \in [0,1]$, the paths $x_j^t = S_t(x_j)$, $x_j^0 = x_j$, $1 \leq j \leq k$, satisfy the following properties:

  (i) $(a_t, x_1^t, \ldots, x_k^t)$, $t \in [0,1]$, is a path in $\mathrm{Pr}_k(\mathcal{R})$. Indeed for each $t \in [0,1]$, $\rho(a_t) - x_1^t = \rho(a) - x_1$, $x_j^t - x_{j-1}^t = x_j - x_{j-1}$; hence $\rho(a_t), x_1^t$, respectively x_j^t, x_{j-1}^t, lie in principal subspaces, $2 \leq j \leq k$.

  (ii) The path x_k^t, $t \in [0,1]$, lies in R_{k+1} and surrounds x_{k+1}. Indeed, $g(R_{k+1}) = R$ and $g(x_k^t)) = g \circ S_t(x_k) = \rho(a_t)$. Since $g(x_{k+1}) = u$ and also the path $\rho(a_t)$ surrounds u, it follows that the path $x_k^t \in R_{k+1}$ surrounds x_{k+1}.

Furthermore, let P_k^t be the parallelopiped associated to $(a_t, x_1^t, \ldots, x_k^t) \in \mathrm{Pr}_k(\mathcal{R})$ $(P_k^0 = P_k)$. For all $t \in [0,1]$, $P_k^t = S_t(P_k)$. For sufficiently small neighbourhoods $U = \mathfrak{Op}\, J \subset W$, the path $\rho(a_t)$ in U lies close to the segment joining $\rho(a), u$ and hence the path $x_j^t = S_t(x_j)$ lies close to the segment joining x_j to $x_j + (u - \rho(a))$, $1 \leq j \leq k$. Since $u - \rho(a) = x_{k+1} - x_k$ (the direction of the translation T_t) it follows that one may choose $U = \mathfrak{Op}\, J \subset W$ sufficiently small so that each $P_k^t \subset W$ i.e. the union $\bigcup_{t \in [0,1]} P_k^t$ lies in W. Since P_k^t is the parallelopiped associated to $y_t = (a_t, x_1^t, \ldots, x_k^t) \in \mathrm{Pr}_k(\mathcal{R}|\mathcal{R}_0)$, it follows from the inductive hypothesis that for all $t \in [0,1]$,

  (iii) $y_t \in \mathrm{Conv}_k(\rho^{-1}(W)|W)$ $(y_0 = (a, x_1, \ldots, x_k))$.

  (iv) $P_k^t \subset \overline{\rho(\mathcal{R})}$.

Applying (iv) to a convergent sequence of neighbourhoods $U_i = \mathfrak{Op}_i J$, it follows that $P_{k+1} \subset \overline{\rho(\mathcal{R})}$. Furthermore, let $W_1 \subset W$ be an open convex neighbourhood of the segment that joins x_k, x_{k+1}. One may assume that the path x_k^t lies in W_1. Hence for each $t \in [0,1]$, the segment that joins x_k^t, x_{k+1} lies in W. Employing (iii) it follows that,

$$y \in \mathrm{Conv}_1(\mathrm{Conv}_k(\rho^{-1}(W)|W)) = \mathrm{Conv}_{k+1}(\rho^{-1}(W)|W),$$

which completes the inductive step and the lemma is proved. $\square$

Let $y = (a, x_1, \ldots, x_m) \in \mathrm{Pr}_m(\mathcal{R}|\mathcal{R}_0)$, whose corresponding parallelopiped is P_m. Employing the analytic caracterization Proposition 8.5 of convex hull extensions, the element y comes equipped with an m-parameter family of paths $a^{t_1,\ldots,t_m} \in \mathcal{R}$, $(t_1, \ldots, t_m) \in [0,1]^m$. Assuming that $\mathcal{R}$ is relatively dense in $\mathcal{R}_0$, the inductive arguments for the proof of Lemma 8.39 show also that the union of all such families of paths $a^{t_1,\ldots,t_m}$ (associated to the given element y) contains the parallelopiped P_m in its closure.

The condition $P_m \subset \mathcal{R}_0$ imposes strong conditions on the affine geometry of the open set $\mathcal{R}_0$. For example, let $y(t)$, $t \in [0,1]$, be a piecewise principal path associated to an element $y \in \mathrm{Pr}_m(\mathcal{R}|\mathcal{R}_0)$. Let N be a neighbourhood of the image $y([0,1])$ in $\mathcal{R}_0$. In general $P_m \not\subset N$. This is illustrated by the following examples in $\mathbf{R}^2$, which generalize easily to $X^{(1)} = J^1(\mathbf{R}^2, \mathbf{R}^2)$. With respect to coordinates $(x,y) \in \mathbf{R}^2$, let $\tau_1 = \ker dx$, $\tau_2 = \ker dy$, be vertical and horizontal tangent line fields in $\mathbf{R}^2$.

Examples. (i) Let $\mathcal{R} \subset \mathbf{R}^2$ be an open set which contains $K = \{(x, \sin x) \mid x \geq 0\}$. Let $z_m = (a, v_1, \ldots, v_m) \in \mathrm{Pr}_m(\mathcal{R})$ (defined with respest to successive principal subspaces (vertical or horizontal lines) with respect to τ_1, τ_2), where $a = (0,0)$, $v_i \in K$, $1 \leq i \leq m$, and $v_m = (m, y_m)$. Let $P_m \subset \mathbf{R}^2$ be the parallelogram that corresponds, as above, to $z_m \in \mathrm{Pr}_m(\mathcal{R})$. One checks that the union $\bigcup_1^\infty P_m$ is the half-space $\{(x,y) \in \mathbf{R}^2 \mid x \geq 0\}$.

(ii) Let $L \subset \mathbf{R}^2$ be the circle of radius r, center the origin. Let P_m be the parallelogram associated to an element $z = (a, v_1, \ldots, v_m) \in \mathrm{Pr}_m(\mathcal{R})$, $a, v_i \in L$, $1 \leq i \leq m$, $a = v_m$, whose associated piecewise principal path $z(t)$ represents a degree one loop in $\mathbf{R}^2 \setminus \{(0,0)\}$. Then P_m is a rectangle of dimensions $3r \times 4r$ (bounded, independent of m).

Theorem 8.40. *Let $\rho \colon \mathcal{R} \to \mathcal{R}_0$ be a relatively dense relation with respect to an open subset $\mathcal{R}_0 \subset X^{(r)}$. Then $\mathcal{R}$ is relatively ample with respect to $\mathcal{R}_0$ i.e. for all $n \geq 1$, $\mathrm{Conv}_n(\mathcal{R}|\mathcal{R}_0) = \mathrm{Pr}_n(\mathcal{R}|\mathcal{R}_0)$.*

Proof. Let $y = (a, x_1, \ldots, x_n) \in \mathrm{Pr}_n(\mathcal{R}|\mathcal{R}_0)$. Let $h \colon [0,1] \to \mathcal{R}_0$ be the associated piecewise principal path, $h(0) = \rho(a)$; $h(1) = x_n$. Let $N \subset \mathcal{R}_0$ be a neighbourhood of the image $h([0,1])$ in $\mathcal{R}_0$. In particular, $y \in \mathrm{Pr}_n(\rho^{-1}(N)|N)$. The relative density hypothesis is quite strong as is shown by the following lemma which proves the theorem, and which contrasts sharply with the the affine geometry of Examples (i), (ii) above.

Lemma 8.41. *Let* $\rho \colon \mathcal{R} \to \mathcal{R}_0$ *be relatively dense with respect to an open subset* $\mathcal{R}_0 \subset X^{(r)}$. *For all n and all data as above,*

$$\mathrm{Pr}_n(\rho^{-1}(N)|N) = \mathrm{Conv}_n(\rho^{-1}(N)|N) \subset \mathrm{Conv}_n(\mathcal{R}|\mathcal{R}_0).$$

Proof. The case $n = 1$ follows from (8.21). Suppose inductively that the lemma is true for all data at $n = m - 1$. Let $y_m = (a, x_1, \ldots, x_m) \in \mathrm{Pr}_m(\rho^{-1}(N)|N)$. Let $W_1 \subset N$, respectively $W_k \subset N$, be an open convex neighbourhood of the segment J_1 that joins $\rho(a), x_1$, respectively an open convex neighbourhood of the segment J_k that joins x_{k-1}, x_k, $2 \leq k \leq m$. Let R_1, respectively R_k denote the principal subspace associated to y that contains $\rho(a), x_1$, respectively the principal subspace associated to y that contains x_{k-1}, x_k, $2 \leq k \leq m$. Let $0 = t_0 < t_1 < \cdots < t_m = 1$. There is a path $a^s \in \rho^{-1}(W_1)$, $a(0) = a$, $s \in [0, t_1]$, such that the path $\rho(a^s) \in R \cap W_1$ surrounds x_1. One may assume $\rho(a^{t_1}) \in W_2$.

Employing relative density one may assume $\rho(a^{t_1})$ is sufficiently close to x_1 so that $\rho(a^{t_1}), x_1, x_2$ are the sides of a parallelogram in W_2 i.e. $y_2 = (a^{t_1}, x_1, x_2) \in \mathrm{Pr}_2(\rho^{-1}(W_2)|W_2)$. Employing Lemma 8.39 (with respect to the interval $[t_1, t_2]$) there are paths $a^t \in \rho^{-1}(W_2)$, $x_1^t \in R_2$, $t \in [t_1, t_2]$, $x^{t_1} = x_1$, such that: (*i*) the path x^t surrounds x_2; (*ii*) for all $t \in [t_1, t_2]$, $\rho(a^t), x_1^t$ lie in a principal subspace parallel to R_1. One may assume also that $\rho(a^{t_2}), x_1^{t_2} \in W_3$. Proceeding inductively in successive W_k on successive intervals $[t_{k-1}, t_k]$, employing relative density at each stage (cf. Lemma 8.39), one constructs paths $a^t \in \rho^{-1}(N)$, $a^0 = a$, $x_k^t \in N$, $x_k^0 = x_k$, $t \in [0, 1]$, $1 \leq k \leq m - 1$, such that the following properties are satisfied:

(i) For all $t \in [t_{k-1}, t_k]$, the subpaths $\rho(a^t), x_j^t$, lie in W_k, $1 \leq j \leq k \leq m$.

(ii) For all $t \in [0, t_k]$, $x_k^t = x_k$, $1 \leq k \leq m$. The path x_k^t, $t \in [t_k, t_{k+1}]$ lies in $R_{k+1} \cap W_{k+1}$ and surrounds x_{k+1}, $1 \leq k \leq m - 1$. One may assume that $\rho(a^{t_j}), x_k^{t_j} \in W_{j+1}$, $1 \leq k \leq j \leq m - 1$.

(iii) For all $t \in [0, 1]$, $z_k^t = (a^t, x_1^t, \ldots, x_k^t) \in \mathrm{Pr}_k(\rho^{-1}(N)|N)$. Furthermore, for all $t \in [t_{k-1}, t_k]$,

$$z_k^t \in \mathrm{Pr}_k(\rho^{-1}(W_k)|W_k), \quad 1 \leq k \leq m.$$

(iv) For all $t \in [t_{k-1}, t_k]$, let P_k^t be the parallelopiped associated to z_k^t. Then $P_k^t \subset W_k$. Consequently, employing Lemma 8.39, for all $t \in [t_{k-1}, t_k]$,

$$z_k^t \in \mathrm{Conv}_k(\rho^{-1}(W_k|W_k) \quad 1 \leq k \leq m.$$

Employing (iii), for all $t \in [0, 1]$, $z_{m-1}^t \in \mathrm{Pr}_{m-1}(\rho^{-1}(N)|N)$. By induction, for all $t \in [0, 1]$, $z_{m-1}^t \in \mathrm{Conv}_{m-1}(\rho^{-1}(N)|N)$. Employing (ii), for each $t \in [0, 1]$ the line segment joining x_{m-1}^t, x_m lies in $W_m \subset N$. Consequently,

$$y_m \in \mathrm{Conv}_1(\mathrm{Conv}_{m-1}(\rho^{-1}(N)|N)|N) = \mathrm{Conv}_m(\rho^{-1}(N)|N),$$

which completes the inductive step and the lemma is proved.   □

Let $a^{t_1,\ldots,t_n}$, $x_j^{t_1,\ldots t_{n-j}}$, $1 \le j \le n-1$, be the n-parameter families of surrounding paths associated to $y = (a, x_1, \ldots, x_n) \in \mathrm{Conv}_n(\rho^{-1}(N)|N)$, in accordance with Proposition 8.5. Employing property (iv) it follows in addition that each surrounding path in these families of surrounding paths is contained at least one of the open convex subsets $W_k \subset N$, $1 \le k \le m$. Accordingly, if each W_k is a small ball with respect to a metric on $X^{(r)}$ then all the surrounding paths take place in these small balls. One therefore obtains the following useful refinement of the lemma.

With respect to a metric d on $X^{(r)}$ ($B_x(r)$ denotes the ball of radius r, center x in this metric), let $\epsilon \in C^0(\mathcal{R}_0, (0, \infty))$ be a continuous positive function such that for all $x \in \mathcal{R}_0$, $d(x, \partial\mathcal{R}_0) > \epsilon(x)$. Let $\rho\colon \mathcal{R} \to \mathcal{R}_0 \subset X^{(r)}$ be a relation. For each $m \ge 1$, let $\mathrm{Pr}_m(\mathcal{R}, \epsilon) \subset \mathrm{Pr}_m(\mathcal{R}|\mathcal{R}_0)$ denote the subspace of all $y = (a, x_1, \ldots, x_m)$ such that each segment J_i lies in a ball $B_{x_i}(\epsilon(x_i))$, $1 \le i \le m$, where J_1, respectively J_i, is the segment in $\mathcal{R}_0$ that joins $\rho(a), x_1$, respectively the segment in $\mathcal{R}_0$ that joins x_{i-1}, x_i, $2 \le i \le m$. Define the corresponding iterated convex hull extensions as follows.

$$\mathrm{Conv}_1(\mathcal{R}, \epsilon) = \mathrm{Conv}_1(\mathcal{R}|\mathcal{R}_0) \cap \mathrm{Pr}_1(\mathcal{R}, \epsilon),$$

and inductively, $\mathrm{Conv}_m(\mathcal{R}, \epsilon) = \mathrm{Conv}_1(\mathrm{Conv}_{m-1}(\mathcal{R}, \epsilon), \epsilon)$. In particular, for all $m \ge 1$, $\mathrm{Conv}_m(\mathcal{R}, \epsilon) \subset \mathrm{Conv}_m(\mathcal{R}|\mathcal{R}_0)$.

Corollary 8.42. *Let $\rho\colon \mathcal{R} \to \mathcal{R}_0$ be relatively dense with respect to an open subset $\mathcal{R}_0 \subset X^{(r)}$. For each $n \ge 1$ (in the above notation), $\mathrm{Conv}_n(\mathcal{R}, \epsilon) = \mathrm{Pr}_n(\mathcal{R}, \epsilon)$.*

Proof. Let $y = (a, x_1 \ldots, x_n) \in \mathrm{Pr}_n(\mathcal{R}, \epsilon)$. Thus (in the above notation) $y \in \mathrm{Pr}_n(\rho^{-1}(N)|N)$ and each segment J_i is contained in a ball $B_{x_i}(\epsilon(x_i)) \subset \mathcal{R}_0$. Thus more particularly, $y \in \mathrm{Pr}_n(\rho^{-1}(N), \epsilon)$. We prove that for all $n \ge 1$,

$$\mathrm{Pr}_n(\rho^{-1}(N), \epsilon) = \mathrm{Conv}_n(\rho^{-1}(N), \epsilon) \subset \mathrm{Conv}_n(\mathcal{R}, \epsilon). \qquad (8.23)$$

The Corollary follows from (8.23). To prove (8.23), let $y \in \mathrm{Pr}_n(\rho^{-1}(N), \epsilon)$ as above. Let W_i be an open convex neighbourhood of the segment J_i in the ball $B_{x_i}(\epsilon(x_i))$, $1 \le i \le n$. Employing Lemma 8.41 and subsequent remarks, it follows that $y \in \mathrm{Conv}_n(\rho^{-1}(N)|N) \subset \mathrm{Conv}_n(\mathcal{R}|\mathcal{R}_0)$; in addition all the surrounding paths in the n-parameter families of surrounding paths associated to y in accordance with Proposition 8.5 are contained in at least one of the subsets $W_i \subset B_{x_i}(\epsilon(x_i))$, $1 \le i \le n$. Consequently $y \in \mathrm{Conv}_n(\rho^{-1}(N), \epsilon)$.   □

8.4.3. Strictly Short Immersions. Let (V, g), (W, h) be Riemannian manifolds; $\dim V = n$, $\dim W = q$. A C^1-map $f: V \to W$ is *strictly short* if for all $v \in V$ and all non-zero vectors $\tau \in T_v(V)$ $(w = f(v) \in W)$,

$$g_v(\tau, \tau) > h_w(df(\tau), df(\tau)). \tag{8.24}$$

Geometrically, (8.24) implies that the C^1-map f strictly decreases Riemannian lengths of smooth curves on V. The map f is *short* if the inequality (8.22) is replaced by $\geq$.

Let $X = V \times W \to V$ be the product bundle. The affine bundle $X^{(1)} \to X$ has fiber $X_b^{(1)} = \mathcal{L}(\mathbf{R}^n, \mathbf{R}^q)$, $b \in X$. Let $\mathcal{R}_0 \subset X^{(1)}$ be the open subset defined by germs of C^1-maps f which are strictly short. Let h' be the quadratic form on $T(V)$ induced by $df: T(V) \to T(W)$. Then f is strictly short if and only if the quadratic form $g - h'$ is positive definite. It follows that for all $b \in X$ the intersection in each fiber $\mathcal{R}_0 \cap X_b^{(1)}$ is open and convex.

Let $\mathcal{R}^k$ be the k-mersion relation in $X^{(1)}$. Then $\mathcal{R}^k$ is open, and in case $k < q$, $\mathcal{R}^k$ is relatively dense in $X^{(1)}$. Indeed, in case $k < q$, $\mathcal{R}^k$ is the complement of a closed stratified subset $\Sigma \subset X^{(1)}$ which satisfies (8.22) (cf. the remarks following (8.22)). Let $\mathcal{R} = \mathcal{R}_0 \setminus \Sigma$. Thus $\mathcal{R} \subset \mathcal{R}_0$ is the open subset that corresponds to germs of C^1-maps that are strictly short and have rank $\geq k$. Thus in case $k < q$, $\mathcal{R}$ is relatively dense in $\mathcal{R}_0$. Employing Theorem 8.40, $\mathcal{R}$ is relatively ample with respect to $\mathcal{R}_0$.

Theorem 8.43 (Gromov). *Let $f \in C^1(V, W)$ be strictly short. Suppose f is homotopic to a C^1 k-mersion $g: V \to W$. Let $\mathcal{N}$ be a neighbourhood of the graph of f in X. If $k < q$ then there is a homotopy $f_t \in C^1(V, W)$, $f_0 = f$, $t \in [0, 1]$, such that,*

(i) *For all $t \in [0, 1]$, $j^1 f_t \in \Gamma(\mathcal{R}_0)$; $j^1 f_1 \in \Gamma(\mathcal{R})$. Hence f is homotopic through strictly short C^1-maps to a strictly short k-mersion.*

(ii) *For all $t \in [0, 1]$, the graph of f_t lies in $\mathcal{N}$. Thus f_1 is a strictly short k-mersion that is a fine C^0-approximation of f.*

Proof. The homotopy connecting f, g induces a bundle map $\alpha: T(V) \to T(W)$ which covers f and is fiberwise of rank $\geq k$ i.e. there is a section (same notation) $\alpha \in \Gamma(\mathcal{R})$ such that $p_0^1 \circ \alpha = f \in \Gamma(X)$. Up to a homotopy obtained by rescaling α one may assume that $\alpha \in \Gamma(\mathcal{R})$ lies in a small neighbourhood $\mathcal{M}$ of the zero-section. In particular, $\mathcal{M} \subset \mathcal{R}_0$. Let $(\alpha, h) \in \Gamma(\mathcal{P}_\infty)$, where h connects (fiberwise) $\alpha, j^1 f$ through contractions to the zero-section. Since $h(1) = j^1 f$ it follows that $(\alpha, h) \in \Gamma(\mathcal{P}_\infty)$ is holonomic. The theorem now follows from applying Theorem 8.37 (ii),(iv) to $(\alpha, h) \in \Gamma(\mathcal{P}_\infty)$. $\qquad\square$

The application of Theorem 8.37(i) yields the stronger conclusion: there is a homotopy $\alpha_t \in \Gamma(\mathcal{R})$, $\alpha_0 = \alpha$, that solves the h-principle with respect to α: $\alpha_1 = j^1 f_1 \in \Gamma(\mathcal{R})$. Consequently Theorem 8.43 is a refinement of the classical k-mersion theorem of Feit [12], and in particular a refinement of the classical immersion theorem of Hirsch [21] in case $k = n < q$ (the immersion case), in the context of strictly short data.

8.4.4. Relative h-Principle for Embeddings. Let $X = V \times W \to V$ be the product bundle, $\dim V = n$, $\dim W = q$, and $n < q$. Let $\mathcal{R}_0 \subset X^{(1)}$ be the open subset defined by the immersion relation. Recall that a bundle map $G : T(V) \to T(W)$ that covers a continuous map $g : V \to W$ is equivalent to a section (same notation) $G \in \Gamma(X^{(1)})$ such that $p_0^1 \circ G = g \in \Gamma(X)$. In particular, if G is a bundle monomorphism, then the induced section $G \in \Gamma(\mathcal{R}_0)$. Let $\Phi \subset \mathcal{R}_0$ denote the open subspace defined by germs of C^1 embeddings from V to W. Note that if $f : V \to W$ is a C^1-embedding and $g : V \to W$ is a C^1-immersion which a fine C^0-approximation to f and such that, in local coordinates, for all $x \in V$ the tangents spaces $df_x, dg_x \in \mathcal{L}(\mathbf{R}^n, \mathbf{R}^q)$ are sufficiently close then g is an embedding that is connected to f by a small C^1-isotopy of embeddings (cf. Hirsch [21] p. 36).

Theorem 8.44. *Let $\rho : \mathcal{R} \to \mathcal{R}_0$ be a microfibration which is relatively dense with respect to the immersion relation $\mathcal{R}_0 \subset X^{(1)}$. Let $f \in C^1(V, W)$ be a C^1-embedding, let $\alpha \in \Gamma(\mathcal{R})$, and suppose there is a homotopy of bundle monomorphisms, $H_t : T(V) \to T(W)$, $H_0 = df$, $t \in [0, 1]$, such that the induced section $H_1 = \rho \circ \alpha \in \Gamma(\mathcal{R}_0)$. Let $\mathcal{N}$ be a neighbourhood of the graph of f in X.*

Then there is a homotopy of formal solutions (α_t, f_t), $(\alpha_0, f_0) = (\alpha, f)$, $t \in [0, 1]$, to the relation $\mathcal{R}_0$ such that the following properties are satisfied:

(i) *The homotopy $f_t : V \to W$ is a C^1-isotopy of embeddings.*

(ii) *The homotopy $\alpha_t \in \Gamma(\mathcal{R})$ solves the h-principle for α: $\rho \circ \alpha_1 = j^1 f_1 \in \Gamma(\Phi)$.*

(iii) *For all $t \in [0, 1]$, the graph of f_t lies in $\mathcal{N}$. In particular, the embedding f_1 is a fine C^0-approximation of f.*

Proof. The homotopy of bundle monomorphisms H induces a holonomic section $\varphi = (\alpha, h_t) \in \Gamma(\mathcal{P}_\infty)$, where $h_t = H_{1-t} \in \Gamma(\mathcal{R}_0)$ $(h_0 = H_1 = \rho \circ \alpha \in \Gamma(\mathcal{R}_0)$; $h_1 = j^1 f)$. Applying Theorem 8.37 to $\varphi \in \Gamma(\mathcal{P}_\infty)$, conclusions (i), (ii) obtain for a C^1-homotopy of immersions f_t rather than for a C^1-isotopy of embeddings. To obtain the refinement of a C^1-isotopy of embeddings, one employs Corollary 8.42 inductively over charts as follows in the proof of Theorem 8.37. With respect to a metric d on $X^{(1)}$, let $\epsilon : \mathcal{R}_0 \to (0, \infty)$ be continuous such that for all $z \in X^{(1)}$, $d(z, \partial \mathcal{R}_0) < \epsilon(z)$. Over a chart U on V let $\psi = (\beta, \ell) \in \Gamma_U(\mathcal{P}_\infty)$ be holonomic such that $\ell(1) = j^1 g \in \Phi_U$, where $g : U \to W$ is a C^1-embedding. As in the proof of Theorem 8.37, employing a sufficiently fine piecewise principal approximation rel $\{0,1\}$ to $\ell(t)$, one may assume that the homotopy $\ell(t)$ induces a holonomic section (same notation) $\ell \in \Gamma_U(\mathrm{Pr}_m(\mathcal{R},\epsilon))$. Employing Corollary 8.42, it follows that $\ell \in \Gamma_U(\mathrm{Conv}_m(\mathcal{R},\epsilon))$. More precisely, employing (8.23), one may assume that for each $x \in U$, $\ell(x) \in \mathrm{Conv}_m(\rho^{-1}(N_x),\epsilon)$, where N_x is a neighbourhood of the associated piecewise principal path $\ell(x)$.

Recall the sequence of auxiliary neighbourhoods $W_i^x \subset N_x$ of successive principal segments in $\ell(x)$, $1 \le i \le m$, $x \in U$. Employing Corollary 8.42, all of the surrounding paths in the m-parameter families of surrounding paths involved in $\ell(x) \in \mathrm{Conv}_m(\rho^{-1}(N_x),\epsilon)$ lie in at least one of the neighbourhoods W_i^x. Consequently, for a sufficiently small system of neighbourhoods W_i^x, the resulting tangent spaces to the successive immersions that are obtained by the m convex integrations of $\ell \in \Gamma_U(\mathrm{Conv}_m(\mathcal{R},\epsilon))$ can be made arbitrarily close in $\mathcal{L}(\mathbf{R}^n, \mathbf{R}^q)$ (in local coordinates). As explained above (employing Hirsch [21] p. 36), these successive convex integrations over U therefore yield C^1-embeddings which are connected by a small isotopy of C^1-embeddings. One concludes that there is a homotopy of holonomic sections $\psi_t = (\beta_t, \ell_t') \in \Gamma_U(\mathrm{Conv}_m(\mathcal{R},\epsilon))$, $\psi_0 = \psi$, $\psi_1 \in \Gamma(\mathcal{R})$, such that for all $t \in [0,1]$, $\ell_t'(1) = j^1 g_t \in \Phi_U$ i.e. g_t is an isotopy of C^1-embeddings on U. In particular, $\rho \circ \beta_1 = j^1 g_1$. An obvious induction over charts of V, (employing the above refinements on C^1-isotopies) proves the theorem. Details are left to the reader. $\square$

Following Gromov, we apply Theorem 8.44 to the study of totally real embeddings. Assume W is a complex manifold, $\dim_{\mathbf{C}} W = q$ (hence W has real dimension $2q$). A bundle homomorphism $H\colon T(V) \to T(W)$ is *totally real* if the complexified bundle homomorphism $H_{\mathbf{C}}\colon \mathbf{C} T(V) \to T(W)$ has complex rank $\ge \min(\dim V, \dim_{\mathbf{C}} W)$. A C^1-map $f\colon V \to W$ is totally real if the complexified differential $d_{\mathbf{C}} f\colon \mathbf{C} T(V) \to W$ is totally real.

Every real analytic map $f\colon V \to W$ extends to a holomorphic map $\mathbf{C} f\colon \mathbf{C} V \to W$ for some (small) complexification $\mathbf{C} V \supset V$ in $\mathbf{C} T(V)$. Totally real maps extend to holomorphic immersions $\mathbf{C} T(V) \to W$ in case $n \le q$ and to totally real submersions in case $n \ge q$.

Let $\mathcal{R} \subset X^{(1)}$ be the total reality condition defined by germs of C^1-totally real maps from V to W. Thus $\mathcal{R} = X^{(1)} \setminus \Sigma$, where Σ is a closed stratified subset such that $R \cap \Sigma$ has codimension ≥ 2 for all principal subspaces $R \subset X^{(1)}$ that are not completely contained in Σ. Consequently, the total reality relation satisfies (8.22). Suppose $n \le q$ and let $\mathcal{R}_0 \subset X^{(1)}$ be the immersion relation. Thus the relation (same notation) $\mathcal{R} - \mathcal{R}_0 \setminus \Sigma$ is open and satisfies (8.21) i.e. the inclusion $i\colon \mathcal{R} \to \mathcal{R}_0$ is relatively dense in $\mathcal{R}_0$. Thus $\mathcal{R}$ represents the subspace of germs of C^1-immersions which are totally real: the corresponding complexified differentials have complex rank n. Applying Theorem 8.44 to the total reality relation $\mathcal{R}$, one obtains the following result.

Theorem 8.45 (Gromov). *Suppose $n \leq q$ and let $f \colon V \to W$ be a C^1-embedding. Let $H_t \colon T(V) \to T(W)$ be a homotopy of bundle monomorphisms, $H_0 = df$, $t \in [0, 1]$, such that the induced section $H_1 \in \Gamma(\mathcal{R})$. Let $\mathcal{N}$ be a neighbourhood of the graph of f in X. Then f is isotopic through C^1-embeddings to a C^1-totally real embedding $f_1 \colon V \to W$ such that the graph of f_1 is contained in $\mathcal{N}$.*

Corollary 8.46. *There is a domain of holomorphy in $\mathbf{C}^3$ which is diffeomorphic to $S^3 \times \mathbf{R}^3$.*

Proof. Following Gromov [18] p. 193, employing homotopy theory (S^3 is parallelizable) one shows that the tangent map of the standard embedding $f \colon S^3 \to \mathbf{C}^3$ is homotopic through bundle monomorphisms to the trivial totally real monomorphism $h \colon T(S^3) \to \mathbf{T}(\mathbf{C}^3)$ ($\mathbf{C}^3 = \mathbf{R}^3 \oplus \sqrt{-1}\mathbf{R}^3$). Applying Theorem 8.45, f is isotopic to a totally real embedding $g \colon S^3 \to \mathbf{C}^3$, which one may assume to be real analytic. One then analytically extends f to a small complexification $\mathbf{C}\, S^3 \equiv S^3 \times \mathbf{R}^3$ whose boundary is a S^2-bundle of sufficiently small radius so that the boundary is pseudo-convex.   □

CHAPTER 9

SYSTEMS OF PARTIAL DIFFERENTIAL EQUATIONS

§1. Underdetermined Systems

9.1.1. Let $p\colon X \to V$ be a smooth bundle, fiber dimension q, $\dim V = n > 1$, and let $\mathcal{R} = F^{-1}(0) \subset X^{(r)}$, where $F\colon X^{(r)} \to \mathbf{R}^k$ is continuous. A C^r-section $f \in \Gamma^r(X)$ is a solution to the relation $\mathcal{R}$ if the image $j^r f(V) \subset \mathcal{R}$: for all $x \in V$, $F(j^r f(x)) = 0 \in \mathbf{R}^k$. The system of equations, $F = (F_1, F_2, \ldots F_k)\colon X^{(r)} \to \mathbf{R}^k$,

$$F_i(j^r f(x)) = 0 \quad 1 \le i \le k, \tag{9.1}$$

is a system of k PDEs in the unknown C^r section $f \in \Gamma^r(X)$. In case $p\colon X = V \times \mathbf{R}^q \to V$ is projection onto an open set $V \subset \mathbf{R}^n$, then (9.1) is a system of k PDEs in the unknown function $f \in C^r(V, \mathbf{R}^q)$; $\Gamma^r(X) \equiv C^r(V, \mathbf{R}^q)$. Evidently, in this generality nothing much can be said about the existence of C^r solutions to (9.1). In this chapter the techniques of convex integration are applied to solve relations $\mathcal{R}$ which satisfy simple convex hull conditions related to the theory of convex hull extensions in Chapter VII. Typically, systems of PDEs which can be solved by these methods are underdetermined i.e. systems (9.1) for which $k \le q - 1$ (more unknown functions than equations) and are non-linear. Generically determined systems and all linear systems are systematically *excluded* i.e. these important systems are beyond the scope of the results and methods of this chapter.

We emphasize here that a solution to a closed relation $\mathcal{R} \subset X^{(r)}$ is a C^r section $f \in \Gamma^r(X)$, not a generalized solution (distribution etc.). With the exception of the Nash C^1-isometric immersion theorem, the systems of PDEs that are solvable by the methods of Convex Integration Theory are not the kinds of systems studied in differential geometry and in applied mathematics. On the other hand, as explained in Chapter X, Convex Integration Theory does contribute to Relaxation Theory and related results in Optimal Control theory.

9.1.2. The general approach to solving closed relations $\mathcal{R} \subset X^{(r)}$ was initiated by Gromov [17], described also in Spring [37] and in Gromov [18] for the general theory in the case of ample relations. The general idea can be seen from Theorem 9.1 where one replaces the closed relation $\mathcal{R}$ by a nearby system of open metaneighbourhoods $\mathrm{Met}_\epsilon\, \mathcal{R} \subset X^{(r)}$ of "radius" $\epsilon > 0$. These nearby open relations are

D. Spring, *Convex Integration Theory: Solutions to the h-principle in geometry and topology*, Modern Birkhäuser classics, DOI 10.1007/978-3-0348-0060-0_9, © Springer Basel AG 2010

called metaneighbourhoods since $\mathcal{R} \subset \overline{\text{Met}_\epsilon} \, \mathcal{R}$, but in general $\mathcal{R} \nsubseteq \text{Met}_\epsilon \, \mathcal{R}$. These metaneighbourhoods are constructed from the geometry of $\mathcal{R}$, in particular from the nowhere flatness conditions imposed on the relation $\mathcal{R}$ (cf. §9.3), and satisfy: $\mathcal{R} = \bigcap_{\epsilon > 0} \overline{\text{Met}_\epsilon} \, \mathcal{R}$.

Let $(\epsilon_n)_{n \geq 1}$ be a sequence of positive reals such that $\lim_{n \to \infty} \epsilon_n = 0$. Beginning with a short map f for $\mathcal{R}$, employing the h-Stability Theorem VII 7.2, a sequence of C^r-sections $f_n \in \Gamma^r(X)$ is constructed such that the image $j^r f_n(V) \subset \text{Met}_{\epsilon_n} \, \mathcal{R}$ i.e. f_n solves the open relation $\text{Met}_{\epsilon_n} \, \mathcal{R}$, $n \geq 1$. In case $\sum_n \epsilon_n < \infty$ one proves $\lim_{n \to \infty} f_n = f$ exists, $f \in \Gamma^r(X)$, and that f solves the relation $\mathcal{R}$. The main results Theorems 9.1, 9.16, are stated in terms of spaces of strictly short maps, thus generalizing the corresponding theorems in Gromov [18] which are based on ampleness assumptions.

We remark that prior to Convex Integration Theory there were no general methods available in differential topology for solving closed relations (i.e. systems of PDEs) in r-jet spaces. The covering homotopy methods introduced by Smale [36] and employed in subsequent papers by Gromov [16], Hirsch [21], Phillips [31] and others are designed to solve open relations in r-jet spaces but these methods do not control derivatives of order $\leq r - 1$. It is this control of all lower order derivatives (in fact all $\perp$-derivatives) in Convex Integration Theory that allows one to prove the convergence of solutions $f_n \in \Gamma^r(X)$ of the open relations $\text{Met}_\epsilon \, \mathcal{R}$, $n \geq 1$, discussed above, to a solution in the limit of the relation $\mathcal{R}$. The control of lower order derivatives is the analytic strength of Convex Integration theory.

9.1.3. Let $p_\perp^r : X^{(r)} \to X^\perp$ be the affine $\mathbf{R}^q$-bundle associated to a codimension 1 tangent hyperplane field $\tau \subset T(V)$. Let $\rho \colon \mathcal{R} \to X^{(r)}$ be continuous. For each base point $b \in X^\perp$, $X_b^{(r)}$ is the $\mathbf{R}^q$-fiber over b. Recall that $z \in \text{Conv}_\tau(\mathcal{R}_b, a)$ if there is a path $a_t \in \mathcal{R}_b$, $a_0 = a$, $t \in [0,1]$, such that the image path $\rho(a_t)$, $t \in [0,1]$, surrounds the r-jet z. In the case of microfibrations $\rho \colon \mathcal{R} \to X^{(r)}$, $\text{Conv}_\tau(\mathcal{R}_b, a)$ is open in the fiber $X_b^{(r)}$, hence the property of surrounding z is equivalent to strictly surrounding z. In general however (for example closed relations $\mathcal{R} \subset X^{(r)}$) the condition of strictly surrounding z is a further requirement that is denoted as follows:

$$z \in \text{IntConv}_\tau(\mathcal{R}_b, a),$$

if and only if there is a path $a_t \in \mathcal{R}_b$, $a_0 = a$, $t \in [0,1]$, whose ρ-image $\rho(a_t)$, $t \in [0,1]$, *strictly surrounds* the r-jet z. Let $\text{IntConv}_\tau(\mathcal{R}) \subset \mathcal{R} \times X^{(r)}$ consist of all pairs (a, z) such that $z \in \text{IntConv}_\tau(\mathcal{R}_b; a)$, $b = p_\perp^r \circ \rho(a) \in X^\perp$. There is a natural projection map $\rho \colon \text{IntConv}(\mathcal{R}) \to X^{(r)}$, $(a, z) \mapsto z \in X^{(r)}$. The relation $\text{IntConv}_\tau(\mathcal{R})$ is called the *interior* convex hull extension of $\mathcal{R}$ with respect to τ. Evidently, $\text{IntConv}_\tau(\mathcal{R}) \subset \text{Conv}_\tau(\mathcal{R})$ i.e. a path that strictly surrounds an r-jet *a fortiori* also surrounds that r-jet.

Analogous to the iterated convex hull extensions of VIII §8.1, we define iterated interior convex hull extensions of a relation $\rho \colon \mathcal{R} \to X^{(r)}$. Let $\tau_i \subset T(V)$ be codimension 1 tangent hyperplane fields on V, $1 \leq i \leq m$, $m \geq 2$. Let $p_i^r \colon X^{(r)} \to X_i^\perp$ be the affine $\mathbf{R}^q$-bundle associated to τ_i, $1 \leq i \leq m$. Define $\mathrm{IntConv}_0(\mathcal{R}) = \mathcal{R}$; $\rho_0 = \rho \colon \mathcal{R} \to X^{(r)}$, and inductively,

$$\mathrm{IntConv}_j(\mathcal{R}) = \mathrm{IntConv}_{\tau_j}(\mathrm{IntConv}_{j-1}(\mathcal{R})), \quad 1 \leq j \leq m.$$

Thus $\mathrm{IntConv}_j(\mathcal{R}) \subset \mathcal{R} \times X^{(r)} \times \cdots \times X^{(r)}$ $((j+1)$ factors). There is a natural projection $\rho_j \colon \mathrm{IntConv}_j(\mathcal{R}) \to X^{(r)}$, $(a, z_1, z_2, \ldots, z_j) \mapsto z_j$ (projection onto the last factor), $1 \leq j \leq m$. Evidently $\mathrm{IntConv}_j(\mathcal{R}) \subset \mathrm{Conv}_j(\mathcal{R})$, $1 \leq j \leq m$.

Assume that $\mathcal{R}$ is a metric space; $B(a; r)$ denotes the open ball at $a \in \mathcal{R}$, radius $r > 0$. Let $p_\perp^r \colon X^{(r)} \to X^\perp$ be the affine $\mathbf{R}^q$-bundle associated to a codimension 1 tangent hyperplane field $\tau \subset T(V)$. Let $\mathrm{Conv}_\tau^\epsilon(\mathcal{R}) \subset \mathrm{Conv}_\tau(\mathcal{R})$ consist of all pairs $(a, z) \subset \mathcal{R} \times X^{(r)}$ such that there is a path $a_t \in \mathcal{R}_b \cap B(a; \epsilon(a))$, $a_0 = a$, $t \in [0, 1]$, whose image $\rho(a_t)$ surrounds z in the fiber $X_b^{(r)}$, $b = p_\perp^r \circ \rho(a) \in X^\perp$. Thus the path a_t is confined to a ball of radius $\epsilon(a)$. Similarly $\mathrm{IntConv}_\tau^\epsilon(\mathcal{R})$ is defined.

Let $\mathcal{R} \subset X^{(r)}$ be closed; $\rho \colon \mathcal{R} \to X^{(r)}$ is the inclusion map. Let $\tau \subset T(V)$ be a codimension 1 tangent hyperplane field on V. For the purposes of this chapter a *formal solution* to the closed relation $\mathcal{R} \subset X^{(r)}$, with respect to the bundle $p_\perp^r \colon X^{(r)} \to X^\perp$ associated to τ, is a pair $(\alpha, f) \in \Gamma(\mathcal{R}) \times \Gamma^r(X)$ such that $p_\perp^r \circ \alpha = j^\perp f \in \Gamma(X^\perp)$. Thus α, f agree on all $\perp$-derivatives of order $\leq r$ of the section f (in Chapter VIII a formal solution (α, f) to $\mathcal{R}$ satisfies $p_{r-1}^r \circ \rho \circ \alpha = j^{r-1} f \in \Gamma(X^{(r-1)})$). In particular, for each $x \in V$, $\alpha(x), j^r f(x)$ both are in the principal subspace over $b(x) = j^\perp f(x) \in X^\perp$ of the bundle $p_\perp^r \colon X^{(r)} \to X^\perp$. A formal solution (α, f) is holonomic if in addition $\alpha = j^r f$ i.e. $j^r f \in \Gamma(\mathcal{R})$, hence f solves the relation $\mathcal{R}$. Following the notation of Chapter VIII, $\mathcal{F} \subset \Gamma(\mathcal{R}) \times \Gamma^r(X)$ is the subspace of all formal solutions to the relation $\mathcal{R}$. $\mathcal{H} \subset \mathcal{F}$ denotes the subspace of holonomic solutions.

A formal solution (α, f) of the relation $\mathcal{R}$ is *strictly short* if in addition $(\alpha, f) \in \Gamma(\mathrm{IntConv}_\tau(\mathcal{R}))$: for all $x \in V$,

$$j^r f(x) \in \mathrm{IntConv}_\tau(\mathcal{R}_{b(x)}, \alpha(x)), \quad b(x) = p_\perp^r \circ \alpha(x) \in X^\perp. \tag{9.2}$$

Thus for each $x \in V$ the convex hull of the arc-component of $\alpha(x)$ in $\mathcal{R}_{b(x)}$ strictly surrounds $j^r f(x)$ in the $\mathbf{R}^q$-fiber $X_{b(x)}^{(r)}$. Note that in case $\mathcal{R}$ is ample with respect to τ: for all $(a, b) \in \mathcal{R}_b \times X_b^{(r)}$, $\mathrm{Conv}_\tau(\mathcal{R}_b, a) = X_b^{(r)} \equiv \mathbf{R}^q$, then every formal solution (α, f) of the relation $\mathcal{R}$ is strictly short. Furthermore if (α, f) is strictly short then of course $(\alpha, f) \in \Gamma(\mathrm{Conv}_\tau(\mathcal{R}))$ is short (strictly surrounding implies surrounding).

Recall II §1 that a subset A of $\mathbf{R}^q$ is nowhere flat if A meets each hyperplane H, $\dim H \leq q - 1$, in a subset $A \cap H$ that is nowhere dense in A. Nowhere flat is an affine notion. The first main result of this chapter is as follows.

Theorem 9.1 $C^{\perp}$-dense h-Principle. *Let $\tau \subset T(V)$ be a smooth, integrable, codimension 1 tangent hyperplane field on V. Let $\mathcal{R} \subset X^{(r)}$ be closed and let $(\alpha, f) \in \mathcal{F}$ be a formal solution to $\mathcal{R}$ which is strictly short i.e. condition (9.2) is satisfied. Let $\mathcal{N}$ be a neighbourhood of the image $j^{\perp}f(V) \subset X^{\perp}$.*

Suppose that $\mathcal{R}$ satisfies the following geometrical properties with respect to the affine $\mathbf{R}^q$-bundle $p_{\perp}^r : X^{(r)} \to X^{\perp}$:

(i) *The projection map $p_{\perp}^r : \mathcal{R} \to X^{\perp}$ is a topological fiber bundle whose fiber is locally path connected (for example $p_{\perp}^r$ is proper and $\mathcal{R} \subset X^{(r)}$ is a smooth submanifold that is transverse to the $\mathbf{R}^q$-fibers of $p_{\perp}^r : X^{(r)} \to X^{\perp}$).*

(ii) *For each $b \in X^{\perp}$ the fiber $\mathcal{R}_b = \mathcal{R} \cap X_b^{(r)}$ is nowhere flat in $X_b^{(r)} \equiv \mathbf{R}^q$.*

Then there is a homotopy of formal solutions $H : [0, 1] \to \mathcal{F}$, $H_t = (\alpha_t, f_t)$, $t \in [0, 1]$, such that the following properties obtain:

(iii) $H_0 = (\alpha, f)$; $H_1 \in \mathcal{H}$ *i.e.* $\alpha_1 = j^r f_1 \in \Gamma(X^{(r)})$.

(iv) *($C^{\perp}$-dense principle) : For all $t \in [0, 1]$ the image $j^{\perp}f_t(V) \subset \mathcal{N}$.*

(v) *(Local Solvability) : For each $x \in V$ there is a local C^r-solution to $\mathcal{R}$ over $\mathfrak{Op}\, x$ i.e. there is a C^r-section $g : \mathfrak{Op}\, x \to X$ such that $j^r g : \mathfrak{Op}\, x \to \mathcal{R}$.*

Conclusion (v) follows from the existence locally of strictly short formal solutions. The proof of the theorem will be given after the requisite notion of metaneighbourhood has been developed. We illustrate the theorem with an example based on the system (9.1) where V is open in $\mathbf{R}^n$ and $p : X = V \times \mathbf{R}^q \to V$ is the projection map; $\Gamma^r(X) \equiv C^r(V, \mathbf{R}^q)$. For all $x = (u_1, \ldots, u_{n-1}, t) \in V$,

$$F(j^r f(x)) = F(j^{\perp} f(x), \partial_t^r f(x)) = 0 \in \mathbf{R}^p,$$

where $\partial_t^r = \partial^r / \partial t^r$ is the pure derivative of order r in the t-coordinate; $F : X^{(r)} \to \mathbf{R}^p$ is a continuous function, and $f \in C^r(V, \mathbf{R}^q)$ is the unknown C^r function. In this example $\tau = \ker dt$, an integrable codimension 1 tangent hyperplane field on V. Thus $X^{(r)} = X^{\perp} \times \mathbf{R}^q$, the product bundle over $X^{\perp}$.

Let $\mathcal{R} = F^{-1}(0) \subset X^{(r)}$. The pair $(\alpha, f) \in \mathcal{R} \times X^{(r)}$ is a formal solution to the relation $\mathcal{R}$ if $\alpha = (j^{\perp} f, \beta) \in \Gamma(\mathcal{R})$, where $\beta \in C^0(V, \mathbf{R}^q)$: for all $x \in V$,

$$F(j^{\perp} f(x), \beta(x)) = 0 \in \mathbf{R}^p.$$

A formal solution (α, f) is holonomic if in addition $\alpha = j^r f \in \Gamma(\mathcal{R})$ i.e. $\beta = \partial_t^r f \in C^0(V, \mathbf{R}^q)$. Generically, in case F is a smooth function, $\mathcal{R} = F^{-1}(0)$

locally is a smooth submanifold of codimension p which is transverse to the $\mathbf{R}^q$-fibers of the bundle $p_\perp^r : X^{(r)} \to X^\perp$. In this case the geometrical hypotheses of the theorem on $\mathcal{R}$ require that the fibers of the bundle $\mathcal{R} \to X^\perp$ are $(q - p)$-manifolds of dimension ≥ 1. Hence $q \geq p + 1$ i.e. the above system of PDEs is underdetermined: there are q unknown functions $f = (f_1, \ldots f_q) \in C^r(V, \mathbf{R}^q)$ and p equations $F = 0 \in \mathbf{R}^p$, such that $q \geq p+1$; in particular $q \geq 2$. Furthermore the geometrical hypotheses on $\mathcal{R}$ (local path connectedness and nowhere flatness) are satisfied locally in the above generic situation. It follows from Theorem 9.1(v) that generically an underdetermined rth order system of PDEs is solvable locally, a remarkable result in non-linear PDE theory.

To illustrate the theorem, we choose F so that the above generic bundle properties are satisfied globally on V. To simplify the presentation, suppose $n = q = 2$ and that $V \subset \mathbf{R}^2$ is open. We consider the system: for all $x = (u, t) \in V$ $(f = (f_1, f_2) \in C^r(V, \mathbf{R}^2))$,

$$(\partial_t^r f_1(x))^2 + (\partial_t^r f_2(x))^2 = A(j^\perp f(x)), \tag{9.3}$$

where $A \colon X^\perp \to (0, \infty)$ is a continuous function. In this example, $X^r = X^\perp \times \mathbf{R}^2$, where the $\mathbf{R}^2$-factor of this product corresponds to the derivative ∂_t^r; $F = A - (p^2 + q^2)$, $(p, q) \in \mathbf{R}^2$ (coordinates for the ∂_t^r-component in X^r), and $\mathcal{R} = F^{-1}(0) \subset X^{(r)}$. Evidently $p_\perp^r \colon \mathcal{R} \to X^\perp$ is a circle bundle whose fiber over $b = j^\perp f(x) \in X^\perp$ is the circle: $\{(p, q) \in \mathbf{R}^2 \mid p^2 + q^2 = A(b)\}$. Thus the bundle hypothesis (i) is satisfied. In addition, for all $b \in X^\perp$ the circle fibers $\mathcal{R}_b \subset \mathbf{R}^2$ are locally path connected and nowhere flat, hence hypothesis (ii) is satisfied. Thus example (9.3) satisfies the hypotheses of the theorem.

A formal solution $(\alpha, f) \in \mathcal{F}$ consists of a map $f \in C^r(V, \mathbf{R}^q)$, and a section $\alpha \in \Gamma(\mathcal{R})$, $\alpha(x) = (j^\perp f(x), (p(x), q(x)) \in X^\perp \times \mathbf{R}^2$ such that for all $x \in V$, $p^2 + q^2 = A(j^\perp f(x))$. Evidently the formal solution (α, f) is strictly short if and only if for all $x \in V$,

$$(\partial_t^r f_1(x))^2 + (\partial_t^r f_2(x))^2 < A(j^\perp f(x)). \tag{9.4}$$

Indeed, for all $x \in V$, (9.4) is necessary and sufficient for the existence of a path in $\mathcal{R}_b$, $b = j^\perp f(x)$, which strictly surrounds the derivative $\partial_t^r f(x) \in \mathbf{R}^2$. Note that strictly short formal solutions to the relation $\mathcal{R}$ exist (for example f is a constant map).

Let (α, f) be a strictly short formal solution to the PDE (9.3). Applying Theorem 9.1, there exists a map $g \in C^r(V, \mathbf{R}^2)$ which solves (9.3) globally on V and such that g, f are $C^\perp$-close.

9.1.4. Metaneighbourhoods. In this section the geometry of metaneighbourhoods is developed in order to prove Theorem 9.1. Strictly speaking, as developed in

Gromov [18], only families of in-paths (cf. §9.1.5) are required for the general theory. However, metaneighbourhoods are part of the underlying geometry of in-paths. We therefore employ both in-paths and metaneighbourhoods in the development of the general theory. We begin with the geometry of metaneighbourhoods in $\mathbf{R}^q$ and then extend to affine $\mathbf{R}^q$-bundles. Let $A \subset \mathbf{R}^q$ and let $\epsilon > 0$; $B(x; r)$ is the open ball of radius r at $x \in \mathbf{R}^q$. For each $x \in A$, $\mathrm{Arc}_\epsilon(A; x)$ denotes the path component of x in $A \cap B(x; \epsilon)$; $\mathrm{Env}_\epsilon(A; x)$ denotes the *interior* in $\mathbf{R}^q$ of the convex hull of $\mathrm{Arc}_\epsilon(A; x)$ (the notation "Env" indicates "enveloppe convexe" in French). In terms of the interior convex hull extensions introduced previously, $\mathrm{Env}_\epsilon(A; a) = \mathrm{IntConv}^\epsilon(A; a)$ with respect to the bundle over a point $\mathbf{R}^q \to *$.

In general $\mathrm{Env}_\epsilon(A; x)$ may be empty. However, in case $A \subset \mathbf{R}^q$ is locally path connected and nowhere flat then for all $(x, \epsilon) \in A \times (0, \infty)$,

$$\mathrm{Env}_\epsilon(A; x) \neq \emptyset; \quad x \in \overline{\mathrm{Env}_\epsilon(A; x)}.$$

Remark 9.2. The following observation is useful for the proofs in this section. Let $y \in \mathrm{Env}_\epsilon(A; x)$. There is a finite number of points $z_i \in \mathrm{Arc}_\epsilon(A; x)$, $1 \leq i \leq m$, such that the convex hull of the set $\{z_1, \ldots, z_m\}$ is a neighbourhood of y in $\mathbf{R}^q$. Indeed, y is in the interior of a q-simplex Δ^q such that $\Delta^q \subset \mathrm{Env}_\epsilon(A; x)$. Since the vertices of Δ^q are each in the convex hull of $\mathrm{Arc}_\epsilon(A; x)$ it follows that there is a finite number of points in $\mathrm{Arc}_\epsilon(A; x)$ whose convex hull contains Δ^q, and thus is a neighbourhood of y. In particular, and this is central for what follows, for each $y \in \mathrm{Env}_\epsilon(A; x)$ there is a continuous map $g \colon [0, 1] \to \mathrm{Arc}_\epsilon(A; x)$, $g(0) = x$, such that the path $g(t)$ strictly surrounds y.

Suppose $A \subset \mathbf{R}^q$ is closed, locally path connected and nowhere flat. Let $\epsilon \colon A \to (0, \infty)$ be continuous. The ϵ-metaneighbourhood of A in $\mathbf{R}^q$ is the subset,

$$\mathrm{Met}_\epsilon A = \bigcup_{x \in A} \mathrm{Env}_{\epsilon(x)}(A; x).$$

Employing Remark 9.2, $\mathrm{Met}_\epsilon A$ is open in $\mathbf{R}^q$. $A \subset \overline{\mathrm{Met}_\epsilon A}$, but in general $\mathrm{Met}_\epsilon A$ is not a neighbourhood of A. For example if $A = S^{q-1} \subset \mathbf{R}^q$ is the unit round sphere, $q \geq 2$, then A is locally path connected, nowhere flat; $\mathrm{Met}_\epsilon A \subset D^q$ is an open annulus and $(\mathrm{Met}_\epsilon A) \cap S^{q-1} = \emptyset$. Evidently,

$$A = \bigcap_\epsilon \overline{\mathrm{Met}_\epsilon A}.$$

Let $\mathcal{N}$ be a neighbourhood of A in $\mathbf{R}^q$. There is a continuous function $\epsilon \colon A \to (0, \infty)$ such that $\mathrm{Met}_\mu A \subset \mathcal{N}$ for all continuous functions $\mu \colon A \to (0, \infty)$ such that $\mu \leq \epsilon$. Furthermore if $\mu \leq \epsilon$ then $\mathrm{Met}_\mu A \subset \mathrm{Met}_\epsilon A$.

9.1.5. In-Paths. Throughout this section $A \subset \mathbf{R}^q$ is closed, locally path connected and nowhere flat. A continuous map $x \colon [0,1] \to \mathbf{R}^q$ is an *in-path* at $a \in A$ in $\mathrm{Met}_\epsilon A$ if: (i) $x(0) = a$; (ii) for all $t \in (0,1]$, $x(t) \in \mathrm{Env}_\epsilon(A;a) \subset \mathrm{Met}_\epsilon A$. Let $x_i(t)$, $i \in \{1,2\}$ be in-paths at $a \in A$ in $\mathrm{Met}_\epsilon A$. Employing the convexity of $\mathrm{Env}_\epsilon(A;a)$ it follows that $x_s(t) = (1-s)x_1(t) + sx_2(t)$, $s \in [0,1]$, is a homotopy of in-paths in $\mathrm{Met}_\epsilon A$.

Lemma 9.3. *Let $\epsilon \colon A \to (0,\infty)$ be continuous. There is a continuous map $\alpha \colon A \to C^0([0,1], \mathbf{R}^q)$ such that for each $a \in A$, $\alpha(a)$ is an in-path at a in $\mathrm{Met}_\epsilon A$.*

Proof. For each $a \in A$ let Δ^q be a q-simplex with vertices $\{a, z_1, z_2, \ldots, z_q\}$ in $\mathbf{R}^q$ where,

$$z_i \in \mathrm{Arc}_{\epsilon(a)/2}(A;a) \quad 1 \le i \le q.$$

Since A is locally path connected there is a neighbourhood $\mathfrak{Op}\, a \subset A$ such that for all $b \in \mathfrak{Op}\, a$, $\{b, z_1, \ldots, z_q\}$ are the vertices of a non-degenerate q-simplex and,

$$z_i \in \mathrm{Arc}_{\epsilon(b)}(A;b) \quad 1 \le i \le q.$$

Let $y \in \Delta^q$ be the barycentre. For all $b \in \mathfrak{Op}\, a$, $y \in \mathrm{Env}_{\epsilon(b)}(A;b)$. Let $\alpha_a \colon \mathfrak{Op}\, a \to C^0([0,1], \mathbf{R}^q)$ be the continuous map: for all $(b,t) \in \mathfrak{Op}\, a \times [0,1]$, $\alpha_a(b)(t) = (1-t)b + ty \in \mathrm{Env}_{\epsilon(b)}(A;b)$. Thus for all $b \in \mathfrak{Op}\, a$, $\alpha_a(b)$ is an in-path at b in $\mathrm{Met}_\epsilon A$.

Let $(p_i \colon A \to [0,1])_{i \ge 1}$ be a partition of unity subordinate to a countable, locally finite cover $(\mathfrak{Op}\, a_i)_{i \ge 1}$ of A such that each $\mathfrak{Op}\, a_i$ satisfies the above properties. Let $\alpha \colon A \to C^0([0,1], \mathbf{R}^q)$ be the continuous map: for all $x \in A$,

$$\alpha(x) = \sum_{i=1}^{\infty} p_i(x)\alpha_{a_i}(x).$$

Thus for all $x \in A$, $\alpha(x) = x$, and since $\mathrm{Env}_{\epsilon(x)}(A;x)$ is a convex set, the convex combination of in-paths is an in-path i.e. $\alpha(x)$ is an in-path at x in $\mathrm{Met}_\epsilon A$. $\quad\square$

For each $a \in A$ the end point of the in-path $\alpha(a)(t)$ satisfies $\alpha(a)(1) \in \mathrm{Env}_{\epsilon(a)}(A;a)$ at $t = 1$. Setting $I(a) = \alpha(a)(1)$, the following corollary is proved.

Corollary 9.4. *Let $\epsilon \colon A \to (0,\infty)$ be continuous. There is a continuous map $I \colon A \to \mathrm{Met}_\epsilon A$ such that for all $a \in A$, $I(a) \subset \mathrm{Env}_{\epsilon(a)}(A;a)$.*

9.1.6. The geometry of metaneighbourhoods in §9.3 is now extended to affine $\mathbf{R}^q$ bundles. Let $p \colon E \to B$ be an affine $\mathbf{R}^q$-bundle over a base space B where (E,d) is paracompact with metric d. Let $\mathcal{R} \subset E$ be a closed subspace that satisfies the following properties (cf. the hypotheses on $\mathcal{R}$ in Theorem 9.1 with respect to the affine bundle $p_\perp^r \colon X^{(r)} \to X^\perp$):

(p_1) The restriction (same notation) $p\colon \mathcal{R} \to B$ is a topological subbundle whose fiber is locally path connected.

(p_2) For each $b \in B$ the fiber $\mathcal{R}_b = \mathcal{R} \cap E_b$ is nowhere flat in $E_b \equiv R^q$.

Let $\epsilon\colon \mathcal{R} \to (0, \infty)$ be continuous. The ϵ-metaneighbourhood of $\mathcal{R}$ in E is the union of all the fiberwise metaneighbourhoods:

$$\mathrm{Met}_\epsilon(\mathcal{R}) = \bigcup_{b \in B} \mathrm{Met}_\epsilon \mathcal{R}_b$$

$$= \bigcup_{b \in B} \bigcup_{x \in \mathcal{R}_b} \mathrm{Env}_{\epsilon(x)}(\mathcal{R}_b; x).$$

Employing the local product structure of the bundle $p\colon \mathcal{R} \to B$, Remark 9.2 generalizes to show that if the vertices of a q-simplex $\Delta^q \subset E_b$ are in the interior of the convex hull of $\mathrm{Arc}_{\epsilon(x)}(\mathcal{R}_b; x)$, then $\Delta^q \times \mathfrak{Op}\, b \subset \mathrm{Met}_\epsilon \mathcal{R}$. The following lemma is easily proved.

Lemma 9.5. *The ϵ-metaneighbourhoods of $\mathcal{R}$ in E satisfy the following properties:*

(i) *For each $\epsilon\colon \mathcal{R} \to (0, \infty)$ $\mathrm{Met}_\epsilon \mathcal{R}$ is open in E. $\mathrm{Met}_\mu \mathcal{R} \subset \mathrm{Met}_\epsilon \mathcal{R}$ if $\mu \le \epsilon$.*

(ii) *$\mathcal{R} = \bigcap_\epsilon \overline{\mathrm{Met}_\epsilon \mathcal{R}}$.*

(iii) *Let $\mathcal{N}$ be a neighbourhood of $\mathcal{R}$ in E. there is a continuous function $\epsilon\colon \mathcal{R} \to (0, \infty)$ such that $\mathrm{Met}_\epsilon \mathcal{R} \subset \mathcal{N}$.*

A continuous map $x\colon [0, 1] \to E$ is defined to be an in-path at $a \in \mathcal{R}$ in $\mathrm{Met}_\epsilon \mathcal{R}$ if $x(t)$ is a fiberwise in-path: $x(0) = a$ and for all $t \in (0, 1]$, $x(t) \in \mathrm{Env}_{\epsilon(a)}(\mathcal{R}_b; a)$, $b = p(a) \in B$ (the convex hull is in the fiber $E_b \equiv \mathbf{R}^q$). Employing the local product structure of $p\colon \mathcal{R} \to B$, Lemma 9.3 extends to prove the following result.

Selection Lemma 9.6. *Let $\epsilon\colon \mathcal{R} \to (0, \infty)$ be continuous. There is a continuous map $\alpha\colon \mathcal{R} \to C^0([0, 1], E)$ such that for all $a \in \mathcal{R}$, $\alpha(a)\colon [0, 1] \to E$ is an in-path at a in $\mathrm{Met}_\epsilon \mathcal{R}$.*

Proof. Let $a \in \mathcal{R}$, $b = p(a) \in E$. Employing the local product structure, there is a bundle neighbourhood $\mathfrak{Op}\, a \equiv N_a \times \mathfrak{Op}\, b \subset \mathcal{R}$, N_a is path connected, and sections $z_i \in \Gamma_{\mathfrak{Op}\, b}(E)$, $1 \le i \le q$, such that for each $u \in \mathfrak{Op}\, a$, $p(u) = x \in \mathfrak{Op}\, b$, $\{u, z_1(x), \ldots, z_q(x)\}$ are the vertices of a q-simplex $\Delta_x^q \subset E_x$ and (cf. Lemma 9.3),

$$z_i(x) \in \mathrm{Arc}_{\epsilon(u)}(\mathcal{R}_x; u) \quad 1 \le i \le q. \tag{9.5}$$

Let $y \in \Gamma_{\mathfrak{Op}\, b}(E)$ such that for all $x \in \mathfrak{Op}\, b$, $y(x)$ is the barycentre of Δ_x^q. Let $\alpha_a\colon \mathfrak{Op}\, a \to C^0([0, 1], E)$ be the continuous map: for all $u \in \mathfrak{Op}\, a$, $p(u) = x \in$

$\mathfrak{Op}\, b$, $\alpha_a(u)(t) = (1 - t)u + ty(x) \in \mathrm{Env}_{\epsilon(u)}(\mathcal{R}_x; u)$. Thus for all $u \in \mathfrak{Op}\, a$, $\alpha_a(u)$ is an in-path at u in $\mathrm{Met}_\epsilon\, \mathcal{R}$.

Let $(p_i \colon \mathcal{R} \rightarrow [0, 1])_{i \geq 1}$ be a partition of unity subordinate to a locally finite cover $(\mathfrak{Op}\, a_i)_{i \in I}$ of $\mathcal{R}$, where each $\mathfrak{Op}\, a_i$ satisfies the above properties. Let $\alpha \colon \mathcal{R} \rightarrow C^0([0, 1], E)$ be the map: for all $u \in \mathcal{R}$,

$$\alpha(u) = \sum_{i \geq 1} p_i(u)\alpha_{a_i}(u).$$

Thus α is continuous and for all $u \in \mathcal{R}$, $\alpha(u) = u$. Since $\mathrm{Env}_{\epsilon(u)}(\mathcal{R}_x; u)$ is a convex set, the convex combination of in-paths is an in-path i.e. for all $u \in \mathcal{R}$, $\alpha(u)$ is an in-path at u in $\mathrm{Met}_\epsilon(\mathcal{R})$. $\square$

For each $u \in \mathcal{R}$ the end point of the in-path $\alpha(a)(1) \in \mathrm{Env}_{\epsilon(u)}(\mathcal{R}_x; u)$ at $t = 1$, $x = p(u) \in B$. Setting $I(u) = \alpha(u)(1)$, the first part of the following corollary is proved.

Corollary 9.7. *Let* $\epsilon \colon \mathcal{R} \rightarrow (0, \infty)$ *be continuous. There is a continuous map* $I \colon \mathcal{R} \rightarrow \mathrm{Met}_\epsilon\, \mathcal{R}$ *such that for all* $u \in \mathcal{R}$, $I(u) \in \mathrm{Env}_{\epsilon(u)}(\mathcal{R}_{p(u)}; u)$. *Furthermore, there is a continuous family of paths* $h \colon [0, 1] \times \mathcal{R} \rightarrow \mathcal{R}$ *such that the following properties are satisfied: for all* $u \in \mathcal{R}$,

(i) $h(0, u) = u$; *the path* $h_t(u)$, $t \in [0, 1]$, *lies in the* ϵ-*ball* $\mathcal{R}_b \cap B(u; \epsilon(u))$, $b = p(u) \in B$.

(ii) *The path* $h_t(u)$, $t \in [0, 1]$, *strictly surrounds* $I(u)$.

Proof. For each $u \in \mathcal{R}$ there is a C-structure (h, H) at u such that the path $h_t(u)$, $t \in [0, 1]$, lies in the ϵ-ball $\mathcal{R}_b \cap B(u; \epsilon(u))$, $b = p(u) \in B$, and strictly surrounds $I(u)$. Employing the bundle $p \colon \mathcal{R} \rightarrow B$, a C-structure at a point u obviously extends to a C-structure over $\mathfrak{Op}\, u \subset \mathcal{R}$. Since the space of C-structures is contractible, arguing as in II Proposition 2.2, it follows that there is a global C-structure (h, H) on $\mathcal{R}$, which proves the lemma. Details are left to the reader. $\square$

Let $f \in \Gamma(\mathcal{R})$ and let $g \in C^0(V, (0, 1])$ be a continuous map. Let $I_g \colon V \rightarrow E$ be the map, $I_g(x) = \alpha(f(x))(g(x))$, the point on the in-path $\alpha(f(x))(t)$ at the parameter $t = g(x)$. In particular for all $x \in V$,

$$I_g(x) \in \mathrm{Env}_{\epsilon(u)}(\mathcal{R}_{p(u)}; u); \quad u = f(x) \in \mathcal{R}.$$

Let $\epsilon \colon \mathcal{R} \rightarrow (0, \infty)$ be continuous. Let $M_\epsilon \subset \mathcal{R} \times \mathrm{Met}_\epsilon\, \mathcal{R}$ consist of all pairs (a, y) such that $y \in \mathrm{Env}_{\epsilon(a)}(\mathcal{R}_b; a)$, $b = p(a) = p(y) \in B$. There is a projection map $\pi \colon M_\epsilon \rightarrow \mathcal{R}$, $(a, y) \mapsto a$, and also a projection map $\rho \colon M_\epsilon \rightarrow E$, $(a, y) \mapsto y \in E$.

The point of introducing the relation $\rho\colon M_\epsilon \to E$ is that it is a microfibration, proved below, hence this relation is subject to the theory of convex hull extensions, whereas (since $\mathcal{R} \subset E$ is closed) the inclusion $i\colon \mathcal{R} \to E$ is not a microfibration (except in the trivial case that $\mathcal{R} = E$).

Lemma 9.8. *The projection map $\rho\colon M_\epsilon \to E$ is a microfibration.*

Proof. Let $f = (g, y)\colon P \to M_\epsilon$, $g \in C^0(P, \mathcal{R})$, $y \in C^0(P, E)$, and $F\colon [0, 1] \times P \to E$ be continuous maps, P a compact polyhedron, such that for all $u \in P$, $F_0(u) = y(u) \in E$. Let $b = p \circ F\colon [0, 1] \times P \to B$ be the projection of the map F into the base space B. Hence for all $u \in P$,

$$F_0(u) \in \mathrm{Env}_{\epsilon(g(u))}(\mathcal{R}_{b(0,u)}; g(u)). \tag{9.6}$$

The bundle map $p\colon \mathcal{R} \to B$ is a Serre fibration. Hence there is a continuous lift $H\colon [0, 1] \times P \to \mathcal{R}$ such that: (i) $H_0 = g$; (ii) $p \circ H = b$. For each $u \in P$, $F_0(u)$ is in the interior of a q-simplex $\Delta^q(u)$ which is a subset of the convex hull (9.6). Since P is compact and $H_0 = g$, employing the arguments for (9.5) in the Selection Lemma (applied to $\mathfrak{Op}\, g(u)$ for each $u \in P$), it follows from the convex hull condition (9.6) that there is a $\delta \in (0, 1]$ such that for all $(s, u) \in [0, \delta] \times P$,

$$F(s, u) \in \mathrm{Env}_{\epsilon(H(s,u))}(\mathcal{R}_{b(s,u)}; H(s, u)).$$

Setting $G = (H, F)\colon [0, \delta] \times P \to M_\epsilon$, the lemma is proved. $\square$

Proof of Theorem 9.1. The constructions and results of §9.1.4 to §9.1.6 are applied to the affine $\mathbf{R}^q$-bundle $p_\perp^r\colon X^{(r)} \to X^\perp$ and to the closed relation $\mathcal{R} \subset X^{(r)}$. $(X^{(r)}, d)$ is a complete metric space. $B(z; r)$ denotes the open ball at $z \in X^{(r)}$, radius $r > 0$ in the metric space $(X^{(r)}, d)$. Let $\mathcal{N}_1$ be a (smaller) neighbourhood of the image $j^\perp(V) \subset X^\perp$ such that $\overline{\mathcal{N}_1} \subset \mathcal{N}$. Also $(X^\perp, d')$ is a complete metric space.

Let $\epsilon\colon \mathcal{R} \to (0, \infty)$ be continuous. Associated to the closed relation $\mathcal{R} \subset X^{(r)}$ is the microfibration $\rho\colon M_\epsilon \to X^{(r)}$ (cf. Lemma 9.8). A section $\varphi = (\alpha, g) \in \Gamma(M_\epsilon)$, $\alpha \in \Gamma(\mathcal{R})$, $g \in \Gamma(X^{(r)})$, satisfies the property that for all $x \in V$,

$$g(x) \in \mathrm{Env}_{\epsilon(\alpha(x))}(\mathcal{R}_{b(x)}, \alpha(x)), \quad b(x) = p_\perp^r \circ \alpha(x) = p_\perp^r \circ g(x) \in X^\perp.$$

The projection $\pi\colon M_\epsilon \to \mathcal{R}$, $(a, y) \mapsto a$, induces a projection (same notation) $\pi\colon \Gamma(M_\epsilon) \to \Gamma(\mathcal{R})$, $(\alpha, g) \mapsto \alpha \in \Gamma(\mathcal{R})$. A section $\varphi = (\alpha, g)$ is holonomic if the projection $\rho \circ \varphi = g$ is holonomic: there is a C^r-section $f \in \Gamma^r(X)$ such that $g = j^r f \in \Gamma(X^{(r)})$. Employing Corollary 9.7, there is a continuous lift $\gamma\colon \Gamma(\mathcal{R}) \to \Gamma(M_\epsilon)$, $\alpha \mapsto (\alpha, I \circ \alpha)$, where for each $x \in V$, $I(\alpha(x))$ is the end point at $t = 1$ of an in-path at $\alpha(x)$ in $\mathrm{Met}_\epsilon\, \mathcal{R}$.

Let $(\alpha, f) \in \mathcal{F}$ be a formal solution to $\mathcal{R}$ that is strictly short i.e. (9.2) is satisfied. Employing Corollary 9.7 there is a continuous family of paths $p_s(x)$, $p_0(x) = \alpha(x)$, $s \in [0, 1]$, in the fiber $\mathcal{R}_{b(x)}$, $b(x) = j^\perp f(x) \in X^\perp$, that strictly surrounds the r-jet $j^r f(x)$ for all $x \in V$.

Let $h \in C^0(V, (0, 1])$ be continuous. Applying the lift I_h to the strictly surrounding path $p_s(x)$ it follows that for each $x \in V$, the path $I_h(p_s(x)) \in \mathrm{Met}_\epsilon \mathcal{R}$, $s \in [0, 1]$, also (strictly) surrounds the r-jet $j^r f(x)$, provided $h > 0$ is sufficiently small. Let $\beta = (\alpha, I_h \circ \alpha) \in \Gamma(M_\epsilon)$. Thus for each $x \in V$ there is a path $\gamma_s(x) = (p_s(x), I_h(p_s(x))) \in M_\epsilon$, $\gamma_0(x) = \beta(x)$, $s \in [0, 1]$, such that the projection $\rho(\gamma_s(x)) = I_h(p_s(x))$ surrounds $j^r f(x)$ in the fiber $X_{b(x)}^{(r)}$, provided $h > 0$ is sufficiently small. Consequently, the formal solution $(\alpha, f) \in \mathcal{F}$ for the closed relation $\mathcal{R}$ induces a holonomic section $(\beta, j^r f) \in \Gamma(\mathrm{Conv}_\tau M_\epsilon)$ for the microfibration $\rho \colon M_\epsilon \to X^{(r)}$, where $\mathrm{Conv}_\tau M_\epsilon$ is the convex hull extension of the relation M_ϵ with respect to the hyperplane field τ. In this way we pass from a formal solution to the relation $\mathcal{R}$ to a holonomic solution of the convex hull extension of the associated relation $\rho \colon M_\epsilon \to X^{(r)}$, provided $h > 0$ is sufficiently small.

Applying the h-stability theorem VII 7.2 to the holonomic section $\varphi = (\beta, j^r f) \in \Gamma(\mathrm{Conv}_\tau M_\epsilon)$, there is a homotopy of holonomic sections $H \colon [0, 1] \to \Gamma(\mathrm{Conv}_\tau M_\epsilon)$, $H_t = (\beta_t, j^r f_t)$, $t \in [0, 1]$, such that the following properties are satisfied:

(1) $H_0 = \varphi$; $H_1 \in \Gamma(M_\epsilon)$ i.e. β_1 is holonomic: $\rho \circ \beta_1 = j^r f_1 \in \Gamma(X^{(r)})$.

(2) For all $t \in [0, 1]$, the image $j^\perp f_t(V) \subset \mathcal{N}_1$.

Let $\beta_t = (\alpha_t, g_t) \in \Gamma(M_\epsilon)$, $t \in [0, 1]$. Employing (1), it follows that: (i) $\alpha_0 = \alpha$; (ii) $g_1 = j^r f_1 \in \Gamma(X^{(r)})$. Consequently, for all $x \in V$ ($\beta_1 \in \Gamma(M_\epsilon)$),

$$j^r f_1(x) \in \mathrm{Env}_{\epsilon(\alpha_1(x))}(\mathcal{R}_{b(x)}; \alpha_1(x)), \quad b(x) = j^\perp f_1(x) \in X^\perp. \qquad (9.7)$$

It follows from (9.7) that for each $x \in V$ there is a path $p_s(x)$, $p_0(x) = \alpha_1(x)$, $s \in [0, 1]$, in $\mathcal{R}_{b(x)} \cap B(p_0(x); \epsilon(p_0(x)))$ i.e. in a small ball of radius $\epsilon(p_0(x))$, which strictly surrounds the r-jet $j^r f_1(x)$. In this way the homotopy of formal solutions $(\alpha_t, j^r f_t)$ to the relation $\mathcal{R}$ improves (α, f) at $t = 0$ which satisfies (9.2) to the formal solution $(\alpha_1, j^r f_1)$ at $t = 1$ which satisfies (9.7) with respect to the continuous function $\epsilon \colon \mathcal{R} \to (0, \infty)$. In particular, employing (9.7),

(3) For all $x \in V$, $d(\alpha_1(x), j^r f_1(x)) \leq \epsilon(\alpha_1(x))$.

The construction of the formal solution (α_1, f_1) to the relation $\mathcal{R}$ for which (9.7) is satisfied is the first step of an iterative procedure of successive improvements of formal solutions by means of the h-Stability Theorem VII 7.2, in order to produce in the limit a C^r-solution of the relation $\mathcal{R}$. The main inductive step in this procedure is the following lemma. In what follows $X^{(r)} \times X^{(r)}$ is the product metric space, for which the subspace $M_\epsilon \subset \mathcal{R} \times \mathrm{Met}_\epsilon \, \mathcal{R} \subset X^{(r)} \times X^{(r)}$ has the induced metric.

Inductive Lemma 9.9. *Let $\epsilon > 0$. Suppose (α, g) is a formal solution to $\mathcal{R}$ such that the image $j^\perp g(V) \subset \mathcal{N}_1$, and such that for all $x \in V$,*

$$ j^r g(x) \in \mathrm{Env}_\epsilon(\mathcal{R}_{b(x)}, \alpha(x)), \quad b(x) = j^\perp g(x) \in X^\perp. \tag{9.7'} $$

Let $\mu \in (0, \epsilon]$. There is a homotopy of formal solutions $H_t = (\alpha_t, g_t)$ of the relation $\mathcal{R}$, $H_0 = (\alpha, g)$, $t \in [0,1]$, such that the following properties are satisfied:

(1) *For all $x \in V$, $j^r g_1(x) \in \mathrm{Env}_\mu(\mathcal{R}_{b(x)}, \alpha_1(x))$; $b(x) = j^\perp g_1(x) \in X^\perp$.*

(2) *The image $j^\perp g_1(V) \subset \mathcal{N}_1$.*

(3) *For all $(t, x) \in [0,1] \times V$, $d(\alpha(x), \alpha_t(x)) \leq (n+1)\epsilon(\alpha(x))$.*

Proof. Let $h \in C^0(V, (0,1])$. Employing Lemma 9.7, there is a continuous lift $I \colon \mathcal{R} \to \mathrm{Met}_\mu \, \mathcal{R}$ and a continuous family of paths $p_s(x) \in \mathcal{R}_{b(x)}$, $p_0(x) = \alpha(x)$, $s \in [0,1]$, that strictly surrounds the r-jet $j^r g(x)$ for all $x \in V$. Let $\beta = (\alpha, I_h \circ \alpha) \in \Gamma(M_\mu)$. For $h > 0$ sufficiently small the lifted path $I_h(p_s(x))$ also (strictly) surrounds $j^r g(x)$. Hence for each $x \in V$, the path $\gamma_s(x) = (p_s(x), I_h(p_s(x))) \in M_\mu$, $\gamma_0(x) = \beta(x)$, $s \in [0,1]$, satisfies the property that the projection $\rho(\gamma_s(x)) = I(p_s(x))$ surrounds $j^r g(x)$ in the fiber $X^{(r)}_{b(x)}$, provided $h > 0$ is sufficiently small. Consequently, $\varphi = (\beta, j^r g) \in \Gamma(\mathrm{Conv}_\tau M_\mu)$.

Employing Lemma 9.8, the projection map $\rho \colon M_\mu \to X^{(r)}$ is a microfibration. Applying the h-stability theorem VII 7.2 to the holonomic section $\varphi = (\beta, j^r g) \in \Gamma(\mathrm{Conv}_\tau M_\mu)$, there is a homotopy of holonomic sections $G \colon [0,1] \to \Gamma(\mathrm{Conv}_\tau M_\mu)$, $G_t = (\beta_t, j^r g_t)$, $t \in [0,1]$, such that the following properties obtain:

(4) $G_0 = \varphi$; $G_1 \in \Gamma(M_\mu)$ is holonomic: $\rho \circ \beta_1 = j^r g_1 \in \Gamma(\mathrm{Met}_\mu \, \mathcal{R})$.

(5) For all $t \in [0,1]$, the image $j^\perp g_t(V) \subset \mathcal{N}_1$.

Let $\beta_t = (\alpha_t, h_t) \in \Gamma(M_\mu)$, $t \in [0,1]$. Employing (4), the following properties obtain: (i) $\alpha_0 = \alpha$; (ii) $h_1 = j^r g_1 \in \Gamma(X^{(r)})$. Since $\beta_1 = (\alpha_1, j^r g_1) \in \Gamma(M_\mu)$, conclusion (1) is proved. Also (5) proves conclusion (2).

Employing (9.7'), for all $(s, x) \in [0,1] \times V$, the above path $p_s(x)$ that surrounds $j^r f(x)$ satisfies $d(p_s(x), \alpha(x)) < \epsilon$. Consequently for $h > 0$ sufficiently small, the lifted path $I_h(p_s(x))$ satisfies $d(I_h(p_s(x)), I_h(\alpha(x))) < \epsilon$. Employing the product metric, it follows that the surrounding path $\gamma_s(x) \in M_\epsilon$ satisfies $d(\gamma_s(x), \beta(x)) < \epsilon$ for all $(s, x) \in [0,1] \times V$. Thus (cf. VII §7.4) for all $x \in V$,

$$ j^r g(x) \in \mathrm{Conv}^\epsilon((M_\mu)_{b(x)}; \beta(x)), \quad b(x) = j^\perp g(x) \in X^\perp. $$

Applying VII Complement 7.16, one can assume that the homotopy of sections $(\beta_t, j^r g_t)$, $t \in [0, 1]$, satisfies the approximation: for all $(t, x) \in [0, 1] \times V$, $d(\beta_t(x), \beta(x)) < (n + 1)\epsilon$. It follows from the product metric on M_μ that for all $x \in V$, $d(\alpha_s(x), \alpha(x)) < (n + 1)\epsilon$, which proves conclusion (3).　□

Returning to the proof of Theorem 9.1, let $\sum_i a_i < \infty$ be a convergent series of positive numbers. Starting with the formal solution (α_1, f_1) which satisfies (9.7) for $\epsilon < a_1$, employing successively the Inductive Lemma 9.9, there is a sequence of formal solutions $(\alpha_m, f_m)_{m \geq 1}$ to the relation $\mathcal{R}$ such that the following properties are satisfied for all $m \geq 1$ and all $x \in V$:

(1) The image $j^\perp f_m(V) \subset \mathcal{N}_1$; $d'(j^\perp f_m(x), j^\perp f_{m+1}(x)) \leq a_m$.

(2) There is a homotopy of formal solutions, $H_m^t = (\alpha_m^t, f_m^t)$ to $\mathcal{R}$, $t \in [0, 1]$, such that $(\alpha_m^0, f_m^0) = (\alpha_m, f_m)$; $(\alpha_m^1, f_m^1) = (\alpha_{m+1}, f_{m+1})$; for all $t \in [0, 1]$, $j^\perp f_m^t(V) \subset \mathcal{N}_1$.

(3) For all $t \in [0, 1]$, $d(\alpha_m(x), \alpha_m^t(x)) < a_m$.

(4) $j^r f_m(x) \in \mathrm{Env}_{a_m}(\mathcal{R}_{b_m(x)}; \alpha_m(x))$, $b_m(x) = j^\perp f_m(x) \in X^\perp$.

Employing properties (2), (4), it follows in particular that,

(5) For all $x \in V$, $d(j^r f_m(x), j^r f_{m+1}(x)) < 2a_m + a_{m+1}$.

It follows from properties (1),(2),(5) above that $(j^\perp f_m)_{m \geq 1}$ is a Cauchy sequence in $\Gamma(X^\perp)$; $(j^r f_m)_{m \geq 1}$ is a Cauchy sequence in $\Gamma(X^{(r)})$; $(\alpha_m)_{m \geq 1}$ is a Cauchy sequence in $\Gamma(\mathcal{R})$. Consequently, $\lim_{m \to \infty} f_m = g \in \Gamma^r(X)$ exists. Employing (1), the image $j^\perp g(V) \subset \overline{\mathcal{N}_1} \subset \mathcal{N}$.

Furthermore from (4), for all $x \in V$ and all $m \geq 1$, $d(j^r f_m(x), \alpha_m(x)) < a_m$. Since $\mathcal{R}$ is closed it follows that,

$$j^r g = \lim_{m \to \infty} j^r f_m = \lim_{m \to \infty} \alpha_m (= \alpha') \in \Gamma(\mathcal{R}).$$

Concatenating the sequence of homotopies $(H_m^t)_{m \geq 1}$, it follows that there is a homotopy of formal solutions $H_t = (\alpha_t, f_t)$ to the relation $\mathcal{R}$, $H_0 = (\alpha, f)$, $t \in [0, 1]$, such that $H_1 = (\alpha', g) \in \Gamma(\mathcal{R})$ is holonomic: $\alpha' = j^r g \in \Gamma(\mathcal{R})$, which completes the proof of the theorem.　□

Corollary 9.10. *Suppose in addition* $\rho: \mathcal{R} \to X^{(r)}$ *is ample. Then* $\mathcal{F}$ *consists of strictly short solutions to* $\mathcal{R}$. *Hence Theorem 9.1 applies to each formal solution in* $\mathcal{F}$. *In particular if* $(\alpha, f) \in \mathcal{F}$, *then there is a homotopy of formal solutions* $(\alpha_t, f_t) \in \mathcal{F}$, $(\alpha_0 f_0) = (\alpha, f)$, $t \in [0, 1]$, *such that* $(\alpha_1, f_1) \in \mathcal{H}$ *i.e.* $\alpha_1 = j^r f_1 \in \Gamma(X^{(r)})$

The point of the corollary is that a formal solution (α, f) does satisfy the strictly short condition (9.2) in case $\mathcal{R}$ is ample. Indeed, for all $x \in V$, $\mathrm{Conv}(\mathcal{R}_{b(x)}; \alpha(x)) = X_{b(x)}^{(r)} \equiv \mathbf{R}^q$.

Example 9.11. To conclude this section we illustrate Corollary 9.10 with an example for which the relation $\mathcal{R}$ is ample. Let $U \subset \mathbf{R}^2$ be open. (x, y) are coordinates in $\mathbf{R}^2$. Consider the underdetermined PDE in the unknown C^r-map $f = (f_1, f_2) \in C^r(U, \mathbf{R}^2)$ $(\partial_x = \partial/\partial x)$,

$$F(\partial_x^r f_1(x, y), \partial_x^r f_2(x, y)) = A(j^{r-1} f(x, y)), \qquad (9.8)$$

where $F \in C^a(U, \mathbf{R})$ is real analytic, A is continuous, and such that F satisfies the following properties:

(1) F has no critical points.

(2) Each level curve $L_c = \{(x, y) \in \mathbf{R}^2 \mid F(x, y) = c\}$ is connected and $\text{Conv}(L_c) = \mathbf{R}^2$. For example $F(x, y) = x + y^3$.

Since F is real analytic without critical points it follows that each level curve L_c is nowhere flat in $\mathbf{R}^2$. Let $X = U \times \mathbf{R}^2 \to U$ be the product bundle. Let $\mathcal{R} \subset X^{(2)}$ be the closed relation defined by the equation, $F(u, v) - A = 0$, where $(u, v) \in \mathbf{R}^2$ are coordinates in the ∂_x^r-component in $X^{(2)}$. Evidently, $p_\perp^r : \mathcal{R} \to X^\perp$ is a trivial bundle whose fibers are the level curves L_c. Hence there is a section φ of the bundle $p_\perp^r : \mathcal{R} \to X^\perp$. Since the convex hull of each level curve L_c is $\mathbf{R}^2$, it follows that $\mathcal{R}$ is ample.

Let $f \in C^r(U, \mathbf{R}^2)$. Employing the section φ there is a formal solution (α, f) to the relation $\mathcal{R}$. Let $\mathcal{N}$ be a neighbourhood of the image $j^\perp f(U) \subset X^\perp$. Applying Corollary 9.10, there is a homotopy of formal solutions (α_t, f_t) to the relation $\mathcal{R}$, $(\alpha_0, f_0) = (\alpha, f)$, $t \in [0, 1]$, such that: (i) for all $t \in [0, 1]$ the image $j^\perp f_t(U) \subset \mathcal{N}$; (ii) (α_1, f_1) is holonomic: $\alpha_1 = j^r f_1 \in \Gamma(\mathcal{R})$ i.e. $f_1 \in C^r(U, \mathbf{R}^2)$ solves the equation (9.8) globally over U.

§2. Triangular Systems

9.2.1. The applications in the previous section §9.1.3 of convex integration to the solving of underdetermined systems of PDEs may be generalized somewhat to include special systems of non-linear PDEs, denoted here as triangular systems of PDEs. Triangular systems have the merit that they may be determined or overdetermined systems of PDEs. However, generic determined systems of PDEs *cannot* be triangular. Hence generic determined systems of PDEs are not solvable by the methods of convex integration theory. This contrasts sharply with the results of the previous section §9.1.3 where the geometrical hypotheses on the closed relation $\mathcal{R} \subset X^{(r)}$ of Theorem 9.1, i.e. the bundle hypothesis and the nowhere flat hypothesis on $\mathcal{R}$, generically are satisfied locally over small charts in the base manifold V. Thus, generically, underdetermined systems are solvable locally (and are solvable globally if the geometrical hypotheses are valid globally), such that the solutions satisfy approximation results on all $\perp$-derivatives.

In this section we study the rth order triangular system (9.9) below (cf. the example VIII §8.3.1). This system is typical of general triangular systems in jet spaces for which the methods of Convex Integration apply. Let $V \subset \mathbf{R}^n$ be open; $x = (x_1, x_2, \ldots, x_n)$ are coordinates in $\mathbf{R}^n$. Let $p\colon X = V \times \mathbf{R}^q \to V$ be the product $\mathbf{R}^q$-bundle. We identify $\Gamma^r(X) \equiv C^r(V, \mathbf{R}^q)$. $\partial_i = \partial/\partial x_i$, $1 \leq i \leq n$. Fix an integer m, $1 \leq m \leq n$. Associated to $f \in C^r(V, \mathbf{R}^q)$ is the r-jet extension of f: for all $x \in V$,

$$j^r f(x) = (j^c f(x), \partial_m^r f(x), \ldots, \partial_1^r f(x)) \in X^{(r)},$$

where the (complementary) jet $j^c f(x)$ includes all the partial derivatives of f of order $\leq r$ except the pure derivatives $\partial_i^r f(x)$, $1 \leq i \leq m$. In particular, $j^c f(x)$ includes $j^{r-1} f(x)$. We write $X^{(r)} = X^c \times \mathbf{R}^{mq}$, where X^c denotes the space of j^c-jets, and the jth factor $\mathbf{R}^q$ corresponds to the derivative ∂_j^r, $1 \leq j \leq m$. We consider the following system of PDEs on V for an unknown C^r-map $f \in C^r(V, \mathbf{R}^q)$. For all $x \in V$,

$$
\begin{aligned}
&F_1(j^c f(x), \partial_1^r f(x)) = 0 \\
&F_2(j^c f(x), \partial_2^r f(x), \partial_1^r f(x)) = 0 \\
&\quad\vdots \\
&F_m(j^c f(x), \partial_m^r f(x), \ldots, \partial_1^r f(x)) = 0.
\end{aligned}
\tag{9.9}
$$

Here $F_j\colon X^j = X^c \times \mathbf{R}^{jq} \to \mathbf{R}^{s_j}$ is continuous, $1 \leq j \leq m$, and for each j, $1 \leq s_j \leq q - 1$. Each subsystem $F_j = 0$ constitutes an underdetermined system of $s_j < q$ PDEs in q unknown functions $f = (f_1, f_2, \ldots, f_q)$. The system is triangular in the sense that the derivative ∂_j^r occurs first in the jth equation $F_j = 0$, $1 \leq j \leq m$.

Let $S^j = F_j^{-1}(0) \subset X^j$, $1 \leq j \leq m$. In case the closed relation S^j satisfies the geometrical hypotheses of Theorem 9.1, it follows that each subsystem $F_j = 0$ admits a global C^r-solution provided there is a strictly short formal solution to the relation S^j, $1 \leq j \leq m$. Theorem 9.11 below solves all the m subsystems of (9.9) simultaneously for a global C^r solution $f \in C^r(V, \mathbf{R}^q)$, provided there is a strictly short formal solution to the system (9.9) as defined below in §9.2.2.

Note that in case $s_j = q - 1$, $1 \leq j \leq m$ (the maximal case), then (9.9) consists of $(q-1)m$ PDEs in q unknown functions $f = (f_1, \ldots f_q) \in C^r(V, \mathbf{R}^q)$. If $m, q \geq 2$, then $(q-1)m \geq q$ with strict inequality if in addition $q \geq 3$. Consequently in case $m \geq 2$ the system (9.9) includes cases of determined and underdetermined systems of PDEs.

9.2.2. Formal and Strictly Short Solutions. Topologically, the analysis of the system (9.9) is as follows. $X^j = X^{j-1} \times \mathbf{R}^q$, $1 \leq j \leq m$; $X^m = X^{(r)}$, $X^0 = X^c$, where the $\mathbf{R}^q$-factor corresponds to the derivative ∂_j^r, $1 \leq j \leq m$. Let $\pi \colon X^j \to X^{j-1}$ be projection onto the first factor, $1 \leq j \leq m$. The composition of these projections induces a projection $\pi_j \colon X^{(r)} \to X^j$, $0 \leq j \leq m-1$.

Starting with an open set $\mathcal{R}^0 \subset X^0$, let $\mathcal{R}^j = S^j \cap \pi^{-1}(\mathcal{R}^{j-1})$, $1 \leq j \leq m$. Thus $\mathcal{R}^j \subset X^j$ is a locally closed relation in the product $\mathbf{R}^q$-bundle $\pi \colon X^j \to X^{j-1}$, $1 \leq j \leq m$. Let $\mathcal{R} = \mathcal{R}^m \subset X^{(r)}$. In case $\mathcal{R}^0 = X^0$, then $\mathcal{R}^j = S^j \subset X^j$, $1 \leq j \leq m$. The case that $\mathcal{R}^0 \subset X^0$ is a small open set is required for the study of local C^r-solutions to the system (9.9).

There is a commutative diagram of projections, $1 \leq j \leq m-1$:

$$
\begin{array}{ccc}
\mathcal{R} & \xrightarrow{\;i\;} & X^{(r)} \\
{\scriptstyle \pi_j}\downarrow & & \downarrow{\scriptstyle \pi_j} \\
\mathcal{R}^j & \xrightarrow[\;i\;]{} & X^j
\end{array}
$$

A pair $(\varphi, f) \in \Gamma(\mathcal{R}) \times \Gamma^r(X)$ such that $\pi_0 \circ \varphi = j^c f \in \Gamma(\mathcal{R}^0)$ is called a *formal solution* to the system (9.9) with respect to the relation $\mathcal{R}$. Thus φ, f agree on all j^c-derivatives globally on V. Explicitly, $\varphi = (j^c f, \alpha_m, \ldots, \alpha_1) \in \Gamma(\mathcal{R})$, where $\alpha_j \in C^0(V, \mathbf{R}^q)$, $1 \leq j \leq m$, such that for all $x \in V$,

$$
\begin{aligned}
& F_1(j^c f(x), \alpha_1(x)) = 0 \\
& F_2(j^c f(x), \alpha_2(x), \alpha_1(x)) = 0 \\
& \qquad\vdots \\
& F_m(j^c f(x), \alpha_m(x), \ldots, \alpha_1(x)) = 0.
\end{aligned}
$$

A formal solution (φ, f) is holonomic if in addition $\varphi = j^r f \in \Gamma(X^{(r)})$ i.e. $\alpha_j = \partial_j^r f \in C^0(V, \mathbf{R}^q)$, $1 \leq j \leq m$. $\mathcal{F} \subset \Gamma(\mathcal{R}) \times \Gamma^r(X)$ denotes the subspace of formal solutions; $\mathcal{H} \subset \mathcal{F}$ denotes the subspace of holonomic solutions, with respect to the relation $\mathcal{R}$.

Let $\tau_j = \ker dx_j \subset T(V)$, $1 \leq j \leq m$. Thus τ_j is an integrable, codimension 1, tangent hyperplane field on V, $1 \leq j \leq m$. The bundle $p_\perp^r \colon X^{(r)} \to X_j^\perp$ associated to τ_j is a product $X^{(r)} = X_j^\perp \times \mathbf{R}^q$, where the $\mathbf{R}^q$-factor corresponds to the derivative ∂_j^r, $1 \leq j \leq m$. The bundle projection $\pi_j \colon X^{(r)} \to X^j$ induces a bundle map which is an isomorphism on the $\mathbf{R}^q$-fibers (which correspond to the derivative ∂_j^r), $1 \leq j \leq m$,

$$
\begin{array}{ccc}
X^{(r)} & \xrightarrow{\;\pi_j\;} & X^j \\
{\scriptstyle p_\perp^r}\downarrow & & \downarrow{\scriptstyle \pi} \\
X_j^\perp & \xrightarrow[\;\pi_j\;]{} & X^{j-1}
\end{array}
\qquad (9.10)
$$

In particular, the principal subspaces through points of $\mathcal{R}$ with respect to the hyperplane field τ_j are identified with the fibers of the bundle $\pi \colon \mathcal{R}^j \to \mathcal{R}^{j-1}$, $1 \leq j \leq m$.

Recall the interior convex hull extensions of $\mathcal{R}$ associated to the sequence (in reverse order) of hyperplane tangent fields $\tau_{m-j} \subset T(V)$, $0 \leq j \leq m - 1$:

$$\mathrm{IntConv}_1(\mathcal{R}) = \mathrm{IntConv}_{\tau_m}(\mathcal{R}) \subset \mathcal{R} \times X^{(r)}.$$

$$\mathrm{IntConv}_j(\mathcal{R}) = \mathrm{IntConv}_{\tau_{m-j+1}}(\mathrm{IntConv}_{j-1}(\mathcal{R}))$$

$$\subset \mathcal{R} \times \prod_{i=1}^{j} X^{(r)}, \quad 1 \leq j \leq m. \tag{9.11}$$

The order of the hyperplane fields in (9.11) reflects the fact that the corresponding families of surrounding paths in $\mathrm{IntConv}_m(\mathcal{R})$ begin at the top of the tower with respect to the fibration $\pi \colon \mathcal{R} = \mathcal{R}^m \to \mathcal{R}^{m-1}$ (associated to the first equation $F_1 = 0$ in the system (9.9)). Subsequent families of surrounding paths correspond to the fibrations $\pi \colon \mathcal{R}^{m-j+1} \to \mathcal{R}^{m-j}$ associated to the equations $F_j = 0$, $1 \leq j \leq m$.

Let $a = (w, \alpha_m, \ldots, \alpha_1) \in X^{(r)}$, $z = (w, y_m, \ldots, y_1) \in X^{(r)}$ have the same X^0-component: $\pi_0(a) = \pi_0(z) = w \in X^0$. Associated to the pair (a, z) is the $(m+1)$-tuple,

$$\gamma = (a, z_1, z_2, \ldots, z_{m-1}, z) \in \prod_1^{m+1} X^{(r)}, \tag{9.12}$$

such that $z_j = (w, y_m, \ldots, y_{m-j+1}, \alpha_{m-j}, \ldots, \alpha_1) \in X^{(r)}$, $1 \leq j \leq m$; $z_m = z$.

For example, let $(\varphi, f) \in \mathcal{F}$ be a formal solution of the closed relation $\mathcal{R} \subset X^{(r)}$. Setting $a = \varphi(x)$, $z = j^r f(x)$, $w = j^0 f(x)$, for each $x \in V$ the above construction associates to each pair $(\varphi(x), j^r f(x))$, the $(m+1)$-tuple,

$$\gamma = (\varphi, z_1, z_2, \ldots z_m) \in \Gamma(\mathcal{R}) \times \prod_{i=1}^{m} \Gamma(X^{(r)}),$$

where the auxiliary sections $z_j \in \Gamma(X^{(r)})$, $1 \leq j \leq m$, are defined as follows: for all $x \in V$,

$$z_j(x) = (j^c f(x), \partial_m^r f(x), \ldots, \partial_{m-j+1}^r f(x), \alpha_{m-j}(x), \ldots, \alpha_1(x)) \in X^{(r)}.$$

Thus $z_m = j^r f \in \Gamma(X^{(r)})$. The sections φ, z_1 differ only in the $\mathbf{R}^q$-factor corresponding to the derivative ∂_m^r. Successive sections z_j, z_{j+1} differ only in the $\mathbf{R}^q$-factor corresponding to the derivative ∂_{m-j}^r, $1 \leq j \leq m - 1$.

A formal solution $(\varphi, f) \in \mathcal{F}$ to the relation $\mathcal{R}$ is called *strictly short* with respect to the sequence of tangent hyperplane fields τ_{m-j}, $0 \le j \le m-1$, if in addition the following condition obtains: $\gamma = (\varphi, z_1, z_2, \ldots z_m) \in \Gamma(\text{IntConv}_m(\mathcal{R}))$. This strictly short condition is clarified in Lemma 9.12 below.

Let $\epsilon > 0$. Employing the product metric on $\mathcal{R} \times \prod_{i=1}^{j} X^{(r)}$, $1 \le j \le m$, the interior convex hull extensions $\text{IntConv}_j^\epsilon(\mathcal{R})$ are inductively defined with respect to ϵ-balls as in (9.12). An explicit analytic characterization of $\text{IntConv}_m(\mathcal{R})$, useful for analyzing the triangular system (9.9), is as follows.

Lemma 9.12. *Let the pair $(a, z) \in \mathcal{R} \times X^{(r)}$ have the same X^0-component as above. An $(m+1)$-tuple $\gamma = (a, z_1, \ldots, z_m) \in \text{IntConv}_m(\mathcal{R})$, $z_m = z$, if and only if γ is of the form (9.12) with respect to the pair (a, z) and there is a parametrized sequence of paths $\alpha_j^{t_1, \ldots, t_k} \in \mathbf{R}^q$, $1 \le k \le j \le m$, $(t_1, \ldots, t_m) \in [0, 1]^m$, (continuous in each variable, not necessarily jointly continuous) such that the following properties obtain: for all $(t_1, t_2, \ldots, t_m) \in [0, 1]^m$,*

(i) $\varphi^{t_1, \ldots, t_m} = (w, \alpha_m^{t_1, \ldots, t_m}, \ldots, \alpha_j^{t_1, \ldots, t_j}, \ldots, \alpha_1^{t_1}) \in \mathcal{R}$.

(ii) $\alpha_j^{t_1, \ldots, t_{k-1}, 0} = \alpha_j^{t_1, \ldots, t_{k-1}}$, $1 \le k \le j \le m-1$; $\alpha_j^{0, \ldots, 0} = \alpha_j$, $1 \le j \le m$.

(iii) *The path $\alpha_j^{t_1, \ldots, t_{j-1}, s}$, $s \in [0, 1]$, strictly surrounds y_j, $1 \le j \le m$.*

Proof. The element $\gamma = (c, z_m) \in \text{IntConv}_m(\mathcal{R})$ if and only if there is a path $p_s \in \text{IntConv}_{m-1}(\mathcal{R})$, $s \in [0, 1]$,

$$p_s = (a^s, z_1^s, \ldots, z_{m-1}^s) \in \text{IntConv}_{m-1}(\mathcal{R}),$$

such that $p_0 = (a, z_1, \ldots z_{m-1})$ and whose projection $\rho_{m-1}(p_s) = z_{m-1}^s$, $s \in [0, 1]$, strictly surrounds $z_m = z$ in a principal subspace associated to τ_1. Hence the r-jets z_{m-1}^s, z_m differ only in the ∂_1^r-component:

$$z_{m-1}^s = (w, y_m, \ldots, y_2, \alpha_1^s) \quad s \in [0, 1],$$

where the path α_1^s, $\alpha_1^0 = \alpha_1$, $s \in [0, 1]$, strictly surrounds $y_1 \in \mathbf{R}^q$.

The path $p_s = (c^s, z_{m-1}^s) \in \text{IntConv}_{\tau_2}(\text{IntConv}_{m-2}(\mathcal{R}))$ if and only if for all $s \in [0, 1]$, there is a path $p_{s,t} \in \text{IntConv}_{m-2}(\mathcal{R})$, $t \in [0, 1]$,

$$p_{s,t} = (a^{s,t}, z_1^{s,t}, \ldots, z_{m-2}^{s,t}) \in \text{IntConv}_{m-2}(\mathcal{R}),$$

such that $p^{s,0} = c^s \in \text{IntConv}_{m-2}(\mathcal{R})$, and whose projection $\rho_{m-2}(p_{s,t}) = z_{m-2}^{s,t}$, $t \in [0, 1]$, strictly surrounds the r-jet z_{m-1}^s in a principal subspace associated to τ_2. Thus the r-jets $z_{m-2}^{s,t}, z_{m-1}^s$ differ only in the ∂_2^r-component:

$$z_{m-2}^{s,t} = (w, y_m, \ldots, y_3, \alpha_2^{s,t}, \alpha_1^s) \quad t \in [0, 1],$$

where the path $\alpha_2^{s,t}$, $\alpha_2^{s,0} = \alpha_2^s$, $t \in [0, 1]$, strictly surrounds y_2. An obvious induction proves the lemma. □

Remark 9.13. Let $\epsilon > 0$. Then $\gamma \in \mathrm{IntConv}_m^\epsilon(\mathcal{R})$ if in addition the following conditions obtain with respect to the above parametrized family of surrounding paths: $d(p_s, p_0) < \epsilon$ for all $s \in [0,1]$; $d(p_{s,t}, p_s) < \epsilon$, for all $(s,t) \in [0,1]^2$, etc. These conditions obtain if the parametrized family of paths in conclusion (iii) above strictly surround y_j in sufficiently small balls in $\mathbf{R}^q$. Suppose $\gamma = (a, z_1, \ldots, z_m) \in \mathrm{IntConv}_m^\epsilon(\mathcal{R})$; $z_m = z$. With respect to the product metric on $\mathcal{R} \times \prod_1^m X^{(r)}$, employing (iii), it follows that $d(a, z) \leq C_m \epsilon$, where the constant C_m depends only on m and the metric d.

It is instructive to write the conclusions of the lemma in terms of the system (9.9). Let (φ, f) be a formal solution to the system (9.9) which is strictly short: $\gamma = (\varphi, z_1, \ldots, z_m) \in \Gamma(\mathrm{IntConv}_m(\mathcal{R}))$; $z_m = j^r f \in \Gamma(X^{(r)})$. Fix $x \in V$. Employing conclusion (i) of Lemma 9.12, there is an m-parameter family $\varphi^{t_1, \ldots, t_m} \in \mathcal{R}$. Thus for all $(t_1, \ldots, t_j) \in [0,1]^j$, $1 \leq j \leq m$,

$$F_1(j^c f(x), \alpha_1^{t_1}) = 0$$
$$F_2(j^c f(x), \alpha_2^{t_1, t_2}, \alpha_1^{t_1}) = 0$$
$$\vdots$$
$$F_m(j^c f(x), \alpha_m^{t_1, \ldots, t_m}, \ldots, \alpha_1^{t_1}) = 0.$$

Employing (iii), the path α_1^s, $s \in [0,1]$, strictly surrounds $\partial_1^r f(x)$ in a principal subspace associated to τ_1; for each $t_1 \in [0,1]$, the path $\alpha_2^{t_1, s}$, $s \in [0,1]$, strictly surrounds $\partial_2^r f(x)$ in a principal subspace associated to τ_2, and so on for m steps to the statement that the path $\alpha_m^{t_1, \ldots, t_{m-1}, s}$, $s \in [0,1]$, strictly surrounds $\partial_m^r f(x)$ in a principal subspace associated to τ_m.

It is clear that triangular systems (9.9) are the natural class of systems of PDEs for which the strictly short condition $\gamma \in \Gamma(\mathrm{IntConv}_m(\mathcal{R}))$ applies. Furthermore if (φ, f) is a formal solution to the system (9.9) it is a geometrical question whether the strictly short condition obtains i.e. whether the requisite family of strictly surrounding paths in Lemma 9.12 exist for each $x \in V$. Thus with respect to the system VIII Example 8.3.1, a small neighbourhood of the constant functions in $C^0(U, \mathbf{R}^2)$ consists of strictly short maps.

Let $v = (v_0, w_m, \ldots, w_1) \in X^0 \times \mathbf{R}^{mq} = X^{(r)}$. The following notation is employed throughout this chapter. With respect to the projection $\pi_j \colon X^{(r)} \to X^j$, let $\pi_j(v) = v_j \in X^j$, $1 \leq j \leq m$. Thus v_j lies in the fiber over v_{j-1}, $1 \leq j \leq m$:

$$v_j = (v_{j-1}, w_j) \in X^{j-1} \times \mathbf{R}^q = X^j. \tag{9.13}$$

Corollary 9.14. *Suppose the projection map $\pi \colon \mathcal{R}^j \to \mathcal{R}^{j-1}$ is a topological bundle whose fibers are ample: for each $b \in \mathcal{R}^{j-1}$, the convex hull of each path component* of the fiber $\mathcal{R}_b^j$ is $X_b^j \equiv \mathbf{R}^q$, $1 \leq j \leq m$. Then each formal solution (φ, f) with respect to the sequence (τ_{m-i}), $0 \leq i \leq m-1$ is strictly short i.e. the required strictly surrounding families of paths in Lemma 9.12 always exist.

Proof. Let $\varphi = (j^c f, \alpha_1, \ldots, \alpha_m) \in \Gamma(\mathcal{R})$. Since the fibers of $\pi \colon \mathcal{R}^1 \to \mathcal{R}^0$ are ample, it follows that for each $x \in V$ there is a path $(\varphi_0(x), \alpha_1^s(x)) \in \mathcal{R}_b^1$, $b = j^c f(x) (= \varphi_0(x))$, $s \in [0, 1]$, such that $\alpha_1^0(x) = \alpha_1(x)$, and such that $\alpha_1^s(x) \in \mathbf{R}^q$, $s \in [0, 1]$, strictly surrounds $\partial_1^r f(x)$. Employing successively the bundles $\pi \colon \mathcal{R}^j \to \mathcal{R}^{j-1}$, $2 \leq j \leq m$, the path $\alpha_1^s(x)$ lifts to paths $\alpha_j^s(x)$, $s \in [0, 1]$, $\alpha_j^0(x) = \alpha_j(x)$, $1 \leq j \leq m$, such that for all $s \in [0, 1]$:

$$\varphi^s(x) = (j^c f(x), \alpha_m^s(x), \ldots, \alpha_1^s(x)) \in \mathcal{R}.$$

Similarly for each $(s, x) \in [0, 1] \times V$, since the fibers of $\pi \colon \mathcal{R}^2 \to \mathcal{R}^1$ are ample, there is a path $(\varphi_1^s(x), \alpha_2^{s,t}(x)) \in \mathcal{R}_b^2$, $b = \varphi_1^s(x)$, $t \in [0, 1]$, such that $\alpha_2^{s,0}(x) = \alpha_2^s(x)$, and such that $\alpha_2^{s,t}(x)$, $t \in [0, 1]$, strictly surrounds $\partial_2^r f(x)$. One lifts the path $\alpha_2^{s,t}(x)$, $t \in [0, 1]$, up to the successive bundles as before to obtain a 2-parameter family of paths,

$$\varphi^{s,t}(x) = (j^c f(x), \alpha_m^{s,t}(x), \ldots, \alpha_2^{s,t}(x), \alpha_1^s(x)) \in \mathcal{R},$$

such that $\varphi^{s,0}(x) = \varphi^s(x)$. An obvious induction proves the existence of a family of parametrized paths $\alpha_j^{t_1, \ldots, t_j}(x) \in \mathbf{R}^q$, $1 \leq j \leq m$, such that the path $\alpha_j^{t_1, \ldots, t_{j-1}, s}(x)$, $s \in [0, 1]$, strictly surrounds $\partial_j^r f(x)$; $\alpha_j^{t_1, \ldots, t_{j-1}, 0}(x) = \alpha_j^{t_1, \ldots, t_{j-1}}(x)$, and such that for all $x \in V$,

$$\varphi^{t_1, \ldots, t_m}(x) = (j^c f(x), \alpha_m^{t_1, \ldots, t_m}, \ldots, \alpha_1^{t_1}) \in \mathcal{R}.$$

Consequently properties (i), (ii), (iii) of Lemma 9.12 are satisfied, from which it follows that (φ, f) is strictly short. □

Lemma 9.15. *For each $\epsilon > 0$ the projection $\rho_m \colon \mathrm{IntConv}_m^\epsilon(\mathcal{R}) \to X^{(r)}$ is a microfibration.*

Proof. Let $f \colon P \to \mathrm{IntConv}_m^\epsilon(\mathcal{R})$, $F \colon [0, 1] \times P \to X^{(r)}$ be continuous such that $\rho_m \circ f = F_0$, where P is a compact polyhedron. Let $f = (\varphi, z_1, \ldots, z_m)$, where $\varphi \in C^0(P, \mathcal{R})$,

$$\varphi(p) = (y(p), \alpha_m(p), \ldots, \alpha_1(p)) \in \mathcal{R}; \quad y(p) = \pi_0 \circ \varphi(p) \in X^0,$$

and $z_i \in C^0(P, X^{(r)})$, $1 \leq i \leq m$; $z_m = F_0$. Since $f(p) \in \mathrm{IntConv}_m^\epsilon(\mathcal{R})$, employing Lemma 9.12, it follows that for each $p \in P$, $f(p)$ is the $(m+1)$-tuple associated to the pair $(\varphi(p), F(0, p))$.

Let also $b = \pi_0 \circ F : [0,1] \times P \to X^0$ be the projection of F into the base space X^0. In particular, $b(0,p) = y(p)$ for all $p \in P$. Since $\mathcal{R}^0 \subset X^0$ is open, employing the tower of bundle maps $\pi \colon \mathcal{R}^j \to \mathcal{R}^{j-1}$, $1 \le j \le m$, there is a $\delta \in (0,1]$ and a continuous map $G \colon [0,\delta] \times P \to \mathcal{R}$, $G_0 = \varphi$, such that $\pi_0 \circ G = b$ on $[0,\delta] \times P$:

$$G(u,p) = (y(u,p), \alpha_m(u,p), \dots, \alpha_1(u,p)) \in \mathcal{R}; \quad G(0,p) = \varphi(p).$$

Employing (9.12), for all $(u,p) \in [0,\delta] \times P$ let $f(u,p)$ be the $(m+1)$-tuple associated to the pair $(G(u,p), F(u,p))$. Since $G(0,p) = \varphi(p)$ it follows that $f(0,p) = f(p)$ for all $p \in P$. We show that for $\delta > 0$ sufficiently small, $f(u,p) \in \mathrm{IntConv}_m^\epsilon(\mathcal{R})$ for all $(u,p) \in [0,\delta] \times P$.

To this end, let $F(u,p) = (b(u,p), w_m(u,p), \dots, w_1(u,p)) \subset X^{(r)}$ for all $(u,p) \in [0,1] \times P$. Applying Lemma 9.12 to $f(p) \in \mathrm{IntConv}_m^\epsilon(\mathcal{R})$, there is a parametrized family of paths $\alpha_j^{t_1,\dots,t_k}(p)$, such that:

(1) The path $\alpha_j^{t_1,\dots,t_{j-1},s}(p)$, $s \in [0,1]$, strictly surrounds $w_j(0,p)$.
(2) $\alpha_j^{t_1,\dots,t_{k-1},0}(p) = \alpha_j^{t_1,\dots,t_{k-1}}(p)$, $1 \le k \le j \le m$.
(3) For all $(t_1, \dots, t_m) \in [0,1]^m$,

$$\varphi^{t_1,\dots,t_m}(p) = (y(p), \alpha_m^{t_1,\dots,t_m}(p), \dots \alpha_1^{t_1}(p)) \in \mathcal{R}; \quad \varphi^{0,\dots,0}(p) = \varphi(p).$$

In what follows the family of paths $\varphi^{t_1,\dots,t_m}(p)$ is extended over $[0,\delta_0] \times P \times [0,1]^m$ for $\delta_0 > 0$ sufficiently small. This extension requires some care since the component j-parameter family of paths $\alpha_j^{t_1,\dots,t_j}(p)$, $1 \le j \le m$, is not jointly continuous on $P \times [0,1]^m$.

Fix $(p, t_1, \dots, t_{m-1}) \in P \times [0,1]^{m-1}$. Employing the tower of fibrations $\pi_j \colon \mathcal{R}^j \to \mathcal{R}^{j-1}$, $1 \le j \le m$, there is a $\delta_1 \in (0,\delta]$ such that for all $(u,q) \in [0,\delta_1] \times \mathfrak{Op}\, p$, there is a family of paths $\varphi^s(u,q) \in \mathcal{R}$, $s \in [0,1]$,

$$\varphi^s(u,q) = (b(u,q), \alpha_m^s(u,q), \dots, \alpha_1^s(u,q)) \in \mathcal{R},$$

such that the following properties are satisfied for all (u,q):

(i) $\varphi^0(u,q) = G(u,q)$; $\varphi^s(0,q) = \varphi^s(q)$, $s \in [0,1]$.
(ii) The path $\alpha_1^s(u,q) \in \mathbf{R}^q$, $s \in [0,1]$, strictly surrounds the coordinate $w_1(u,q)$ of $F(u,q) \in X^{(r)}$.

Similarly, there is a $\delta_2 \in (0,\delta_1]$ such that for all $(u,q,r_1) \in [0,\delta_2] \times \mathfrak{Op}\,(p,t_1)$ there is a family of paths $\varphi^{r_1,s}(u,q) \in \mathcal{R}$, $s \in [0,1]$,

$$\varphi^{r_1,s}(u,q) = (b(u,q), \alpha_m^{r_1,s}(u,q), \dots, \alpha_2^{r_1,s}(u,q), \alpha_1^{r_1}(u,q)) \in \mathcal{R},$$

such that the following properties are satisfied for all (u,q,r_1):

(iii) $\varphi^{r_1,0}(u,q) = \varphi^{r_1}(u,q); \; \varphi^{r_1,s}(0,q) = \varphi^{r_1,s}(q), \; s \in [0,1].$

(iv) The path $\alpha_2^{r_1,s}(u,q) \in \mathbf{R}^q, \; s \in [0,1],$ strictly surrounds the coordinate $w_2(u,q)$ of $F(u,q) \in X^{(r)}.$

An obvious induction proves the existence of a $\delta_j > 0$ such that for all $(u,q,r_1, \ldots,r_{j-1}) \in [0,\delta_j] \times \mathfrak{Op}\,(p,t_1,\ldots,t_{j-1}), \; 1 \leq j \leq m,$ there is a family of paths $\varphi^{r_1,\ldots,r_{j-1},s}(u,q) \in \mathcal{R},$

$$\varphi^{r_1,\ldots,r_{j-1},s}(u,q) = (b(u,q), \alpha_m^{r_1,\ldots,r_{j-1},s}(u,q), \ldots, \alpha_j^{r_1,\ldots,r_{j-1},s}(u,q),$$
$$\alpha_{j-1}^{r_1,\ldots,r_{j-1}}(u,q), \ldots, \alpha_1^{r_1}(u,q)) \in \mathcal{R}; \quad 1 \leq j \leq m,$$

such that the following properties are satisfied for all $(u,q,r_1,\ldots,r_{j-1}), 1 \leq j \leq m$:

(v) $\varphi^{r_1,\ldots,r_{j-1},0}(u,q) = \varphi^{r_1,\ldots,r_{j-1}}(u,q).$

(vi) For all $s \in [0,1], \; \varphi^{r_1,\ldots,r_{j-1},s}(0,q) = \varphi^{r_1,\ldots,r_{j-1},s}(q).$

(vii) The path $\alpha_j^{r_1,\ldots,r_{j-1},s}(u,q) \in \mathbf{R}^q, \; s \in [0,1],$ strictly surrounds the coordinate $w_j(u,q)$ of $F(u,q) \in X^{(r)}.$

Since $P \times [0,1]^{m-1}$ is compact, it follows that there is a $\delta_0 > 0$ such that properties (v), (vii) are satisfied for all $u \in [0,\delta_0]$ and all $(q,r_1,\ldots,r_{m-1}) \in P \times [0,1]^{m-1}.$ Employing Lemma 9.12, one concludes that the $(m+1)$-tuple $f(u,p) \in \mathrm{IntConv}_m(\mathcal{R})$ for all $(u,p) \in [0,\delta_0].$ Furthermore, since $f \colon P \to \mathrm{IntConv}_m^\epsilon(\mathcal{R}),$ it follows that $f(u,p) \in \mathrm{IntConv}_m^\epsilon(\mathcal{R})$ for all $(u,p) \in [0,\delta_0] \times P,$ provided that $\delta_0 > 0$ is sufficiently small, which completes the proof of the lemma. $\qquad\qquad\square$

With these preliminaries, the main result about triangular systems is the following theorem.

Theorem 9.16. *Let $(\alpha, f) \in \mathcal{F}$ be a formal solution to the triangular system (9.9) that is strictly short. Let $\mathcal{N}$ be an open neighbourhood of the image $j^c f(V) \subset X^c.$ Setting $\mathcal{R}^0 = \mathcal{N} \subset X^0,$ let $\mathcal{R}^j = \pi^{-1}(\mathcal{R}_{j-1}) \cap S^j, \; 1 \leq j \leq m; \; \mathcal{R} = \mathcal{R}^m \subset X^{(r)}.$ Suppose the locally closed relations $\mathcal{R}^j \subset F_j^{-1}(0)$ satisfy the following geometrical properties, $1 \leq j \leq m.$*

(p_1) *The projection $\pi \colon \mathcal{R}^j \to \mathcal{R}^{j-1}$ is a topological bundle whose fibers are locally path connected.*

(p_2) *For each $b \in \mathcal{R}^{j-1}$ the fiber $\mathcal{R}_b^j = \mathcal{R}^j \cap X_b^j$ is nowhere flat in $X_b^j \equiv \mathbf{R}^q.$*

Then there is a homotopy of formal solutions $H \colon [0,1] \to \mathcal{F}, \; H_t = (\alpha_t, f_t), \; t \in [0,1],$ such that the following properties obtain:

(i) *$H_0 = (\alpha, f); \; H_1 \in \mathcal{H}$ i.e. $\alpha_1 = j^r f_1 \in \Gamma(X^{(r)}).$*

(ii) *For all $t \in [0,1]$ the image $j^c f_t(V) \subset \mathcal{N}.$*

The proof employs convex hull extension theory and is similar to the proof of Theorem 9.1, with the additional technical complication that the closed relations $\mathcal{R}^j$ are subsets of X^j instead of $X^{(r)}$, $1 \leq j \leq m$. The proof of the theorem will be given after the development of some preliminaries about interior convex hull extensions.

9.2.3. Throughout this section $(X^{(r)}, d)$ is a complete metric space and the locally closed relations $\mathcal{R}^j \subset F_j^{-1}(0)$ associated to the system (9.9) are assumed to satisfy the geometrical properties (p_1), (p_2) of Theorem 9.16.

Employing the nowhere flatness properties (p_2), the relation $\mathcal{R}^j \subset X^j$ admits the system of metaneighbourhoods $\mathrm{Met}_\epsilon \, \mathcal{R}^j$ and a continuous map $I^j : \mathcal{R}^j \to \mathrm{Met}_\epsilon \, \mathcal{R}^j$, $1 < j \leq m$ (cf. Corollary 9.4). In this section these results are extended to the relation $\mathcal{R} \subset X^{(r)}$ associated to the system (9.9).

Let $\epsilon > 0$. Let $M_\epsilon = \mathrm{IntConv}_m^\epsilon(\mathcal{R})$, the mth iterated interior convex hull extension defined with respect to ϵ-balls. In case $m = 1$ i.e. the system (9.9) reduces to the system (9.1), then $\mathrm{IntConv}_1^\epsilon(\mathcal{R})$ coincides with the relation $M_\epsilon \subset \mathcal{R} \times \mathrm{Met}_\epsilon \, \mathcal{R}$ employed in Lemma 9.8. Thus Lemma 9.15 generalizes the microfibration property of M_ϵ that obtains for the system (9.1), to the corresponding microfibration property of $M_\epsilon = \mathrm{IntConv}_m(\mathcal{R})$ associated to the system (9.9).

Employing Corollary 9.7 to $\mathcal{R}^j \subset X^j$ and to the bundle $\pi \colon X^j \to X^{j-1}$, there is a continuous map $I^j \colon \mathcal{R}^j \to \mathrm{Met}_\epsilon \, \mathcal{R}^j$, such that for all $u \in \mathcal{R}^j$, $I^j(u) \in \mathrm{Env}_\epsilon(\mathcal{R}_b^j; u)$, $b = \pi(u) \in \mathcal{R}^{j-1}$, $1 \leq j \leq m$. With respect to the notation (9.13), let $I \colon \mathcal{R} \to X^{(r)}$ be the continuous map, $I(v) = (v_0, x_m, \ldots, x_1) \in X^{(r)}$ where,

$$I^j(v_j) = (v_{j-1}, x_j) \in \mathrm{Env}_\epsilon(\mathcal{R}_{v_{j-1}}^j, v_j), \quad 1 \leq j \leq m. \qquad (9.14)$$

Thus for each j there is a path p_j^s, $p_j^0 = v_j$, $s \in [0,1]$, that strictly surrounds $I^j(v_j)$ in a ball of radius ϵ at v_j in $\mathcal{R}^j$, $1 \leq j \leq m$.

Let $g = (g_j \in C^0(V, (0,1]))_{1 \leq j \leq m}$ be an auxiliary sequence of continuous maps. Let $\varphi \in \Gamma(\mathcal{R})$. Thus $\varphi_j \in \Gamma(\mathcal{R}^j)$, $1 \leq j \leq m$. Recall that $I_{g_j}^j(\varphi_j(x)) \in \mathrm{Env}_\epsilon(\mathcal{R}^j)$ is the point on the in-path at $\varphi_j(x) \in \mathcal{R}^j$ at the parameter $g_j(x) \in (0,1]$. Employing (9.14), associated to g is the section $I_g \circ \varphi \in \Gamma(X^{(r)})$, such that for all $x \in V$,

$$\begin{aligned} I_g \circ \varphi(x) &= (\varphi_0(x), y_m(x), \ldots, y_1(x)) \in X^{(r)}; \\ I_{g_j}^j(\varphi_j(x)) &= (\varphi_{j-1}(x), y_j(x)) \in \mathrm{Env}_\epsilon(\mathcal{R}^j), \quad 1 \leq j \leq m. \end{aligned} \qquad (9.15)$$

Lemma 9.17. *Let $\epsilon > 0$ and let $\varphi \in \Gamma(\mathcal{R})$. Employing (9.12) let $\gamma = (\varphi, \ldots, I_g \circ \varphi)$ be the $(m+1)$-tuple associated to the pair of sections $(\varphi, I_g \circ \varphi)$. For $g_m > 0$ sufficiently small and if $g_j > 0$ is sufficiently small with respect to $g_{j+1}, \ldots, g_m$, $1 \leq j \leq m-1$, then $\gamma \in \Gamma(\mathrm{IntConv}_m^\epsilon(\mathcal{R}))$.*

Proof. Employing Corollary 9.7, there is a continuous family of paths in ϵ-balls at $\varphi_j(x) \in \mathcal{R}^j$ that strictly surround $I^j_{g_j}(\varphi_j(x)) \in \operatorname{Env}_\epsilon(\mathcal{R}^j)$, $1 \le j \le m$ (continuous in $x \in V$). For each $\delta \in (0, \epsilon]$ this continuous family of strictly surrounding paths can be chosen to be in δ-balls provided $g_j > 0$ is sufficiently small. Furthermore, this family of paths lifts up the tower of fibrations, $\pi \colon \mathcal{R}^k \to \mathcal{R}^{k-1}$, $j+1 \le k \le m$. In what follows the function g_m is fixed. To prove the lemma we indicate the construction of parametrized families of strictly surrounding paths as required by Lemma 9.12.

Starting with the top fibration of the tower, $\pi \colon \mathcal{R} \to \mathcal{R}^{m-1}$ ($\mathcal{R} = \mathcal{R}^m$), for each $x \in V$,
$$I^m_{g_m}(\varphi_m(x)) = (\varphi_{m-1}(x), y_m(x)) \in \operatorname{Met}_\epsilon \mathcal{R}.$$
Since $\operatorname{Met}_\epsilon \mathcal{R}$ is open (in the restriction of the bundle $\pi \colon X^m \to X^{m-1}$ to $\mathcal{R}^{m-1}$), for each $x \in V$ there is a neighbourhood $\mathfrak{Op}\,\varphi_{m-1}(x)$ in $\mathcal{R}^{m-1}$ such that for all $z \in \mathfrak{Op}\,\varphi_{m-1}(x)$, $(z, y_m(x)) \in \operatorname{Met}_\epsilon \mathcal{R}$. In particular, for each $x \in V$ the coordinate $y_m(x)$ is strictly surrounded by paths in the fibers $\mathcal{R}_z$ for all $z \in \mathfrak{Op}\,\varphi_{m-1}(x)$. Thus there is a cover of the image $\varphi_{m-1}(V)$ in $\mathcal{R}^{m-1}$ by open sets $\mathfrak{Op}\,\varphi_{m-1}(x)$ as above, $x \in V$.

For each j, $1 \le j \le m-1$, let $(U_r)_{r \in A}$ be an open cover of the image $\varphi_j(V)$ in $\mathcal{R}^j$. With respect to the fibration $\pi \colon \mathcal{R}^j \to \mathcal{R}^{j-1}$, for each $x \in V$,
$$I^j_{g_j}(\varphi_j(x)) = (\varphi_{j-1}(x), y_j(x)) \in \operatorname{Met}_\epsilon \mathcal{R}^j.$$
Since $\operatorname{Met}_\epsilon \mathcal{R}^j$ is open (in the restriction of the bundle $\pi \colon X^j \to X^{j-1}$ to $\mathcal{R}^{j-1}$), for each $x \in V$ there is a neighbourhood $\mathfrak{Op}\,\varphi_{j-1}(x)$ in $\mathcal{R}^{j-1}$ such that for all $z \in \mathfrak{Op}\,\varphi_{j-1}(x)$, $(z, y_j(x)) \in \operatorname{Met}_\epsilon \mathcal{R}^j$. In particular, for each $x \in V$ the coordinate $y_j(x)$ is strictly surrounded by paths in the fibers $\mathcal{R}^j_z$ for all $z \in \mathfrak{Op}\,\varphi_{j-1}(x)$. Thus there is a cover of the image $\varphi_{j-1}(V)$ in $\mathcal{R}^{j-1}$ of open sets above of the form $\mathfrak{Op}\,\varphi_{j-1}(x)$, $x \in V$. Furthermore, for g_j sufficiently small these strictly surrounding paths lie in an open set of the cover $(U_r)_{r \in A}$. Thus the cover $(U_r)_{r \in A}$ induces a cover $(V_r)_{r \in B}$ of the image $\varphi_{j-1}(V)$ in $\mathcal{R}^{j-1}$ by open sets $\mathfrak{Op}\,\varphi_{j-1}(x)$ as above, $x \in V$.

Employing these open covers of the image $\varphi_j(V)$ in $\mathcal{R}^j$ constructed inductively as above, $1 \le j \le m-1$, if g_j is sufficiently small with respect to succeeding functions $g_{j+1}, \dots, g_m$, $1 \le j \le m-1$, then for each $x \in V$ there are (jointly) continuous parametrized families of paths that strictly surround the coordinates $y_j(x)$ of $I_g \circ \varphi(x)$, as required in Lemma 9.12, whose lifts up the tower of fibrations $\pi \colon \mathcal{R}^j \to \mathcal{R}^{j-1}$ are contained in the corresponding open covers, $1 \le j \le m$. Consequently, $I_g \circ \varphi \in \operatorname{IntConv}_m(\mathcal{R})$. For g sufficiently small, it follows also that $I_g \circ \varphi \in \operatorname{IntConv}^\epsilon_m(\mathcal{R})$. $\qquad\square$

Corollary 9.18 (Parameters). *Let $\epsilon > 0$ and let $\varphi \in C^0(Z, \Gamma(\mathcal{R}))$, where Z is a compact (parameter) space. Employing (9.12), for each $z \in Z$ let $\gamma(z)$*

$= (\varphi(z), \ldots, I_g \circ \varphi(z))$ *be the* $(m+1)$-*tuple associated to the pair of sections* $(\varphi(z), I_g \circ \varphi(z))$. *For* $g_m > 0$ *sufficiently small and if* $g_j > 0$ *is sufficiently small with respect to* $g_{j+1}, \ldots, g_m$, $1 \le j \le m-1$, *then the induced map* $\gamma \in C^0(Z, \Gamma(\text{IntConv}_m^\epsilon(\mathcal{R})))$.

Let $(\varphi, f) \in \Gamma(\mathcal{R}) \times \Gamma^r(X)$ be a strictly short formal solution to the relation $\mathcal{R}$. Thus $\varphi = (j^c f, \alpha_m, \ldots, \alpha_1) \in \Gamma(\mathcal{R})$, where $\alpha_j \in C^0(V, \mathbf{R}^q)$, $1 \le j \le m$. Applying Lemma 9.12, there is a family of parametrized paths $\alpha_j^{t_1, \ldots, t_k}(x)$, $x \in V$, $(t_1, \ldots, t_m) \in [0,1]^m$, $1 \le k \le j \le m$, such that the following properties obtain: for all $x \in V$ and all $(t_1, \ldots, t_m) \in [0,1]^m$,

(i) $\varphi^{t_1, \ldots, t_m}(x) = (j^c f(x), \alpha_m^{t_1, \ldots, t_m}(x), \ldots, \alpha_1^{t_1}(x)) \in \mathcal{R}$.

(ii) $\alpha_j^{t_1, \ldots, t_{k-1}, 0}(x) = \alpha_j^{t_1, \ldots, t_{k-1}}(x)$, $1 \le k \le j$; $\alpha_j^{0, \ldots, 0}(x) = \alpha_j(x)$, $1 \le j \le m$.

(iii) The path $\alpha_j^{t_1, \ldots, t_{j-1}, s}(x) \in \mathbf{R}^q$, $s \in [0,1]$, strictly surrounds $\partial_j^r f(x)$, $1 \le j \le m$.

Lemma 9.19. *The above parametrized families of paths,* $\alpha_j^{t_1, \ldots, t_j}(x)$, *that satisfy properties* (i) *to* (iii) *may be chosen to be continuous functions in* $C^0([0,1]^j \times V, \mathbf{R}^q)$, $1 \le j \le m$.

Furthermore, let $\epsilon > 0$ *and let* γ, *respectively* γ^c, *be the* $(m+1)$-*tuple associated to the strictly short formal solution* (φ, f) *as in* (9.12), *respectively the corresponding* $(m+1)$-*tuple obtained by employing continuous families of strictly surrounding paths. If* $\gamma \in \Gamma(\text{IntConv}_m^\epsilon(\mathcal{R}))$ *then one can choose* $\gamma^c \in \Gamma(\text{IntConv}_m^\epsilon(\mathcal{R}))$.

Proof. For each $x \in V$ the family of paths $\varphi^{t_1, \ldots, t_m}(x)$ in $\mathcal{R}$, satisfies the strictly surrounding properties (iii) with respect to each of its $\mathbf{R}^q$-components, and $\varphi^{0, \ldots, 0}(x) = \varphi(x)$. Employing the successive strong deformation retracts $[0,1]^j \to [0,1]^{j-1}$, $1 \le j \le m$ (along successive coordinate directions) it follows that the tower of strictly surrounding parametrized paths $\varphi^{t_1, \ldots, t_m}(x)$ is canonically homotopic by a contraction to $\varphi(x)$. Hence $\varphi^{t_1, \ldots, t_m}(x)$ defines a (generalized) C-structure based at $\varphi(x)$. Employing the tower of fibrations $\pi \colon \mathcal{R}^j \to \mathcal{R}^{j-1}$, $1 \le j \le m$, for each $x \in V$, this C-structure extends over $\mathfrak{Op}\, x$ in V. Since the space of C-structures is contractible, arguing as in II Proposition 2.2, it follows that there is a global C-structure on V based at the section $\varphi \in \Gamma(\mathcal{R})$. Consequently there is a continuous map (same notation) $\varphi \colon [0,1]^m \to \Gamma(\mathcal{R})$, which satisfies the above properties (i), (ii), (iii). In case $\gamma \in \Gamma(\text{IntConv}_m^\epsilon(\mathcal{R}))$, then the above C-structure extensions over sufficiently small $\mathfrak{Op}\, x$ respect the strictly surrounding properties in ϵ-balls. Hence one may assume also that $\gamma^c \in \Gamma(\text{IntConv}_m^\epsilon(\mathcal{R}))$. $\square$

Proof of Theorem 9.16. Let (φ, f) be strictly short. Let $\mathcal{N}_1$ be a neighbourhood of the image $j^c f(V)$ in X^0 such that $\overline{\mathcal{N}_1} \subset \mathcal{N}$.

Inductive Lemma 9.20. *Let $\epsilon > 0$ and let (φ, f) be a strictly short formal solution to the relation $\mathcal{R}$ whose associated section (cf. (9.12)) $\gamma = (\varphi, z_1, \ldots, z_{m-1}, j^r f)$ $\in \Gamma(M_\epsilon)$.*

Let $\delta \in (0, \epsilon]$. There is a homotopy of formal solutions $H_t = (\varphi_t, f_t)$ of the relation $\mathcal{R}$, $H_0 = (\varphi, f)$, $t \in [0, 1]$, such that the following properties are satisfied:

(1) *The associated section γ_1 to the formal solution (φ_1, f_1) satisfies $\gamma_1 \in \Gamma(M_\delta)$.*

(2) *The image $j^c f_1(V) \subset \mathcal{N}_1$.*

(3) *For all $(t, x) \in [0, 1] \times V$, $d(\varphi(x), \varphi_t(x)) \leq m(n+1)\epsilon$.*

(4) *For all $x \in V$, $d(\varphi_1(x), j^r f_1(x)) \leq C\delta$ (C is an independent constant).*

Proof. Recall the section $I_g \circ \varphi \in \Gamma(X^{(r)})$ associated to a sequence of auxiliary functions $g = (g_j \in C^0(V, (0, 1]))_{1 \leq j \leq m}$. Employing (9.12), let $\sigma = (I_g \circ \varphi, \ldots, j^r f)$, the $(m+1)$-tuple of sections associated to the pair (φ, f). Employing Lemma 9.19, there is a jointly continuous family of parametrized paths $\varphi^{t_1, \ldots, t_m}(x) \in \mathcal{R}$, $(t_1, \ldots, t_m) \in [0, 1]^m$, $x \in V$, whose components satisfy the strictly surrounding properties (iii) of the lemma with respect to the derivatives $\partial_j^r f(x)$. Since these families of strictly surrounding paths are continuous, it follows that for g sufficiently small, the continuous lift $I_g \circ \varphi^{t_1, \ldots, t_m}(x)$ also satisfies the strictly surrounding properties (iii) with respect to the derivatives $\partial_j^r f(x)$ for all $x \in V$, $1 \leq j \leq m$. Since $\gamma \in \Gamma(M_\epsilon)$, for g sufficiently small, one may assume that the above strictly surrounding properties associated to $I_g \circ \varphi^{t_1, \ldots, t_m}$ take place in corresponding ϵ-balls.

Employing (9.12), for each $(t_1, \ldots, t_m) \in [0, 1]^m$ let $\rho(t_1, \ldots, t_m)$ be the $(m+1)$-tuple associated to the pair of sections $(\varphi^{t_1, \ldots, t_m}, I_g \circ \varphi^{t_1, \ldots, t_m})$. Employing Corollary 9.18 to the parameter space $[0, 1]^m$, it follows that for g sufficiently small $\rho \in C^0([0, 1]^m, \Gamma(M_\delta))$.

Let $\lambda = (\varphi, \ldots, I_g \circ \varphi, \ldots, j^r f)$ be the $(2m+1)$-tuple of sections obtained by concatenating $\rho(0, \ldots, 0)$ with σ. It follows from the above strictly surrounding properties of σ, ρ that for g sufficiently small,

$$\lambda \in \Gamma(\mathrm{Conv}_m^\epsilon(\mathrm{IntConv}_m^\delta(\mathcal{R}))) = \Gamma(\mathrm{Conv}_m^\epsilon(M_\delta)).$$

One now applies the proof of VIII Theorem 8.4 to the microfibration $\rho_m \colon M_\delta \to X^{(r)}$ and to the section $f \in \Gamma^r(X)$ which is short with respect to $\lambda \in \Gamma(\mathrm{Conv}_m^\epsilon(M_\delta))$. Thus applying the h-stability theorem VII Theorem 7.2 (cf. also Complement 7.16 for the ϵ-ball restrictions) m-times, there is a homotopy of holonomic sections $\lambda_t \in \Gamma(\mathrm{Conv}_m^\epsilon(M_\delta))$ of the form $\lambda_t = (\varphi_t, \ldots, j^r f_t)$ (a $(2m+1)$-tuple), $\varphi_t \in \Gamma(\mathcal{R})$, $t \in [0, 1]$, such that the following properties obtain:

(i) $\lambda_0 = \lambda$; $\lambda_1 \in \Gamma(M_\delta)$.

(ii) For all $t \in [0,1]$, the image $j^c f_t(V) \subset \mathcal{N}_1$.

(iii) For all $(t,x) \in [0,1] \times V$, $d(\varphi(x), \varphi_t(x)) \leq m(n+1)\epsilon$.

(iv) For all $t \in [0,1]$, $\pi_0 \circ \varphi_t = j^c f_t \in \Gamma(X^0)$ i.e. $\varphi_t, j^r f_t$ agree on all j^c-jets.
Indeed, each of the m-applications above of the h-stability theorem provides
a homotopy of holonomic sections with respect to all $j^\perp$-jets (hence all j^c-
jets) for the bundle $p_\perp^r : X^{(r)} \to X_j^\perp$ associated to τ_j, $1 \leq j \leq m$.

Employing (i), (φ_1, f_1) is a formal solution to $\mathcal{R}$ whose associated section $\gamma_1 \in$
$\Gamma(M_\delta)$. Furthermore, from Remark 9.13, $d(\varphi_1(x), j^r f_1(x)) \leq C\delta$ for all $x \in V$,
where C is an independent constant. Thus conclusion (4) is proved, and the proof
of the lemma is complete. □

Returning to the proof of Theorem 9.16, let $\sum_i a_i < \infty$ be a convergent series of positive numbers. The initial strictly short formal solution (φ, f) in general
does not satisfy the ϵ-hypothesis of the Lemma 9.20. However the ϵ-hypothesis
is required only to obtain the estimates in conclusions (3). One therefore applies
Lemma 9.20 to (φ, f) to obtain a homotopy of formal solutions to $\mathcal{R}$ such that
(φ_1, f_1) satisfies conclusions (1), (2), (4). In particular, for sufficiently small $\delta > 0$
one may assume that the formal solution (φ_1, f_1) satisfies the property that the
associated section $\gamma_1 \in \Gamma(M_\delta)$, where in addition $C\delta < a_1$.

Starting with the formal solution (φ_1, f_1) to the relation $\mathcal{R}$ and employing
successively the Inductive Lemma 9.20, there is a sequence of formal solutions
$(\varphi_m, f_m)_{m \geq 1}$ to the relation $\mathcal{R}$ such that the following properties are satisfied for
all $m \geq 1$ and all $x \in V$ (the metric d on $X^{(r)}$ induces a metric (same notation)
on the subspace $X^0 = X^0 \times \{0\} \subset X^{(r)}$):

(5) The image $j^c f_m(V) \subset \mathcal{N}_1$; $d(j^c f_m(x), j^c f_{m+1}(x)) < a_m$.

(6) There is a homotopy of formal solutions $H_m^t = (\varphi_m^t, f_m^t)$ to $\mathcal{R}$, $t \in [0,1]$,
such that $(\varphi_m^0, f_m^0) = (\varphi_m, f_m)$; $(\varphi_m^1, f_m^1) = (\varphi_{m+1}, f_{m+1})$; for all $t \in [0,1]$,
$j^c f_m^t(V) \subset \mathcal{N}_1$.

(7) For all $t \in [0,1]$, $d(\varphi_m, \varphi_m^t) < a_m$.

(8) $d(\varphi_m(x), j^r f_m(x)) < a_m$.

Employing properties (7), (8), it follows in particular that,

(9) $d(j^r f_m(x), j^r f_{m+1}(x)) < 2a_m + a_{m+1}$.

It follows from properties (5), (7), (9) above that $(j^c f_m)_{m \geq 1}$ is a Cauchy sequence
in $\Gamma(X^0)$; $(j^r f_m)_{m \geq 1}$ is a Cauchy sequence in $\Gamma(X^{(r)})$; $(\varphi_m)_{m \geq 1}$ is a Cauchy
sequence in $\Gamma(\mathcal{R})$. Consequently $\lim_{m \to \infty} f_m = g \in \Gamma(X^{(r)})$ exists.

Let $\mathcal{A}^0 = \overline{\mathcal{N}_1}$; $\mathcal{A}^j = \pi^{-1}(\mathcal{A}^{j-1} \cap S^j)$, $1 \leq j \leq m$; $\mathcal{A} = \mathcal{A}^m$. Thus $\mathcal{A}^j \subset \mathcal{R}^j$,
and $\mathcal{A}^j$ is closed in X^j, $1 \leq j \leq m$. Employing (6), H_m^t is a homotopy of formal

solutions to $\mathcal{R}$. It follows that for all $m \geq 1$,

$$\pi_0 \circ \varphi_m^1(V) = j^c f_m(V) \subset \mathcal{N}_1 \subset \mathcal{A}^0.$$

Consequently for all $m \geq 1$, $\varphi_m \in \Gamma(\mathcal{A})$. Since $\mathcal{A} \subset X^{(r)}$ is closed it follows that $\varphi' = \lim_{m \to \infty} \varphi_m \in \Gamma(\mathcal{A})$.

Furthermore from (8), for all $x \in V$ and all $m \geq 1$, $d(j^r f_m(x), \varphi_m(x)) < a_m$. It follows that,

$$j^r g = \lim_{m \to \infty} j^r f_m = \lim_{m \to \infty} \varphi_m \in \Gamma(\mathcal{A}) \subset \Gamma(\mathcal{R}).$$

Concatenating the sequence of homotopies $(H_m^t)_{m \geq 1}$, it follows that there is a homotopy of formal solutions $G_t = (\varphi_t, f_t)$ to the relation $\mathcal{R}$, $G_0 = (\varphi, f)$, $t \in [0, 1]$, such that $G_1 = (\varphi', g) \in \Gamma(\mathcal{R})$ is holonomic: $\varphi' = j^r g \in \Gamma(\mathcal{R})$, which completes the proof of the theorem. □

Employing Corollary 9.14, one obtains the following corollary in the ample case.

Corollary 9.21. *Suppose in addition* $\pi \colon S^j = F_j^{-1}(0) \to X^{j-1}$ *is a topological bundle whose fibers are locally path connected and are ample. Let* $\mathcal{R}^0 = X^0$; $\mathcal{R}^j = \pi^{-1}(\mathcal{R}^{j-1}) \cap S^j$, $1 \leq j \leq m$; $\mathcal{R} = \mathcal{R}^m \subset X^{(r)}$. *Then the closed relation* $\mathcal{R} \subset X^{(r)}$ *satisfies the* C^s-*dense h-principle,* $0 \leq s \leq r - 1$: *Explicitly, let* $(\varphi, f) \in \Gamma(\mathcal{R}) \times \Gamma^s(X)$ *such that* $p_s^r \circ \varphi = j^s f \in \Gamma(X^{(s)})$. *Let* $\mathcal{N}$ *be a neighbourhood of the image* $j^s f(V)$ *in* $X^{(s)}$.

There is a homotopy $H_t = (\varphi_t, f_t) \in \Gamma(\mathcal{R}) \times \Gamma^r(X)$, $t \in [0, 1]$, *such that the following properties are satisfied.*

(1) $H_0 = (\varphi, f)$; $\varphi_1 = j^r f_1 \in \Gamma(\mathcal{R})$.
(2) *For all* $t \in [0, 1]$, *the image* $j^s f_t(V) \subset \mathcal{N}$.

Proof. Since the bundles $p_s^{r-1} \colon X^{(r-1)} \to X^{(s)}$, $p_{r-1}^c \colon X^c \to X^{(r-1)}$ have contractible fibers, up to a preliminary homotopy of the pair (φ, f) (employing also the bundles $\pi \colon S^j \to X^{j-1}$, $1 \leq j \leq m$) one may assume $f \in \Gamma^r(X)$ and in addition that $\pi_0 \circ \varphi = j^c f \in \Gamma(X^c)$ i.e. $(\varphi, f) \in \mathcal{F}$ is a formal solution to the relation $\mathcal{R}$. Since the fibers of the bundle $\pi \colon \mathcal{R}^j \to \mathcal{R}^{j-1}$ are ample, $1 \leq j \leq m$, employing Corollary 9.14, it follows in addition that (φ, f) is strictly short. Let $\mathcal{M}$ be an open neighbourhood of the image $j^c f(V)$ in X^c such that the projection $p_s^{r-1} \circ p_{r-1}^c(\mathcal{M}) \subset \mathcal{N}$. Applying Theorem 9.16 to the strictly short formal solution (φ, f) and to the neighbourhood $\mathcal{M}$, the corollary is proved. □

9.2.4. Regularity of Solutions. If $\mathcal{R} \subset X^{(r)}$ is a generic C^s-smooth submanifold whose intersections with principal subspaces $R \subset X^{(r)}$ ($\dim R = q$) have positive dimension (this corresponds to underdetermined systems of PDEs with

C^s-smooth coefficients), then genericity implies nowhere flatness of the intersections with R and one expects $\mathcal{R}$ to admit C^s-smooth holonomic sections $V \to \mathcal{R}$ that correspond to C^{r+s}-solutions $V \to X$ of $\mathcal{R}$. However convex integration yields only C^r-solutions that satisfy C^{r-1}-density properties. Furthermore, no regularity assumption on $\mathcal{R}$ improves the regularity of solutions in the above sense. Indeed, Spring [38] provides counterexamples of C^1-triangular systems of order r which are solved by solutions of class C^r but not by solutions of class C^{r+1}, subject to the required C^{r-1}-density properties. The simplest example is as follows ($r = 1$). Consider the following 1st order system ($\partial_t = \partial/\partial t$):

$$(\partial_x f_1)^2 + (\partial_x f_2)^2 = A(x, y, f)$$
$$(\partial_y f_1)^2 + (\partial_y f_2)^2 = A(x, y, f) \tag{9.16}$$

where $f = (f_1, f_2): K \to \mathbf{R}^2$, K a rectangle in $\mathbf{R}^2$, and A is a C^1-positive function. The system (9.16) is triangular. Since A is a positive function, the corresponding tower of bundles (cf. §9.2.2) consists of circle bundle of "radius" $\sqrt{A}$. Furthermore these circle fibers are nowhere flat in their principal subspaces $R \equiv \mathbf{R}^2$. Associated to the system (9.16) is the space of short maps $\mathcal{S} \subset C^1(K, \mathbf{R}^2)$: $h = (h_1, h_2) \in \mathcal{S}$ if,

$$(\partial_x h_1)^2 + (\partial_x h_2)^2 < A(x, y, h)$$
$$(\partial_y h_1)^2 + (\partial_y h_2)^2 < A(x, y, h) \tag{9.17}$$

Let Sol_p, $p \geq 1$, be the subspace of $C^1(K, \mathbf{R}^2)$ consisting of C^p-solutions to the 1st order system (9.16). The following result is proved in Spring [38].

Theorem 9.22. *Suppose the C^1-function A satisfies the inequalities: there is a constant $m > 0$ such that for all $p \in K \times \mathbf{R}^2$,*

$$A(p) \geq m; \quad (\partial_x A(p) \pm \partial_y A(p))^2 \geq m. \tag{9.18}$$

Then there is an open subspace $W \subset \mathcal{S}$ in the C^0-topology such that $W \cap \overline{\mathrm{Sol}_2} = \emptyset$ i.e. short maps in W cannot be C^0-approximated arbitrarily closely by C^2-solutions to the system (9.16).

The idea of the proof is as follows. The inequalities (9.18) are employed to prove that a C^2-solution to (9.16) must be an immersion of the rectangle K into $\mathbf{R}^2$. One chooses W to be a suitably small neighbourhood of a C^1-short map h that restricts to an immersion with induced opposite orientations on two subrectangles K_1, K_2. For example h is an immersion with folds. The theorem follows from the fact that h cannot be approximated arbitrarily closely by immersions of K into $\mathbf{R}^2$.

This theorem contrasts sharply with the case $p = 1$. Indeed, let $h \in \mathcal{S}$ and let $\mathcal{N}$ be a neighbourhood of the graph of h in $X = K \times \mathbf{R}^2$. Since K is contractible the tower of S^1-bundles is trivial. Hence there exists a formal solution (α, h) to the system (9.16). Applying Theorem 9.16(ii) $(X^0 = X; j^c h = j^0 h)$ to the formal solution (α, h), one obtains the result that $\mathcal{N} \cap \mathrm{Sol}_1 \neq \emptyset$. Consequently, $\mathcal{S} = \overline{\mathrm{Sol}_1}$. Analogous counterexamples are presented in Spring [38] in the case of rth order systems, for all $r \geq 1$.

§3. C^1-Isometric Immersions

9.3.1. Let (V, g), (W, h) be Riemannian manifolds, $\dim V = n$, $\dim W = q$, $q \geq n + 1$, where g, h are Riemannian C^0-metrics on V, respectively W. Thus g, h are continuous fields of Euclidean structures g_x, respectively h_y, on $T_x(V)$, $x \in V$, respectively $T_y(W)$, $y \in W$. A C^1-map $f \colon V \to W$ is *isometric* if $f^* h = g$ i.e. f pulls back the metric h on W to the metric g on V: for all $x \in V$ and all $\tau \in T_x(V)$,

$$g_x(\tau, \tau) = h_y(df(\tau), df(\tau)). \tag{9.19}$$

Note that (9.19) implies that the tangent map df is a fiberwise monomorphism; thus f is a C^1-isometric immersion.

Assume first that $(W, h) = (\mathbf{R}^q, st)$, where st is the standard Euclidean metric on $\mathbf{R}^q$. Thus in local coordinates $(u_1, \ldots, u_n)$ on a chart U of V (9.19) is a system of $\frac{1}{2}n(n+1)$ 1st order non-linear PDEs in $f = (f_1, \ldots, f_q) \in C^1(U, \mathbf{R}^q)$ $(\partial_i = \partial/\partial u_i, 1 \leq i \leq n)$,

$$g_{ij} = \partial_i f \cdot \partial_j f$$

$$= \sum_{k=1}^{k=n} \frac{\partial f_k}{\partial u_i} \frac{\partial f_k}{\partial u_j} \quad 1 \leq i \leq j \leq n. \tag{9.20}$$

where $g = \sum g_{ij} du_i \, du_j$ is the metric tensor on U: $g_{ij} = \langle \partial_i, \partial_j \rangle$. Let $z = j^1 f(x) = (x, y = f(x), v_1, \ldots, v_n) \in J^1(U, \mathbf{R}^q)$. For $v_1, \ldots, v_{n-1} \in \mathbf{R}^q$ fixed, the system (9.20) induces a quadratic algebraic system for the vector $v_n \in \mathbf{R}^q$:

$$v_i \cdot v_n = g_{in} \quad 1 \leq i \leq n - 1,$$

$$v_n \cdot v_n = g_{nn}. \tag{9.21}$$

The solution of (9.21) for $v_n \in \mathbf{R}^q$ is a round sphere S^{q-n} in a codimension $(n-1)$ hyperplane π in $\mathbf{R}^q$. In particular S^{q-n} is everywhere flat in $\mathbf{R}^q$ if and only if $n \geq 2$. Thus in case $n = 1$ i.e. $V = S^1$ or $\mathbf{R}^1$, then S^{q-1} is nowhere flat in $\mathbf{R}^q$.

Returning to the general case of (W, h), let $X = V \times W \to V$ be the product bundle. For each $b = (x, y) \in X$ let $Is \equiv Is_g \subset X_b^{(1)} = \mathcal{L}(\mathbf{R}^n, \mathbf{R}^q)$ be the induced isometric immersion relation with respect to the metric g_x on $\mathbf{R}^n$: $z = (x, y, A) \in Is_g$, $A \in \mathcal{L}(\mathbf{R}^n, \mathbf{R}^q)$, if and only if for all $v \in \mathbf{R}^n$,

$$g_x(v, v) = h_y(A(v), A(v)).$$

In particular A is a linear monomorphism. Let $\mathcal{R} \subset X^{(1)}$ be the closed subspace defined by the isometric immersion relation: for all $b \in X$, $\mathcal{R} \cap X_b^{(1)} = Is_g$. The projection $p_0^1 \colon \mathcal{R} \to X$ is a bundle whose fiber is the manifold of orthonormal n-frames in $\mathbf{R}^q$. Let $\mathcal{S} \subset \Gamma(\mathcal{R}) \times \Gamma^1(X)$ be the subspace of *strictly short* formal solutions to $\mathcal{R}$, in the following sense. $(\alpha, f) \in \mathcal{S}$ if and only if $p_0^1 \circ \alpha = f$ (f is the 0-jet component of α), and the induced quadratic form f^*h on V satisfies $g - f^*h$ is positive definite. Thus locally in a chart U, $g - f^*h = \ell_1^2 + \cdots + \ell_n^2$, where ℓ_i is a 1-form on U, $1 \leq i \leq n$. A geometrical analysis of this decomposition is provided in Lemma 9.24. The principal result is as follows.

Theorem 9.23 (Gromov). *The isometric immersion relation $\mathcal{R} \subset X^{(1)}$ satisfies the C^0-dense h-principle with respect to the space $\mathcal{S}$ of strictly short formal solutions. Explicitly, let $(\alpha, f) \in \mathcal{S}$. Let also $\mathcal{N}$ be a neighbourhood of the graph of f in X.*

Then there is a homotopy $H_t = (\alpha_t, f_t) \in \mathcal{R} \times \Gamma^1(X)$, $H_0 = (\alpha, f)$, $t \in [0, 1]$, such that,

(i) *$\alpha_1 = j^1 f \in \Gamma(\mathcal{R})$ i.e. f_1 is a C^1-isometric immersion. In particular the h-principle is satisfied for $\alpha \in \Gamma(\mathcal{R})$.*

(ii) *(C^0-density) For all $t \in [0, 1]$, the graph of f_t lies in $\mathcal{N}$.*

The original version of this theorem is due to Nash [30] in the case $(W, h) = (\mathbf{R}^q, st)$, $q \geq n + 2$, and f is a C^1-immersion. The refinements that include the case $q = n + 1$ are due to Kuiper [26]. This celebrated Nash-Kuiper Theorem may be viewed as a precursor to Convex Integration theory (cf. Spring [39] for historical remarks). The general formulation stated above is adapted from Gromov [18].

The case of curves ($\dim V = 1$) is a consequence of Theorem 9.1. Indeed, in this case codimension 1 hyperplane fields on V are points. Hence $X^\perp = X = V \times W$; $j^\perp f = j^0 f \in \Gamma(X)$, and principal subspaces $R = X_b^{(1)} \equiv \mathbf{R}^q$. Employing the remarks following (9.21), it follows that for each $b \in X$ the intersection $\mathcal{R} \cap R$ is a sphere S^{q-1} which is nowhere flat in $R = \mathbf{R}^q$. Furthermore, in case $n = 1$, $(g_{11}(t) - f'(t) \cdot f'(t))dt^2$ is positive definite if and only if the derivative $f'(t)$ lies in the interior of the corresponding sphere fiber S^{q-1} above i.e. the strictly short condition (9.1) for (α, f) is equivalent to $(\alpha, f) \in \mathcal{S}$ above. Thus in the case of curves ($\dim V = 1$), Theorem 9.23 is a corollary of Theorem 9.1.

The C^0-dense property of Theorem 9.23(ii) is possible only because of the C^1-smoothness of the constructed isometric immersion $f_1 \colon V \to W$. As explained above in §9.2.4, convex integration of closed relations in spaces of 1-jets $X^{(1)}$ yields solutions only of class C^1 in general. Thus f_1 satisfies no curvature conditions i.e. second order derivative conditions. Indeed, the C^0-density conclusion of Theorem 9.23(ii) is false in general for curvature reasons in case the isometric immersion f_1 is required also to be of class C^2. For example if g is a C^1-metric of constant curvature on S^2 then this metric cannot be realized by a C^2-isometric immersion of S^2 in a small neighbourhood of a round sphere of sufficiently small radius in $\mathbf{R}^3$.

Proof of Theorem 9.23. Since $p_0^1 \colon \mathcal{R} \to X$ is a bundle, up to a small perturbation of f, employing VIII Theorem 8.43, one may suppose f is a C^∞-short immersion. Let $A, B \in X_b^{(1)} = \mathcal{L}(\mathbf{R}^n, \mathbf{R}^q)$, $b = (x, y) \in X$. Let $q_A = A^* h_y$, $q_B = B^* h_y$, be the induced quadratic forms on $T_x(V)$. In what follows we assume that pointwise h_y is the standard Euclidean metric on $T_y(W) = \mathbf{R}^q$ with respect to an orthonormal basis. Let also $z_1 = (b, A)$, $z_2 = (b, B) \in X^{(1)}$ and suppose $z_1, z_2 \in R$, a principal subspace in $X_b^{(1)}$ associated to a codimension 1 hyperplane field $\tau = \ker \ell$, where ℓ is a non-zero 1-form on $\mathfrak{Op}\, x$. The following lemma relates the geometry of z_1, z_2 in R to the algebraic condition $q_A - q_B = \ell^2$ (in particular, $q_A - q_B$ is positive semi-definite of rank 1).

Lemma 9.24. *Suppose $B \in \mathcal{L}(\mathbf{R}^n, \mathbf{R}^q)$ is injective. Then $q_A - q_B = \ell^2$ if and only if there is a $(q - n + 1)$-dimensional hyperplane $\pi \subset R$ such that:*

(i) *π is orthogonal to $A(\tau(x)) = B(\tau(x)) \subset R$.*

(ii) *z_1 lies in a round sphere S^{q-n} in π; $z_2 \in B^{q-n+1}$, the ball bounded by S^{q-n} in π. In particular, $A \in \mathcal{L}(\mathbf{R}^n, \mathbf{R}^q)$ is injective.*

Proof. Since $A(\tau(x)) = B(\tau(x)) \subset \mathbf{R}^q$, it follows that for an adapted basis $\mathcal{B}$ of $\mathbf{R}^n$ with respect to $\tau(x)$, we can write (in matrix form) $A = (v_1, \ldots, v_{n-1}, v)$, $B = (v_1, \ldots, v_{n-1}, w)$ where $v_i, v, w \in \mathbf{R}^q$, $1 \le i \le n - 1$. Let $(u_1, \ldots, u_n)$ be coordinates in the adapted basis $\mathcal{B}$. Thus,

$$q_A - q_B = \sum_{i=1}^n (a_{in} - b_{in}) du_i \, du_n,$$

where $a_{in} = v_i \cdot v$, $b_{in} = v_i \cdot w$, $1 \le i \le n - 1$; $a_{nn} = v \cdot v$, $b_{nn} = w \cdot w$. Employing linear algebra, $q_A - q_B = \ell^2$ is positive semi-definite of rank 1 if and only if $a_{in} = b_{in}$, $1 \le i \le n - 1$, and $a_{nn} > b_{nn} \ge 0$. Let $\pi \subset R$ be the codimension $n - 1$ hyperplane defined by the linear equations ($\xi \in \mathbf{R}^q$), $a_{in} = v_i \cdot \xi$, $1 \le i \le n - 1$. Then z_1 lies in the round sphere $S^{q-1} \subset \pi$ of radius $r = \sqrt{a_{nn}}$, and z_2 lies in the ball $B^{q-n+1} \subset \pi$ bounded by S^{q-n}. $\square$

Let $\mathrm{Conv}_1^+ \subset X^{(1)} \times X^{(1)}$ denote the subspace of pairs (z_1, z_2), $z_1 = (b, A)$, $z_2 = (b, B)$ as above where $z_1, z_2 \in R$, a principal subspace associated to

$\tau = \ker \ell$, B is injective, and $q_A - q_B = \ell^2$. More generally let $\mathrm{Conv}_m^+(\mathcal{R}) \subset \mathrm{Pr}_m(\mathcal{R})$ denote the subspace of m-tuples $(z_0, z_1, \ldots, z_m)$ such that $z_0 \in \mathcal{R}$, and $(z_{i-1}, z_i) \in \mathrm{Conv}_1^+$, $1 \le i \le m$. In particular, if $z_i = (x, y, A_i)$, $0 \le i \le m$, then $g_x = A_0^* h_y$, and the linear maps A_i are all injective.

Lemma 9.25. *For all $m \ge 1$, $\mathrm{Conv}_m^+(\mathcal{R}) \subset Conv_m(\mathcal{R})$.*

Proof. The case $m = 1$ is trivial since there is a path $z_0^t \in S^{q-n}$, $t \in [0, 1]$, that surrounds z_1. Furthermore the metric on $T_x(V)$ induced by z_t is $g(x)$ (constant metric)). Let $(z_0, z_1, z_2) \in \mathrm{Conv}_2^+(\mathcal{R})$. Let $z_1, z_2 \in R_1$, the principal subspace associated to a hyperplane field $\tau_1 = \ker \ell_1$. There is a path z_1^t in the sphere $S_1^{q-n} \subset R_1$ that surrounds z_2. Let $z_0, z_1 \in R$, and let R^t be the translate of R to z_1^t in $X_b^{(1)}$ by the translation $z \mapsto z + (z_1^t - z_1)$, $t \in [0, 1]$. The path z_1^t lifts to a path z_0^t in the corresponding S^{q-n}-bundle such that for all $t \in [0, 1]$, $(z_0^t, z_1^t) \in \mathrm{Conv}_1^+$, and $z_0^t \in \mathcal{R} \cap R^t$. In particular, the path z_0^t induces the constant metric g_x on $T_x(V)$. Consequently for all $t \in [0, 1]$, $(z_0^t, z_1^t) \in \mathrm{Conv}_1^+$, which proves $(z_0, z_1, z_2) \in \mathrm{Conv}_2(\mathcal{R})$. An obvious induction proves the lemma. $\qquad\square$

Remark 9.26. Let $\mathcal{I} \subset X^{(1)}$ be the immersion relation. Note that all the surrounding paths constructed above in $\mathrm{Conv}_m^+(\mathcal{R})$ are in $\mathcal{I}$. Furthermore if $(z_0, \ldots, z_m) \in \mathrm{Conv}_m^+(\mathcal{R})$ then the associated principal path joining z_0, z_m is homotopic in $\mathcal{R} \cup \mathcal{R}_0$ rel the end points to the straight line that joins z_0, z_1.

Following Gromov, a useful description of Conv_1^+ starts with $z = (x, y, A) \in \mathcal{I}$ such that $g_x - A^* h_y = \ell^2$, for a non-zero 1-form on $T_x(V)$. Setting $\tau = \ker \ell$, let R be the associated principal subspace through z. Then one chooses $z_0 \in \mathcal{R} \cap R = S^{q-n}$. Thus z_0 induces the metric g_x and $(z_0, z) \in \mathrm{Conv}_1^+$.

This description generalizes also to construct elements of $\mathrm{Conv}_m^+(\mathcal{R})$. Suppose $g_x - A^* h_y = \ell_1^2 + \cdots + \ell_m^2$, where each ℓ_i is a non-zero 1-form on $T_x(V)$. There is an element,

$$(z_0, z_1, \ldots, z_m = z) \in \mathrm{Conv}_m^+(\mathcal{R}),$$

$z_i = (x, y, A_i) \in \mathcal{I}$, $z_0 \in \mathcal{R}$, such that $A_i^* h_y = g_i$ ($A_0^* h_y = g_x$) satisfies, $g_{i-1} - g_i = \ell_i^2$, $1 \le i \le m$. In particular for each i, $\tau_i = \ker \ell_i$; R_i is the principal subspaces through z_i with respect to τ_i, and $z_{i-1} \in R_i$ lies in the round sphere S_i^{q-n} such that $A_i^* h_y = g_{i-1}$. Furthermore, let $\alpha \in \Gamma(\mathcal{R})$. Since the fibers of $\mathcal{R} \to X$ are Steifel manifolds, there is a path in $\mathcal{R}_x$ that joins $\alpha(x), z_0$. Consequently, $\psi = (\alpha(x), z_1, \ldots z_m) \in \mathrm{Conv}_m^+(\mathcal{R}) \subset \mathrm{Conv}_m(\mathcal{R})$.

Returning to the proof of Theorem 9.23, let $\mathcal{R}_0 \subset X^{(1)}$ be the subspace of germs of C^1-maps which are strictly short with respect to the metric g on $T(V)$ (cf. VIII §8.4.3). Then $\mathcal{R}_0$ is open and for all $b \in X$, $\mathcal{R}_0 \cap X_b^{(1)}$ is convex. With respect to a metric on $X^{(1)}$, let $\mathcal{M}_\epsilon$ be an open ϵ-neighbourhood of $\mathcal{R}$ in $\mathcal{I}$. There is a fiberwise projection $p\colon \mathcal{M}_\epsilon \to \mathcal{R}$. In local coordinates let $z = (x, y, A) \in \mathcal{R}$. Thus $g_x = A^* h(y)$ on $T_x(V)$. For $t > 0$, let $z_t = (x, y, tA) \in X^{(1)}$. The associated quadratic form $g_x^t = (tA)^* h_y$ satisfies $g_x - g_x^t = (1 - t^2) g_x$ is positive definite on $T_x(V)$ for all $t \in (0, 1)$. Hence there is a $\delta > 0$ such that $z_t \in \mathcal{R}_0 \cap \mathcal{M}_\epsilon$ for all $t \in (0, \delta)$. One concludes that for each open $U \subset V$ there is a smooth section $\varphi \in \Gamma_U^\infty(\mathcal{R}_0 \cap \mathcal{M}_\epsilon)$, $p_0^1 \circ \varphi = j^0 f_U \in \Gamma_U(X)$, such that $g - g_\varphi$, respectively $g_\varphi - f^* h$, are positive definite quadratic forms on $T(U)$, where g_φ is the metric on $T(U)$ induced by φ. Up to a small homotopy of α one may assume $p \circ \varphi = \alpha_U$.

Let $U \subset V$ be open and assume $g_\varphi - f^* h = \ell_1^2 + \cdots + \ell_m^2$, where each $\tau_i = \ker \ell_i$ is a smooth integrable codimension 1 hyperplane field on $T(U)$. Lifting α_U to φ and employing the tower of sphere bundles associated to $\mathrm{Conv}_m^+(\mathcal{R}) \subset \mathrm{Pr}_m(\mathcal{R})$, the construction as above of $\psi \in \mathrm{Conv}_m^+(\mathcal{R})$ yields a section $(z_m = j^1 f_U)$, $\psi_1 = (\varphi, z_1, \ldots, z_{m-1}, j^1 f_U) \in \Gamma_U(\mathrm{Conv}_m(\mathcal{M}_\epsilon))$. Applying VIII Theorem 8.4, there is a homotopy of strictly short formal solutions (φ_t, f_t) to $\mathcal{M}_\epsilon \cap \mathcal{R}_0$ over U, $(\varphi_0, f_0) = (\varphi, f_U)$, $t \in [0, 1]$, such that $\varphi_1 = j^1 f_1$ and which satisfies the required C^0-density properties. Projecting φ_t into $\mathcal{R}$, there is a homotopy $(\alpha_t, f_t) \in \mathcal{S}$ over U such that $j^1 f_1 \in \Gamma_U(\mathcal{M}_\epsilon \cap \mathcal{R}_0)$.

Proceeding inductively on a covering of V by charts one constructs a homotopy $(\alpha_t, f_t) \in \mathcal{S}$, $(\alpha_0, f_0) = (\alpha, f)$, $t \in [0, 1]$, such that $j^1 f_1 \in \Gamma(\mathcal{M}_\epsilon \cap \mathcal{R}_0)$. This inductive construction over a covering of V by charts $(W_j)_{j \geq 1}$, $(U_j)_{j \geq 1}$, $\overline{W}_j \subset U_j$ for all j, constructs for each index j a small lift of α to $\varphi_j \in \Gamma_{U_j}(\mathcal{M}_\epsilon)$ in order to maintain the positive definiteness of the induced form $g_{\varphi_j} - f_{j-1}^* h$ on $T(U_j)$, where $f_{j-1} \in \Gamma^1(X)$ is constructed in the previous inductive step. We may assume $g_{\varphi_j} - f_{j-1}^* h = \ell_1^2 + \cdots + \ell_n^2$ as above, $n = \dim V$. The construction of f_j proceeds in n steps (cf. the proof of VIII Lemma 8.13) where the n successive convex integrations of the formal solution (φ_j, f_{j-1}) take place in a nested sequence of subcharts of U_j, and are rel $\mathfrak{Op}\, \partial U_j$ (matching up at the boundary). After these n steps one also applies a relative version of VIII Theorem 8.43 (cf. Remark 9.26 for the construction of paths in $\mathcal{P}_\infty(\mathcal{I} \cap \mathcal{R}_0 | \mathcal{R}_0)$) to obtain a strictly short C^1-immersion. In particular, after n steps f_j is a strictly short C^1 immersion; the image $j^1 f_j(\bigcup_{i=1}^j (W_i)) \subset \mathcal{M}_\epsilon \cap \mathcal{R}_0$; f_j matches up with f_{j-1} on $V \setminus U_j$. Choosing a nested sequence of neighbourhoods $(\mathcal{M}_{\epsilon_i})_{i \geq 1}$ whose intersection is $\mathcal{R}$, and following the proof procedure of Theorem 9.1, one constructs a concatenated homotopy (α_t, f_t) that satisfies all the conclusions of the theorem. Details are left to the reader. $\square$

In case $f\colon V \to W$ is a strictly short C^1-immersion or embedding then one can construct a section $\alpha \in \Gamma(\mathcal{R})$ such that $(\alpha, f) \in \mathcal{S}$. Indeed, the fiber of the bundle $p_0^1\colon \mathcal{R} \to X$ is the Stiefel manifold of all orthonormal n-frames in $\mathbf{R}^q$, which itself is a deformation retract of the manifold of all n-frames in $\mathbf{R}^q$. One chooses α to be the deformation retract of the associated section induced by f on bundle of n-frames in $\mathbf{R}^q$. The following result is left as an exercise.

Theorem 9.27. *Each strictly short C^1-embedding $f\colon V \to W$, $\dim W > \dim V$, admits a C^1-isotopy of embeddings to an isometric embedding which is a small approximation of f in the fine C^0-topology.*

Complement 9.28 (Relative Case). *Suppose in addition there is a closed set $K \subset V$ such the $\alpha_K = j^1 f_K \in \Gamma_K(\mathcal{R})$ and such that over $V \setminus K$, $(\alpha, f) \in \mathcal{S}$. Then there is a homotopy rel K, $(\alpha_t, f_t) \in \Gamma(\mathcal{R}) \times \Gamma^1(X)$, $(\alpha_0, f_0) = (\alpha, f)$, $t \in [0, 1]$, such that $\alpha_1 = j^1 f_1 \in \Gamma(\mathcal{R})$ i.e. f_1 is a C^1-isometric immersion.*

The h-principle for C^1-isometric immersions, including the relative case above, extends to parametric h-principles. In particular, one has the following corollary, Gromov [18] p. 16, concerning the famous Smale inversion of S^2 in $\mathbf{R}^3$.

Corollary 9.29 (Gromov). *The inversion of $S^2 \subset \mathbf{R}^3$ can be realized by a regular homotopy of C^1-isometric immersions.*

CHAPTER 10

RELAXATION THEORY

§1. Filippov's Relaxation Theorem

10.1.1. In this chapter we briefly examine the relationship between Convex Integration theory and the Relaxation Theorem, due to A.F. Filippov [13], in Optimal Control theory, and we prove a general C^r-Relaxation Theorem 10.2. In broadest terms the underlying analytic approximation problem for both the Relaxation Theorem and for Convex Integration theory is the following. Let $A \subset \mathbf{R}^q$ and let $f \colon [0,1] \to \mathbf{R}^q$ be a continuous vector valued function which is differentiable a.e. (almost everywhere), such that the derivative $f'(t) \in \operatorname{Conv} A$ a.e., where $\operatorname{Conv} A$ denotes the convex hull of A in $\mathbf{R}^q$. Let also $\epsilon > 0$. The problem is to construct a continuous map $g \colon [0,1] \to \mathbf{R}^q$, differentiable a.e. such that: (i) the derivative $g'(t) \subset A$ a.e.; (ii) for all $t \in [0,1]$, $\|f - g\| < \epsilon$. Simply put, the problem is to C^0-approximate the continuous map $f \colon [0,1] \to \mathbf{R}^q$, whose derivatives lie in the convex hull of A a.e., by a continuous map g whose derivatives lie in the set A a.e.

In terms of Convex Integration Theory, the topological interest lies in the smooth version of this problem: $A \subset \mathbf{R}^q$ is open and connected, and the functions f, g are required to be of class C^1. This smooth version of the problem is solved in chapter III in various degrees of generality, beginning with III Theorem 1.4, the One Dimensional Theorem, and culminating in III Theorem 2.1, the $C^\perp$-Approximation Theorem, which applies to smooth maps $f \colon [0,1]^n \to \mathbf{R}^q$ and includes a relative theorem and an approximation result in the C^{r-1}-topology, $r \geq 1$. These theorems originated in Gromov's seminal paper Gromov [17] published in 1973. Historically, the Nash C^1-isometric immersion theorem, Nash [30] (1954), may be considered an early topological precursor to the technique of convex integration (cf. Spring [39] for a brief historical account).

This C^1-version of the problem is not discussed in the literature of Optimal Control theory since applications of the Relaxation theorem involve functions with discontinuous derivatives. Thus in the Relaxation Theorem, $A \subset \mathbf{R}^q$ is compact, not necessarily connected, and for technical reasons the functions f, g are assumed to be absolutely continuous. In this form the solution to the problem was obtained by A. F. Filippov [13] (1967). Work on this problem in optimal

D. Spring, *Convex Integration Theory: Solutions to the h-principle in geometry and topology*, Modern Birkhäuser classics, DOI 10.1007/978-3-0348-0060-0_10, © Springer Basel AG 2010

control theory goes back even further to earlier papers by Ważeski [41] and others. More precisely, Filippov considers the following problem of differential inclusions:

$$x'(t) \in F(t, x(t)); \quad t \in I, \tag{10.1}$$

where $I = [c, d]$ is a compact interval; F is a set-valued map whose domain a ball B in $\mathbf{R}^{n+1}$, with values that are compact sets in $\mathbf{R}^n$. The curve $x(t)$ in $\mathbf{R}^n$ is assumed to be absolutely continuous. In particular, $x(t) = x(c) + \int_0^t x'(s)\, ds$ and the above differential inclusion is understood throughout to be satisfied a.e. The theory of differential inclusions has a longer history, including earlier papers by Ważewski [41], A. Marchaud [27], S.K. Zaremba [42], and more recently the books, Aubin and Cellina [2], Aubin and Frankowska [3], and the survey article, Blagodat·Skikh and Filippov [4].

Following Filippov, the set valued map F is assumed to be Lipschitzean with constant k: $\alpha(F(x), F(y)) \leq k\|x - y\|$, for all $x, y \in B$, where α is the Hausdorff distance on subsets of $\mathbf{R}^n$. For the purposes of this chapter we state a somewhat simplified version of Filippov's Relaxation Theorem, proved in Aubin and Cellina [2].

Relaxation Theorem 10.1. *Let F be a set-valued map, domain a closed ball $B = \{x \in \mathbf{R}^n \mid \|x - a\| \leq r\}$, with values the compact subsets of $\mathbf{R}^n$, such that F is Lipschitzean with constant k. Let $I = [-T, T]$ and let $x\colon I \to B$ be a solution (i.e. absolutely continuous) to the differential inclusion :*

$$x'(t) \in \operatorname{Conv} F(x(t)); \quad x(0) = a. \tag{10.2}$$

such that for all $t \in I$, $x(t)$ lies in the interior of B. Let $\epsilon > 0$. Then there is an absolutely continuous function $y\colon I \to B$, such that,

$$y'(t) \in F(y(t)); \quad y(0) = a.$$

and such that for all $t \in I$, $\|x(t) - y(t)\| \leq \epsilon$.

Rather than proving the theorem, we discuss the proof in terms of Convex Integration Theory. The proof of the Relaxation Theorem depends critically on the construction of a *quasi-solution* to the differential relation (10.2). An absolutely continuous function $z\colon I \to B$ is a quasi-solution to the differential inclusion (10.2) if for all $t \in I$, $\|z(t) - x(t)\| \leq \delta$, and such that:

$$d(z'(t), F(z(t))) \leq M\epsilon, \tag{10.3}$$

for a suitable constant M and sufficiently small $\delta > 0$. For the purposes of the proof of the Relaxation Theorem 10.1, $M = k/(2e^{kT} - 1)$. In topological terms, a quasi-solution solves the *open* relation $z'(t) \in \mathfrak{Op}\, F(z(t))$ a.e., where $\mathfrak{Op}\, F(z(t))$ denotes a small open neighbourhood of $F(z(t))$ in $\mathbf{R}^n$. Thus a critical part of the proof of the Relaxation Theorem involves the passage from a differential inclusion into compact sets, $x'(t) \in F(x(t))$, to a differential inclusion into an open set, $z'(t) \in \mathfrak{Op}\, F(z(t))$. This passage to an open set clarifies an important methodological point. From the point of view of Convex Integration Theory, the expected hypothesis in Theorem 10.1 is that $x'(t) \in \mathrm{IntConv}\, F(x(t))$ i.e. $x'(t)$ lies in the interior of the convex hull of the compact set $F(x(t))$. Since the convex hull of an open set in $\mathbf{R}^q$ is open, it follows that the interior convex hull hypothesis is not required for the construction of a quasi-solution to the open differential inclusion $z'(t) \in \mathfrak{Op}\, F(z(t))$.

It is with respect to this open differential inclusion that the proof of the Relaxation Theorem has elements in common with the proof of the One Dimensional Theorem III 1.4. In the terminology of control theory, the One-dimensional Theorem constructs a C^1-quasi-solution to a differential inclusion into an open, connected set. From this point of view the One-Dimensional Theorem is a type of C^1-Relaxation Theorem with parameters. Furthermore, a close examination of the measure-theoretic construction of quasi-solutions in the Relaxation Theorem reveals a close resemblance to the construction of C^1-solutions to open, connected relations in the One Dimensional Theorem. Indeed, the critical use of the rapidly oscillating functions $\theta_\epsilon \colon [0,1] \to [0,1]$ in the proof of the One Dimensional Theorem has its counterpart in the construction of quasi-solutions in the proof of the Relaxation Theorem.

The rest of the proof of the Relaxation Theorem involves the construction of a suitably convergent sequence of quasi-solutions $(z_j(t))_{j\geq 1}$ to a converging sequence of open relations, $z_j'(t) \in \mathfrak{Op}_j F(z_j(t))$, $j \geq 1$. It is here that the Lipschitz properties of the set-valued map F enter in the proof in order to ensure the analytic convergence of the sequence of quasi-solutions. Details are found in Aubin and Cellina [2], based on Filippov [13], who also proves the existence of short time C^1-solutions to differential inclusions in case the compact set-valued map F satisfies some additional geometrical properties.

From point of view of Convex Integration Theory, Filippov's convergence arguments, based on the Lipschitz conditions on the set-valued map F, constitute a completely new convergence technique that is independent of the convergence arguments based on the nowhere flat conditions that are employed in chapter IX to solve closed differential relations in spaces of r-jets.

Another approach in Optimal Control theory to solving differential inclusions (10.1) is to consider set-valued maps whose values at each point are Lie algebras generated by certain families of vector fields on the underlying manifold. The theory of reachable sets is the basic invariant object for these Lie determined systems. Cf. Jurdjevic [25] for this approach.

§2. C^r-Relaxation Theorem

10.2.1. In this section Theorem 10.1 is generalized to rth order partial differential inclusions, $r \geq 1$. The proof of the main result, Theorem 10.2, is based on the PDE theory of IX Theorem 9.1. Let V be a smooth n-dimensional submanifold of $\mathbf{R}^n$, $n \geq 1$, possibly with corners in case $\partial V \neq \emptyset$. For example, V is open in $\mathbf{R}^n$ or $V = \prod_{i=1}^{n}[a_i, b_i]$, an n-rectangle in $\mathbf{R}^n$. Let also $W \subset \mathbf{R}^q$ be a smooth q-dimensional submanifold. Recall the space $J^r(V, W)$ of r-jets of germs of C^r-maps from V to W. In order to avoid complications in case the manifolds V, W have boundary, we recall here the standing assumption in chapter I that (by adding small collars across the boundary) maps in $C^r(V, W)$ will be assumed to extend across ∂V and to lie in the interior of W. Let $(u_1, \ldots, u_{n-1}, t)$ be coordinates on $\mathbf{R}^n$. Let $\tau = \ker dt$ be the integrable codimension 1 tangent field on $\mathbf{R}^n$. Recall also the product decomposition associated to the restriction of τ to V:

$$J^r(V, W) = J^\perp(V, W) \times \mathbf{R}^q,$$

where the $\mathbf{R}^q$-factor in this product decomposition corresponds to the pure rth order derivative ∂_t^r. In what follows the map $p_\perp^r \colon J^r(V, W) \to J^\perp(V, W)$ is projection onto the first factor and $s \colon J^r(V, W) \to V$ is the source map. Let $f \in C^r(V, W)$. Employing the above product decomposition, for all $x \in V$,

$$j^r f(x) = (j^\perp f(x), \partial_t^r f(x)) \in J^r(V, W).$$

where $j^\perp f(x)$ includes all the derivatives of order $\leq r$ except $\partial_t^r f(x)$. Let $F \colon J^\perp(V, W) \to \mathcal{P}(\mathbf{R}^q)$ be a set-valued map whose values are *path connected* closed subsets of $\mathbf{R}^q$. The path connected hypothesis is imposed only to simplify notation so that convex hulls as employed in Theorem 10.1 coincides with convex hulls of path components as employed in Convex Integration Theory. Associated to F is the partial differential inclusion: for all $x \in V$,

$$\partial_t^r f(x) \in F(j^\perp f(x)) \tag{10.4}$$

where $f \in C^r(V, W)$ is an unknown C^r-map. Note that in case $n = 1$, and $V = [a, b]$, a closed interval, then the partial differential inclusion (10.4) assumes the form: for all $t \in [a, b]$ $(x^{(k)}(t) = \dfrac{d^k x}{dt^k}(t))$,

$$x^{(r)}(t) \in F(t, x(t), x'(t), x^{(2)}(t), \ldots, x^{(r-1)}(t)), \tag{10.5}$$

where $x \in C^r([a, b], \mathbf{R}^q)$ is a parametrized C^r-curve in $\mathbf{R}^q$. In case $r = 1$, the differential inclusion (10.5) reduces to the differential inclusion $x'(t) \in F(t, x(t))$, which is related to Filippov's Relaxation Theorem in the case of classical C^1-solutions and the set values of F are path connected compact sets.

Returning to the partial differential inclusion (10.4), let $\mathcal{R} \subset J^r(V,W)$ be the graph of the set-valued map F:

$$\mathcal{R} = \bigcup_{z \in J^\perp(V,W)} \{z\} \times F(z) \subset J^r(V,W). \tag{10.6}$$

Thus $\mathcal{R}$ is a differential relation on the space of r-jets $J^r(V,W)$ whose fibers $\mathcal{R}_z \equiv F(z)$ are path connected and closed for all $z \in J^\perp(V,W)$. Recall that $\Gamma(\mathcal{R})$ is the space of sections of the source map $s\colon \mathcal{R} \to V$. A map $f \in C^r(V,W)$ solves the relation $\mathcal{R}$ if $j^r(f) \in \Gamma(\mathcal{R})$. Equivalently, for all $x \in V$, $\partial_t^r f(x) \in F(j^\perp f(x))$ i.e. f solves the partial differential inclusion (10.4). If F is continuous then $\mathcal{R}$ is a closed differential relation.

Formal Solutions. Employing the notation of IX §9.2.1, a formal solution to the relation $\mathcal{R}$ is a pair (α, f) where $\alpha \in \Gamma(\mathcal{R})$, $f \in C^r(V,W)$, such that for all $x \in V$, $j^\perp f(x) = p_\perp^r \circ \alpha \in J^\perp(V,W)$ i.e. the continuous section α agrees with f on all $\perp$-derivatives. In terms of the partial differential inclusion (10.4), (α, f) is a formal solution if $\alpha = (j^\perp f, \beta)$ where $\beta \in C^0(V,W)$ is such that for all $x \in V$,

$$\beta(x) \in F(j^\perp f(x)).$$

Evidently f solves the differential inclusion (10.4) if and only if $\beta = \partial_t^r f$ i.e. $\alpha = j^r f \in \Gamma(\mathcal{R})$. Thus the existence of a formal solution to the relation $\mathcal{R}$ is a necessary topological condition for the existence of a C^r-solution to the partial differential inclusion (10.4).

A map $f \in C^r(V,W)$ is *strictly short* for F if for all $x \in V$,

$$\partial_t^r f(x) \in \mathrm{IntConv}\, F(j^\perp f(x)). \tag{10.7}$$

Thus f is strictly short if the derivatives $\partial_t^r f(x)$ lie in the interior of the convex hull of the path connected closed set $F(j^\perp f(x))$ for all $x \in V$. Since the set values of F are path connected, the above strictly short condition is equivalent to the strictly short condition in IX §9.2.1 applied to formal solutions (α, f) i.e. the section $\alpha \in \Gamma(\mathcal{R})$ plays no role in determining strictly short maps in the path connected case. Applying IX Theorem 9.1, one has the following solution to the partial differential inclusion (10.4).

C^r-Relaxation Theorem 10.2. *Let $\mathcal{R} \subset J^r(V,W)$ be the relation (10.6) associated to the set-valued map F. Suppose $\mathcal{R}$ satisfies the following geometrical properties with respect to the product $\mathbf{R}^q$-bundle $p_\perp^r\colon J^r(V,W) \to J^\perp(V,W)$:*

(i) *The projection map $p_\perp^r\colon \mathcal{R} \to J^\perp(V,W)$ is a topological fiber bundle whose fiber is locally path connected (for example $\mathcal{R}$ is a smooth submanifold of*

$J^r(V, W)$ *that is transverse to the* $\mathbf{R}^q$-*fibers of the bundle* $p^r_\perp : J^r(V, W) \rightarrow J^\perp(V, W)$).

(ii) *For each* $z \in J^\perp(V, W)$, *the fiber* $\mathcal{R}_z$ *is nowhere flat in* $\{z\} \times \mathbf{R}^q = \mathbf{R}^q$.

Let (α, f) *be a formal solution to the relation* $\mathcal{R}$ *such that* f *is strictly short. Let* $\mathcal{N}$ *be a neighbourhood of the image* $j^\perp f(V)$ *in* $J^\perp(V, W)$.

Then there is a homotopy of formal solutions (α_t, f_t), $t \in [0, 1]$, *to the relation* $\mathcal{R}$, $(\alpha_0, f_0) = (\alpha, f)$, *such that the following properties obtain:*

(iii) $\alpha_1 = j^r f_1 \in \Gamma(\mathcal{R})$. *Hence* f_1 *solves the partial differential inclusion (10.4).*

(iv) ($C^\perp$-*dense principle*) : *For all* $t \in [0, 1]$, *the image* $j^\perp f_t(V) \subset \mathcal{N}$.

Thus the map $f_1 \in C^r(V, W)$ solves the partial differential relation (10.4) and is a $C^\perp$-approximation to the given short map $f \in C^r(V, W)$. The fiber bundle hypothesis on $\mathcal{R}$ implies that the set-values $F(z) \in \mathbf{R}^q$, $z \in J^\perp(V, W)$, are homeomorphic to the fiber.

Let the manifold $V \subset \mathbf{R}^n$ be an n-rectangle. Since V is contractible it follows that bundles over V are trivial. Hence for each $f \in C^r(V, W)$ there is a section $\alpha \in \Gamma(\mathcal{R})$ such that (α, f) is a formal solution i.e. the formal solution hypothesis is redundant in the n-rectangle case. The following corollaries therefore obtain.

Corollary 10.3. *Suppose in addition* $V \subset \mathbf{R}^n$ *is an* n-*rectangle. Let* $f \in C^r(V, W)$ *be strictly short with respect to* F *as in (10.7): for all* $x \in V$,

$$\partial^r_t f(x) \in \text{IntConv}\, F(j^\perp f(x))$$

Let $\mathcal{N}$ *be a neighbourhood of the image* $j^\perp f(V)$ *in* $J^\perp(V, W)$. *There is a map* $g \in C^r(V, W)$ *which solves the partial differential inclusion (10.4) and such that the image* $j^{r-1}g(V) \subset \mathcal{N}$.

Corollary 10.4. *Suppose* $V = [a, b]$. *Let* $x(t)$ *be a* C^r-*parametrized curve in* W, $r \geq 1$, *which is strictly short with respect to* F: *for all* $t \in [a, b]$,

$$x^{(r)}(t) \in \text{IntConv}\, F(t, x(t), x'(t), \ldots, x^{(r-1)}(t)).$$

Let $\epsilon > 0$. *There is a* C^r-*parametrized curve* $y(t)$ *in* W *such that for all* $t \in [0, 1]$:

(i) $\|y^{(s)}(t) - x^{(s)}(t)\| \leq \epsilon$, *for all* s, $0 \leq s \leq r - 1$.

(ii) $y^{(r)}(t) \in F(t, y(t), y'(t), \ldots, y^{(r-1)}(t))$ *i.e.* $y(t)$ *solves the differential inclusion* F.

In case $r \geq 2$, the one-variable Corollary 10.4 is new. The extensive literature on the Relaxation Theorem in Optimal Control theory, [2], [3], [4], does not discuss rth order differential inclusions for $r \geq 2$.

References

[1] Adachi M., *Immersion Theory*; Transl. of Math. Monographs vol. 124, Amer. Math. Soc., 1993.

[2] Aubin J-P., Cellina A., *Differential Inclusions; Set-Valued Maps and Viability Theory*, Springer Verlag, New York, 1984.

[3] Aubin J-P., Frankowska H., *Set-Valued Analysis*, Birkhäuser, Boston, 1990.

[4] Blagodat·Skikh V. I., Filippov A. F., *Differential inclusions and optimal control*, Proc. Steklov Inst. Math. 4 (1986), 199–259.

[5] Bredon G. E., *Topology and Geometry*; Graduate Texts in Mathematics vol. 139, Springer-Verlag, New York, 1993.

[6] Bryant R. L., Chern S. S., Gardner R. B., Goldschmidt H. L., Griffiths P. A., *Exterior Differential Systems* M.S.R.I. Vol. 18, Springer-Verlag, Berlin, New York, 1991.

[7] Eliashberg Y., *On singularities of folding type*, Izv. Akad. Nauk. SSSR **34** (1970), 1111–1127.

[8] Eliashberg Y., Gromov M., *Removal of singularities of smooth mappings*, Math. USSR (Izvestia) **3** (1971), 615–639.

[9] ——— , *Construction of nonsingular isoperimetric films*, Proc. Steklov Inst. Math. **116** (1971), 13–28.

[10] Eliashberg Y., *Surgery of singularities of smooth mappings*, Izv. Akad. Nauk. SSSR **36** (1972), 1321–1347.

[11] Eliashberg Y., Mishachev N. M., *Wrinkling of smooth maps and its applications* I, II, III; preprints and in preparation, 1994.

[12] Feit S., *k-mersions of manifolds*, Acta Mathematica **122** (1969), 173–195.

[13] Filippov A. F., *Classical solutions of differential inclusions with multivalued right-hand side*, Siam Journal of Control **6** (1967), 609–621.

[14] Geiges H., Gonzalo J., *An Application of convex integration to 3-dimensional contact geometry*; preprint (1992).

[15] Ginsburg V. L., *Calculation of contact and symplectic cobordism groups*, Topology **31** 4 (1992), 767–773.

[16] Gromov M., *Stable maps of foliations into manifolds*, Trans. Math. USSR (Izvestia) **3** (1969), 671–693.

[17] ____ , *Convex integration of differential relations* I, Izv. Akad. Nauk SSSR **37** (1973), 329–343.

[18] ____ , *Partial Differential Relations*, Springer-Verlag, New York, 1986.

[19] ____ , *Oka's principle for holomorphic sections of elliptic bundles*, Jour. Amer. Math. Soc. **2** 4 (1989), 851–897.

[20] Haefliger A., *Lectures on the theorem of Gromov*, Proc. Liverpool Singularities Symp. II. Springer Lecture Notes Math., vol. 207, Springer-Verlag, New York, 1970, pp. 128-141.

[21] Hirsch M. W., *Immersions of Manifolds*, Trans. Amer. Math. Soc. **93** (1959), 242–276.

[22] ____ , *Differential Topology*, Graduate Texts in Mathematics vol. 33, Springer-Verlag, New York, 1976.

[23] Hörmander L., *Convex Sets*, North Holland, 1994.

[24] Hurewicz W., Wallman H., *Dimension Theory*, Princeton Univ. Press,, 1948.

[25] Jurdjevic V., *Geometric Control Theory*, Cambridge Studies in Advanced Mathematics vol. 51, Camb. Univ. Press, 1997.

[26] Kuiper, *On C^1-isometric imbeddings* I, II, Indag. Math. **17** (1955), 545–556, 683–689.

[27] Marchaud A., *Sur les champs de demi-cônes et les équations différentielles du premier ordre*, Bull. Soc. Math. France **62** (1934).

[28] McDuff D., *Applications of convex integration to symplectic and contact geometry*, Ann. Inst. Fourier **37** (1987), 107–133.

[29] ____ ; Book Review of [18] Bull. Amer. Math. Soc. **18** 2 (1987), 214–220.

[30] Nash J., *C^1-isometric imbeddings*, Ann. Math. **63** (1954), 384–396.

[31] Phillips A., *Submersions of open manifolds*, Topology **6** 2 (1967), 170–226.

[32] Poenaru V., *Sur la théorie des immersions*, Topology **1** (1962), 81–100.

[33] ____ , *Regular Homotopy in Codimension 1*, Ann. Math. **83** 2 (1966), 257–265.

[34] ____ , *Homotopy theory and differential singularities*, Springer Lecture Notes Math., vol. 197, 1970, pp. 106–133.

[35] Reznick B., *Sums of Even Powers of Real Linear Forms*; Memoirs of the Amer. Math. Soc. vol. 96, No. 463, Amer. Math. Soc., Providence, Rhode Island, 1992.

[36] Smale S., *The classification of immersions of spheres in Euclidean spaces*, Ann. Math. **69** (1959), 327–344.

[37] Spring D., *Convex integration of non-linear systems of partial differential equations*, Ann. Inst. Fourier **33** (1983), 121–177.

[38] ____ (1991), *On the Regularity of Solutions in Convex Integration Theory*, Invent. Math. **104** (1991), 165–178.

[39] ____ , *Notes on the history of immersion theory*, From Topology to Computation: Proceedings of the Smalefest (M. W. Hirsch, J.E. Marsden, M Shub, eds.), Springer Verlag, New York, 1993, pp. 114–116.

[40] Steenrod N., *The Topology of Fiber Bundles*, Princeton University Press, 1951.

[41] Ważewski T., *Sur une généralisation de la notion des solutions d'une équation au contingent*, Bull. Acad. Polon. Sci. Sér. sci. math. astr. et phys. **10** (1962), 11–15.

[42] Zaremba S. K., *Sur les équations aux paratingent*, Bull. des Sci. Math. **60** (1936), 139–160.

INDEX

INDEX OF NOTATION

Breinigsville, PA USA
26 December 2010
252124BV00003B/34/P